Unimodularity in Randomly Generated Graphs

CONTEMPORARY MATHEMATICS

719

Unimodularity in Randomly Generated Graphs

AMS Special Session
Unimodularity in Randomly Generated Graphs
October 8–9, 2016
Denver, Colorado

Florian Sobieczky
Editor

AMERICAN MATHEMATICAL SOCIETY

2010 *Mathematics Subject Classification.* Primary 60K35, 60K37, 60J35, 60C05, 60G55, 37A05, 37A35.

Library of Congress Cataloging-in-Publication Data

Names: Sobieczky, Florian, editor.

Title: Unimodularity in randomly generated graphs / Florian Sobieczky, editor.

Description: Providence, Rhode Island : American Mathematical Society, [2018] | Series: Contemporary mathematics ; volume 719 | "AMS Special Session on Unimodularity in Randomly Generated Graphs, October 8-9, 2016, Denver, Colorado." | Includes bibliographical references.

Identifiers: LCCN 2018027509 | ISBN 9781470439149 (alk. paper)

Subjects: LCSH: Random graphs. | Graph theory. | AMS: Probability theory and stochastic processes – Special processes – Interacting random processes; statistical mechanics type models; percolation theory. msc | Probability theory and stochastic processes – Special processes – Processes in random environments. msc | Probability theory and stochastic processes – Markov processes – Transition functions, generators and resolvents. msc | Probability theory and stochastic processes – Combinatorial probability – Combinatorial probability. msc | Probability theory and stochastic processes – Stochastic processes – Point processes. msc | Dynamical systems and ergodic theory – Ergodic theory – Measure-preserving transformations. msc | Dynamical systems and ergodic theory – Ergodic theory – Entropy and other invariants, isomorphism, classification. msc

Classification: LCC QA166.17 .U55 2018 | DDC 519.2/3–dc23

LC record available at https://lccn.loc.gov/2018027509

Contemporary Mathematics ISSN: 0271-4132 (print); ISSN: 1098-3627 (online)

DOI: https://doi.org/10.1090/conm/719

∞ The paper used in this book is acid-free and falls within the guidelines established to ensure permanence and durability.

Visit the AMS home page at `https://www.ams.org/`

10 9 8 7 6 5 4 3 2 1 23 22 21 20 19 18

Dedicated to Prof. Russell Lyons in admiration of his work
and gratitude for his inspiration.

Contents

Preface

'Well, this soup is inedible, again.'
'Why? This is even a very fine soup, today!'
'Nobody claims this isn't a fine soup.
Just that it is too hot to be edible.'
Karl Valentin, 'Der Hasenbraten'

The special session 'Unimodularity in randomly generated graphs' was held at the Denver AMS sectional meeting in October 2016. At this time of the year the Indian Summer turns the Aspen trees of the the Rocky Mountain Front Range into a golden curtain. It brought together people from different communities turning the discussion into a fruitful event.

Russ Lyons and Yuval Peres with the Rocky Mountains in the background

Unimodularity is a term originally used in the field of topological groups, where it refers to the equality of the left and right Haar measure. 'Unimodularity of graphs' was invented as a concept in percolation theory under the name of the 'Mass Transport Principle' where it was used with great success to generalise results from percolation on the Euclidean lattice to vertex-transitive graphs. It is equivalent to unimodularity of the automorphism group of the graph but has been developed further in terms of measures of rooted random graph valued processes. More generally, for measurable spaces of equivalence relations allowing graphings, unimodularity plays a central role in terms of the invariance and reversibility with respect to these equivalence relations. This is only a very incomplete glimpse of the range of topics in which unimodularity is the main underlying concept. Numerous related terms (such as hyperfiniteness, quasi-invariance, Diestel-Leader graphs and horocyclic products) are currently being developed further, and several of these developments are discussed in the articles of this volume. If different communities value different aspects of the same notion, let us remember the great Karl Valentin, and try to understand the other specialist's view.

This special session and the publication of these proceedings have been supported by the Austrian Ministry for Transport, Innovation and Technology, the Federal Ministry of Science, Research and Economy, and the Province of Upper Austria in the frame of the COMET center SCCH.

Hagenberg, *Florian Sobieczky*, August 2018

Contemporary Mathematics
Volume **719**, 2018
https://doi.org/10.1090/conm/719/14464

Monotonicity of average return probabilities for random walks in random environments

Russell Lyons

ABSTRACT. We extend a result of Lyons (2017) from fractional tiling of finite graphs to a version for infinite random graphs. The most general result is as follows. Let $\mathbf{P}$ be a unimodular probability measure on rooted networks (G, o) with positive weights w_G on its edges and with a percolation subgraph H of G with positive weights w_H on its edges. Let $\mathbf{P}_{(G,o)}$ denote the conditional law of H given (G, o). Assume that $\alpha := \mathbf{P}_{(G,o)}[o \in \mathsf{V}(H)] > 0$ is a constant $\mathbf{P}$-a.s. We show that if $\mathbf{P}$-a.s. whenever $e \in \mathsf{E}(G)$ is adjacent to o,

$$\mathbf{E}_{(G,o)}\big[w_H(e) \mid e \in \mathsf{E}(H)\big]\mathbf{P}_{(G,o)}\big[e \in \mathsf{E}(H) \mid o \in \mathsf{V}(H)\big] \le w_G(e)\,,$$

then

$$\forall t > 0 \quad \mathbf{E}\big[p_t(o; G)\big] \le \mathbf{E}\big[p_t(o; H) \mid o \in \mathsf{V}(H)\big]\,.$$

1. Introduction

Associated to a graph with nonnegative numbers on its edges such that the sum of numbers of edges incident to each given vertex is finite, there is a continuous-time random walk that, when at a vertex x, crosses each edge e incident to x at rate equal to the number on e. When all rates equal 1, this is called continuous-time simple random walk. In general, the rate at which the random walk leaves x equals the sum of the numbers on the edges incident to x.

It is well known and easy to prove that every such (weighted) random walk has the property that the probability of return to the starting vertex is a decreasing function of time. Equivalently, the return probability at any fixed time decreases if all the rates are increased by the same factor. However, the return probability is *not* a decreasing function of the set of rates in general. Indeed, the behavior of the return probabilities is not intuitive; a small example is shown in Figure 1.1. Examples show that the return probability to a vertex x need not be monotonic even when rates are changed only on edges not incident to x. On the other hand, on a finite graph, the *average* of the return probabilities *is* decreasing in the rates, as shown by Benjamini and Schramm (see Theorem 3.1 of [**HH05**]). Recall that on a finite graph, the stationary measure for this continuous-time random walk is uniform on the vertices.

2010 *Mathematics Subject Classification.* Primary 60K37, 60J35; Secondary 05C80, 05C81.
Key words and phrases. Continuous time, Markov chains.
Partially supported by the National Science Foundation under grant DMS-1612363 and by Microsoft Research.

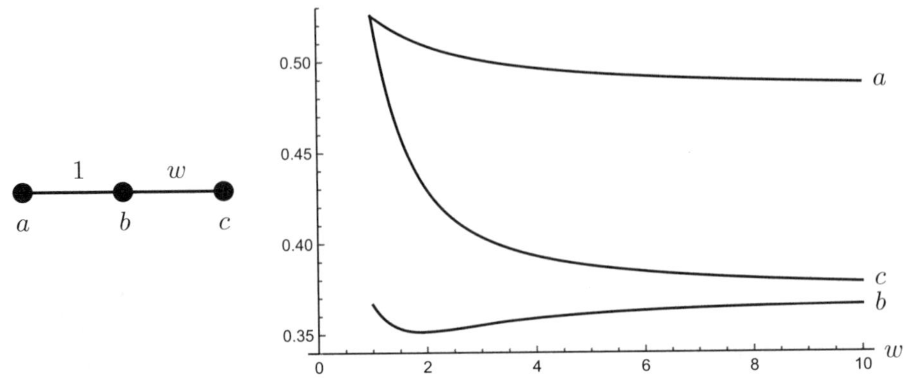

FIGURE 1.1. The return probabilities *at time 1* of vertices a, b, and c in the graph on the left as the rate w varies.

In Theorems 4.1 and 4.2 of [**Lyo17**], we extended and strengthened the theorem of Benjamini and Schramm to the case of graphs of different sizes and even to the case of one graph G that is "fractionally tiled" by a set of subgraphs H_i with a certain condition on the edge weights of G and H_i. Our purpose here is to establish a version of those results for infinite graphs.

For a very simple example of our results here, consider the square lattice $\mathbb{Z}^2$ and the subgraph H formed by deleting every vertex both of whose coordinates are odd; see Figure 1.2. There are four subgraphs of $\mathbb{Z}^2$ that are isomorphic to H. Considering those four copies of H, we find that each vertex of $\mathbb{Z}^2$ is covered three times, once by a vertex of degree 4 and twice by a vertex of degree 2. An appropriate average return probability in H is thus 1/3 that of a vertex of degree 4 plus 2/3 that of a vertex of degree 2. Consider continuous-time simple random walk on each graph, where edges are crossed at rate 1; the return probabilities are shown in Figure 1.3. As illustrated in Figure 1.4, we have for all $t \geq 0$,

$$p_t\big((0,0);\mathbb{Z}^2\big) \leq \frac{1}{3}p_{3t/2}\big((0,0);H\big) + \frac{2}{3}p_{3t/2}\big((0,1);H\big) .$$

Effectively, we have used rates 3/2 on every edge of H. This inequality follows from Corollary 2.3. It is sharp in the following sense: if $3t/2$ is replaced by βt for some $\beta > 3/2$, then the resulting inequality fails for some $t > 0$.

This particular example can be easily derived from Theorem 4.2 of [**Lyo17**]. With some more work, so can all the results here when the unimodular probability measures involved are sofic. However, the general case (which is not known to be sofic) does not follow from earlier work. Nevertheless, our results and proofs are modeled on Theorem 4.2 of [**Lyo17**]. The challenge here was to formulate the proper statements for infinite graphs and to make the appropriate adjustments to the proofs required for using direct integrals instead of direct sums.

For a more complicated example of our results, suppose that G is the usual nearest-neighbor graph on $\mathbb{Z}^d$ ($d \geq 2$) and H is the infinite cluster of supercritical Bernoulli (site or bond) percolation on G. Let $\delta := \mathbf{E}\big[\deg_H(o) \bigm| o \in H\big]/(2d) \in (0,1)$. Then

$$\forall t \geq 0 \quad p_t(o;\mathbb{Z}^d) \leq \mathbf{E}\big[p_{t/\delta}(o;H) \bigm| o \in \mathsf{V}(H)\big] .$$

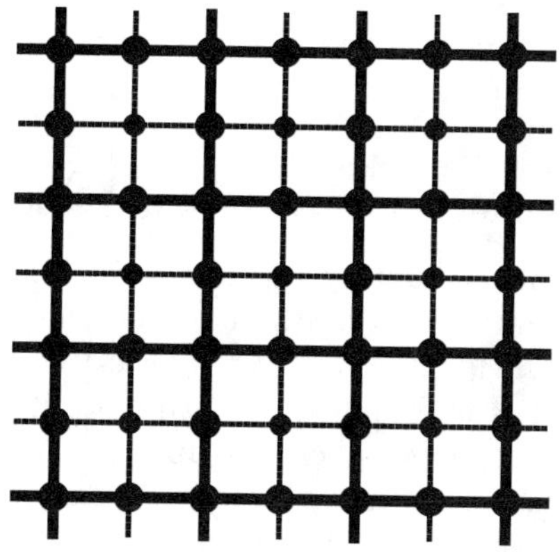

FIGURE 1.2. The square lattice $\mathbb{Z}^2$ and the subgraph H, drawn thicker, formed by deleting every vertex both of whose coordinates are odd.

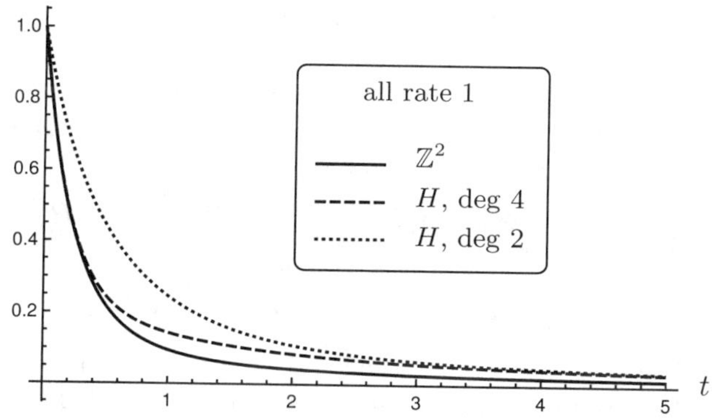

FIGURE 1.3. Return probabilities for continuous-time simple random walk on each graph.

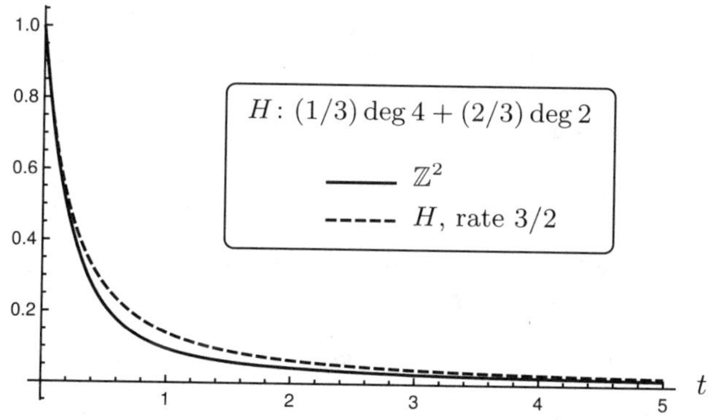

FIGURE 1.4. Comparison of continuous-time random walk with rates 1 on $\mathbb{Z}^2$ and rates 3/2 on H, the latter averaged over the starting point.

This is obtained by using $w_H \equiv 1/\delta$ in Corollary 2.3. The preceding inequality is false for any larger value of δ.

2. Statements of Results and Background

Let G be a simple, locally finite graph with weights $w_G(e) \geq 0$ on the edges e. Consider the continuous-time random walk on G where edge e is crossed at rate $w_G(e)$ when the walk is incident to e. Let $p_t(x; G)$ denote the probability that a random walk started at x is found at x at time t. If Δ_G is the corresponding Laplacian, i.e., $\Delta_G(x, y) := -w(e)$ when e is an edge joining x and y with weight $w(e)$, all other off-diagonal elements of Δ_G are 0, and the row sums are 0, then $p_t(x; G)$ is the (x, x)-entry of $e^{-t\Delta_G}$. If the entries of Δ_G are unbounded, then we take the minimal Markov process, which dies after an explosion. The infinitesimal generator is then the self-adjoint extension of $-\Delta_G$ (for uniqueness of the extension, see [**HKMW13**]). For the definition of unimodular in our context, see [**AL07**].

THEOREM 2.1. *Let* $\mathbf{P}$ *be a unimodular probability measure on rooted networks* (G, o) *with positive weights* w_G *on its edges and with a percolation subgraph* H *of* G *with positive weights* w_H *on its edges. Let* $\mathbf{P}_{(G,o)}$ *denote the conditional law of* H *given* (G, o). *Assume that* $\alpha := \mathbf{P}_{(G,o)}\big[o \in \mathsf{V}(H)\big] > 0$ *is a constant* $\mathbf{P}$*-a.s. If* $\mathbf{P}$*-a.s. whenever* $e \in \mathsf{E}(G)$ *is adjacent to* o,

$$\mathbf{E}_{(G,o)}\big[w_H(e) \bigm| e \in \mathsf{E}(H)\big]\,\mathbf{P}_{(G,o)}\big[e \in \mathsf{E}(H) \bigm| o \in \mathsf{V}(H)\big] \leq w_G(e)\,, \tag{2.1}$$

then

$$\forall t > 0 \quad \mathbf{E}\big[p_t(o; G)\big] \leq \mathbf{E}\big[p_t(o; H) \bigm| o \in \mathsf{V}(H)\big]\,.$$

The case where G is finite is Theorem 4.2 of [**Lyo17**], although it is disguised. The case where $G = H$ and $w_G \geq w_H$ is Theorem 5.1 of [**AL07**].

REMARK 2.2. Theorem 2.1 is sharp in a sense: if

$$\mathbf{E}\Big[\sum_{e\sim o} \mathbf{E}_{(G,o)}\big[w_H(e) \bigm| e \in \mathsf{E}(H)\big]\,\mathbf{P}_{(G,o)}\big[e \in \mathsf{E}(H) \bigm| o \in \mathsf{V}(H)\big]\Big] > \mathbf{E}\Big[\sum_{e\sim o} w_G(e)\Big]\,, \tag{2.2}$$

then for all sufficiently small, positive t,

$$\mathbf{E}\big[p_t(o; G)\big] > \mathbf{E}\big[p_t(o; H) \bigm| o \in \mathsf{V}(H)\big]\,. \tag{2.3}$$

For example, let (G, o) be any unimodular random rooted graph and consider Bernoulli(α) site percolation on G. Let H be the induced subgraph. Then

$$\forall t \geq 0 \qquad \mathbf{E}\big[p_t(o; G)\big] \leq \mathbf{E}\big[p_{t/\alpha}(o; H) \bigm| o \in \mathsf{V}(H)\big]\,.$$

This is obtained by using $w_H \equiv 1/\alpha$. This is sharp: for all $\beta < \alpha$, there is some t such that $\mathbf{E}\big[p_t(o; G)\big] > \mathbf{E}\big[p_{t/\beta}(o; H)\big]$.

The following corollary is immediate from Theorem 2.1.

COROLLARY 2.3. *Let* G *be a unimodular transitive graph and* H *be a random subgraph of* G *with edge weights* w_H *such that the law of* (H, w_H) *is* $\mathrm{Aut}(G)$*-invariant. If*

$$\forall e \sim o \qquad \mathbf{E}\big[w_H(e) \bigm| e \in H\big]\,\mathbf{P}[e \in H \mid o \in H] \leq 1\,,$$

then continuous-time simple random walk on G and the continuous-time network random walk on (H, w_H) satisfy

$$\forall t > 0 \quad p_t(o; G) \le \mathbf{E}\big[p_t(o; H, w_H) \bigm| o \in \mathsf{V}(H)\big] .$$

One might expect also the following as a corollary: Suppose that G is a fixed Cayley graph and w_1, w_2 are two random fields of positive weights on its edges with the properties that each field w_i has an invariant law and a.s. $w_1(e) \ge w_2(e)$ for each edge e. Then $\mathbf{E}\big[p_{1,t}(o; G)\big] \le \mathbf{E}\big[p_{2,t}(o, G)\big]$ for all $t > 0$, where $p_{i,t}$ denotes the return probability to a fixed vertex o at time t with the weights w_i. This is indeed known to be true for amenable G [**FM06**] and also when the pair (w_1, w_2) is invariant [**AL07**]. However, it is open in general and was asked by Fontes and Mathieu (personal communication). Even more generally, the following question is open, even for finite graphs where it was raised by [**Lyo17**]:

QUESTION 2.4. Suppose that $\mathbf{P}_1$ and $\mathbf{P}_2$ are two unimodular probability measures on rooted graphs with positive edge weights for which there is a coupling that is carried by the set of pairs $\big((G, o, w_G), (H, o, w_H)\big)$ with H a subgraph of G and $w_H(e) \le w_G(e)$ for all $e \in \mathsf{E}(H)$. Is $\mathbf{E}_1\big[p_t(o; G)\big] \le \mathbf{E}_2\big[p_t(o; H)\big]$ for all $t > 0$?

We prove Theorem 2.1 and Remark 2.2 in the following section. Here we present the background required, especially regarding von Neumann algebras.

We will use the notation $A \le B$ for self-adjoint operators A and B to mean that $B - A$ is positive semidefinite. Sometimes we regard the edges of a graph as oriented, where we choose one orientation (arbitrarily) for each edge. In particular, we do this whenever we consider the ℓ^2-space of the edge set of a graph. In this case, we denote the tail and the head of e by e^- and e^+. Define $d_G \colon \ell^2\big(\mathsf{V}(G)\big) \to \ell^2\big(\mathsf{E}(G)\big)$ by

$$d_G(a)(e) := \sqrt{w_G(e)}\,\big[a(e^-) - a(e^+)\big] .$$

Then $\Delta_G = d_G^* d_G$.

Consider the Hilbert space $\mathscr{G} := \int^{\oplus} \ell^2\big(\mathsf{V}(G)\big)\, d\mathbf{P}(G, o)$; see Section 5 of [**AL07**] for details of this direct integral. Let Tr denote the normalized trace corresponding to $\mathbf{P}$, as in Section 5 of [**AL07**]. That is, given an equivarant operator $T = \int^{\oplus} T_G \, d\mathbf{P}(G, o)$ on $\mathscr{G}$ in the von Neumann algebra Alg associated by [**AL07**] to $\mathbf{P}$, we define

$$\mathrm{Tr}(T) := \int (T_G \mathbf{1}_{\{o\}}, \mathbf{1}_{\{o\}})\, d\mathbf{P}(G, o) .$$

This trace on Alg is obviously finite. A closed densely defined operator is ***affiliated*** with Alg if it commutes with all unitary operators that commute with Alg. Write AffAlg for the set of all such operators. An operator $T \in \mathsf{AffAlg}$ is called Tr-***measurable*** if for all $\epsilon > 0$, there is an orthogonal projection $E \in \mathsf{Alg}$ whose image lies in the domain of T and $\mathrm{Tr}(E^\perp) < \epsilon$. For example, Δ_G is Tr-measurable because if E_n denotes the orthogonal projection to the space of functions that are nonzero only on those (G, o) where the sum of the edge weights at o is at most n, then $\lim_{n\to\infty} \mathrm{Tr}(E_n^\perp) = 0$ and $\|\Delta_G E_n\| \le 2n$. We will need another representation of the trace. For $s \in [0, 1]$ and a Tr-measurable operator $T \ge 0$ with spectral resolution E_T, define

$$m_s(T) := \inf\big\{\lambda \ge 0 ;\ \mathrm{Tr}\big(E_T(\lambda, \infty)\big) \le 1 - s\big\} ;$$

see Remark 2.3.1 of [**FK86**]. By Lemma 2.5(iii) of [**FK86**], if $0 \le S \le T$ are Tr-measurable, then

$$\forall s \in [0,1] \quad m_s(S) \le m_s(T)\,. \tag{2.4}$$

A proof similar to that of Corollary 2.8 of [**FK86**] shows that for bounded monotone $f\colon \mathbb{R} \to \mathbb{R}$ and $T \in \mathsf{AffAlg}$, we have

$$\mathrm{Tr}\big(f(T)\big) = \int_0^1 f\big(m_s(T)\big)\,ds\,. \tag{2.5}$$

From (2.5) and (2.4), we obtain

$$\mathrm{Tr}\, f(S) \le \mathrm{Tr}\, f(T) \tag{2.6}$$

for bounded increasing $f\colon \mathbb{R} \to \mathbb{R}$ and $0 \le S \le T$ that are Tr-measurable operators in AffAlg. Furthermore, if f is strictly increasing, then equality holds in (2.6) iff $S = T$: if equality holds, then $f(S) = f(T)$ (because Tr is faithful by Lemma 2.3 of [**AL07**]), whence $f^{-1}\big(f(S)\big) = f^{-1}\big(f(T)\big)$.

Let $w_{G,n}$ denote the weights on G when for every $x \sim y$, the edge weight $w_G(x,y)$ is replaced by 0 if the sum of the weights incident to x and y is larger than n. We claim that

$$\int p_t(o; w_{G,n})\,d\mathbf{P}(G,o) = \lim_{n\to\infty} \int p_t(o; w_G)\,d\mathbf{P}(G,o)\,. \tag{2.7}$$

To see this, let E_n denote, as before, the orthogonal projection to the space of functions that are nonzero only on those (G,o) where the sum of the edge weights at o is at most n. Then $\Delta_{G,w_{G,n}} E_n = \Delta_{G,w_G} E_n$ for all n. Since $\lim_{n\to\infty} \mathrm{Tr}(E_n^\perp) = 0$, it follows that $\lim_{n\to\infty} \Delta_{G,w_{G,n}} = \Delta_{G,w_G}$ in the measure topology (Definition 1.5 of [**FK86**]). Since $\Delta_{G,w_{G,n}} \le \Delta_{G,w_G}$, we have $m_s\big(\Delta_{G,w_{G,n}}\big) \le m_s\big(\Delta_{G,w_G}\big)$ by (2.4). Therefore, $\lim_{n\to\infty} m_s\big(\Delta_{G,w_{G,n}}\big) = m_s\big(\Delta_{G,w_G}\big)$ by Lemma 3.4(ii) of [**FK86**]. Now use $f(\lambda) := e^{-t\lambda}$ in (2.5) to obtain $\lim_{n\to\infty} \mathrm{Tr}\big(e^{-t\Delta_{G,w_{G,n}}}\big) = \mathrm{Tr}\big(e^{-t\Delta_{G,w_G}}\big)$, which is the same as (2.7).

Suppose that Φ is a positive, unital, linear map from a unital C^*-algebra A to a von Neumann algebra with finite trace, Tr. The proof of Theorem 3.9 of [**AMS07**] shows that

$$\mathrm{Tr}\, j\big(\Phi(T)\big) \le \mathrm{Tr}\, \Phi\big(j(T)\big) \tag{2.8}$$

for self-adjoint operators $T \in \mathsf{A}$ and functions $j\colon \mathbb{R} \to \mathbb{R}$ that are convex on the convex hull of the spectrum of T. (In fact, those authors show the more general inequality $\mathrm{Tr}\, k\big(j\big(\Phi(T)\big)\big) \le \mathrm{Tr}\, k\big(\Phi\big(j(T)\big)\big)$ for every increasing convex $k\colon \mathbb{R} \to \mathbb{R}$.)

3. Proofs

Proof of Theorem 2.1. Suppose first that the entries of Δ_G and Δ_H are uniformly bounded, so that $\Delta_{\mathscr{G}}$ and $\Delta_{\mathscr{H}}$ are bounded operators in Alg.

In addition to the Hilbert space $\mathscr{G} := \int^\oplus \ell^2\big(\mathsf{V}(G)\big)\,d\mathbf{P}(G,o)$ we considered in the preceding section, also let

$$\mathscr{H} := \int^\oplus \int^\oplus \ell^2\big(\mathsf{V}(H)\big)\,d\mathbf{P}_{(G,o)}(H)\,d\mathbf{P}(G,o)\,.$$

By Lemma 2.3 of [**AL07**], we have that

$$\mathbf{P}\Big[\forall x \in \mathsf{V}(G) \;\; \mathbf{P}_{(G,o)}\big[x \in \mathsf{V}(H)\big] = \alpha\Big] = 1\,. \tag{3.1}$$

Similarly, (2.1) implies that a.s.

$$\forall e \in \mathsf{E}(G) \quad \alpha^{-1} \int_{e \in \mathsf{E}(H)} w_H(e)\, d\mathbf{P}_{(G,o)} \le w_G(e)\,. \tag{3.2}$$

By (3.1), for every $f = \int^{\oplus} f(G,o)\, d\mathbf{P}(G,o) \in \mathscr{G}$, we have that

$$\phi(f) := \alpha^{-1/2} \int^{\oplus}\!\! \int^{\oplus} \sum_{x \in \mathsf{V}(H)} f(G,o)(x) \mathbf{1}_{\{x\}}\, d\mathbf{P}_{(G,o)}(H)\, d\mathbf{P}(G,o) \in \mathscr{H}$$

has the same norm as f. Moreover, $\phi\colon \mathscr{G} \to \mathscr{H}$ defines an isometry, i.e., $\phi^*\phi$ is the identity map. Define $\Phi\colon \mathcal{L}(\mathscr{H}) \to \mathcal{L}(\mathscr{G})$ by $\Phi T := \phi^* T \phi$. Then Φ is a positive unital map.

Consider quadruples (G, H, w_G, w_H) of graphs G and H and weight functions w_G and w_H with H a subgraph of G. An isomorphism of a pair of such quadruples is defined in the obvious way. As before, however, we will generally omit including the weight functions in the notations for networks. Similarly to how Alg is defined, let A be the von Neumann algebra of (equivalence classes of) bounded linear maps $T = \int^{\oplus}\int^{\oplus} T_{(G,o,H)}\, d\mathbf{P}_{(G,o)}(H)\, d\mathbf{P}(G,o) \in \mathcal{L}(\mathscr{H})$ that are equivariant in the sense that for all isomorphisms $\psi\colon (G_1, H_1) \to (G_2, H_2)$, all $o_1 \in \mathsf{V}(G_1)$, $o_2 \in \mathsf{V}(G_2)$, and all $x, y \in \mathsf{V}(H_1)$, we have $(T_{(G_1,o_1,H_1)}x, y) = (T_{(G_2,o_2,H_2)}\psi x, \psi y)$; in particular, $T_{(G,o,H)}$ does not depend on o. Then Φ maps A into Alg.

Let $\Delta_{\mathscr{G}} := \int^{\oplus} \Delta_G\, d\mathbf{P}(G,o) \in \mathsf{Alg}$ and $\Delta_{\mathscr{H}} := \int^{\oplus}\int^{\oplus} \Delta_H\, d\mathbf{P}_{(G,o)}(H)\, d\mathbf{P}(G,o) \in \mathsf{A}$. Then $\Phi(\Delta_{\mathscr{H}}) \in \mathsf{Alg}$ and, therefore, $j(\Delta_{\mathscr{G}}), j\big(\Phi(\Delta_{\mathscr{H}})\big) \in \mathsf{Alg}$ for all bounded Borel $j\colon \mathbb{R} \to \mathbb{R}$.

We claim that

$$\Delta_{\mathscr{G}} \ge \Phi(\Delta_{\mathscr{H}})\,. \tag{3.3}$$

To see this, let $f \in \mathscr{G}$. We have

$$\big(\Delta_{\mathscr{G}}(f),\, f\big) = \mathbf{E}\big[\|d_G f(G,o)\|^2\big] \tag{3.4}$$

and

$$\big(\Phi\Delta_{\mathscr{H}}(f),\, f\big) = \big(\phi^*\Delta_{\mathscr{H}}\phi f,\, f\big) = \big(\Delta_{\mathscr{H}}\phi f,\, \phi f\big)\,. \tag{3.5}$$

Now

$$\begin{aligned}
\big(\Delta_{\mathscr{H}}\phi f, \phi f\big) &= \alpha^{-1} \iint \|d_H f(G,o)\|^2\, d\mathbf{P}_{(G,o)}\, d\mathbf{P}(G,o) \\
&= \alpha^{-1} \iint \sum_{e \in \mathsf{E}(H)} w_H(e)\big(f(G,o)(e^-) - f(G,o)(e^+)\big)^2\, d\mathbf{P}_{(G,o)}\, d\mathbf{P}(G,o) \\
&= \alpha^{-1} \int \sum_{e \in \mathsf{E}(G)} \int_{\mathsf{E}(H) \ni e} w_H(e)\, d\mathbf{P}_{(G,o)} \cdot \big(f(G,o)(e^-) - f(G,o)(e^+)\big)^2\, d\mathbf{P}(G,o) \\
&\le \int \sum_{e \in \mathsf{E}(G)} w_G(e) \cdot \big(f(G,o)(e^-) - f(G,o)(e^+)\big)^2\, d\mathbf{P}(G,o) \\
&= \mathbf{E}\big[\|d_G f(G,o)\|^2\big]
\end{aligned}$$

by (3.2). Combining this with (3.4) and (3.5), we get our claimed inequality (3.3).

By (3.3) and (2.6), we have

$$\operatorname{Tr} j(\Delta_{\mathscr{G}}) \le \operatorname{Tr} j\big(\Phi(\Delta_{\mathscr{H}})\big)$$

for every decreasing function j. (We have strict inequality if j is strictly decreasing and we have strict inequality in (3.3).) Use $j(s) := e^{-ts}$ in this and in (2.8) to obtain

$$\operatorname{Tr} j(\Delta_{\mathscr{G}}) \le \operatorname{Tr} \Phi\big(j(\Delta_{\mathscr{H}})\big)\,. \tag{3.6}$$

The left-hand side equals $\mathbf{E}\big[p_t(o;G)\big]$. The right-hand side equals

$$\begin{aligned}\operatorname{Tr} \Phi\big(j(\Delta_{\mathscr{H}})\big) &= \alpha^{-1} \iint_{o\in \mathsf{V}(H)} \big(j(\Delta_H)\mathbf{1}_{\{o\}}, \mathbf{1}_{\{o\}}\big)\, d\mathbf{P}_{(G,o)}\, d\mathbf{P}(G,o)\\ &= \mathbf{E}\big[p_t(o;H) \mid o \in \mathsf{V}(H)\big]\,,\end{aligned}$$

which completes the proof of the theorem in the case of bounded vertex weights.

We deduce the general case from this by a truncation argument. Recall (2.7) and its notation, which we use also for H. Let μ_n be the law of $(w_{G,n}, w_{H,n})$. Since the diagonal entries of $\Delta_{(G,w_{G,n})}$ and $\Delta_{(H,w_{H,n})}$ are bounded and (2.1) holds μ_n-a.s., we have proved that

$$\forall t > 0 \quad \mathbf{E}\big[p_t\big(o;(G,w_{G,n})\big)\big] \le \mathbf{E}\big[p_t\big(o;(H,w_{H,n})\big) \mid o \in \mathsf{V}(H)\big]\,.$$

Taking $n \to \infty$ and using the bounded convergence theorem, we get the desired result. □

A similar proof shows that (3.6) holds if j is any decreasing convex function.

Proof of Remark 2.2. The right-hand side of (2.2) is equal to $\mathbf{E}\big[\Delta_G(o,o)\big]$ and the left-hand side is $\mathbf{E}\big[\Delta_H(o,o) \mid o \in \mathsf{V}(H)\big]$. Now both sides of (2.3) equal 1 for $t=0$. We claim that the derivative of the left-hand side at $t=0$ is $-\mathbf{E}\big[\Delta_H(o,o) \mid o \in \mathsf{V}(H)\big]$ and the derivative of the right-hand side at $t=0$ is $-\mathbf{E}\big[\Delta_G(o,o)\big]$. This clearly implies the remark. To evaluate these derivatives, note that for every fixed G, the spectral representation

$$p_t(o;G) = \int_0^\infty e^{-\lambda t}\, d\big(E_{\Delta_G}(\lambda)\mathbf{1}_{\{o\}}, \mathbf{1}_{\{o\}}\big)$$

shows that $t \mapsto p_t(o;G)$ is monotone decreasing and convex. By Tonelli's theorem, it follows that for $(G,o) \sim \mathbf{P}$,

$$\mathbf{E}\big[p_t(o;G)\big] - 1 = \mathbf{E}\Big[\int_0^t p_s'(o;G)\, ds\Big] = \int_0^t \mathbf{E}\big[p_s'(o;G)\big]\, ds\,.$$

The fundamental theorem of calculus and the monotone convergence theorem now yield that

$$\frac{d}{dt}\mathbf{E}\big[p_t(o;G)\big]\Big|_{t=0} = \mathbf{E}\big[p_0'(o;G)\big] = -\mathbf{E}\big[\Delta_G(o,o)\big]\,.$$

A similar calculation applied to the distribution of (H,o) given $o \in \mathsf{V}(H)$ yields the derivative of the left-hand side of (2.3). □

References

[AL07] David Aldous and Russell Lyons, *Processes on unimodular random networks*, Electron. J. Probab. **12** (2007), no. 54, 1454–1508, DOI 10.1214/EJP.v12-463. MR2354165

[AMS07] Jorge Antezana, Pedro Massey, and Demetrio Stojanoff, *Jensen's inequality for spectral order and submajorization*, J. Math. Anal. Appl. **331** (2007), no. 1, 297–307, DOI 10.1016/j.jmaa.2006.08.029. MR2306006

[FK86] Thierry Fack and Hideki Kosaki, *Generalized s-numbers of τ-measurable operators*, Pacific J. Math. **123** (1986), no. 2, 269–300. MR840845

[FM06] L. R. G. Fontes and P. Mathieu, *On symmetric random walks with random conductances on $\mathbb{Z}^d$*, Probab. Theory Related Fields **134** (2006), no. 4, 565–602, DOI 10.1007/s00440-005-0448-1. MR2214905

[HH05] Deborah Heicklen and Christopher Hoffman, *Return probabilities of a simple random walk on percolation clusters*, Electron. J. Probab. **10** (2005), no. 8, 250–302, DOI 10.1214/EJP.v10-240. MR2120245

[HKMW13] Xueping Huang, Matthias Keller, Jun Masamune, and Radosław K. Wojciechowski, *A note on self-adjoint extensions of the Laplacian on weighted graphs*, J. Funct. Anal. **265** (2013), no. 8, 1556–1578, DOI 10.1016/j.jfa.2013.06.004. MR3079229

[Lyo17] Russell Lyons, *Comparing graphs of different sizes*, Combin. Probab. Comput. **26** (2017), no. 5, 681–696, DOI 10.1017/S096354831700013X. MR3681977

Department of Mathematics, Indiana University, 831 E. 3rd Street, Bloomington, Indiana 47405-7106

Email address: rdlyons@indiana.edu

URL: http://pages.iu.edu/~rdlyons/

Contemporary Mathematics
Volume **719**, 2018
https://doi.org/10.1090/conm/719/14465

Counterexamples for percolation on unimodular random graphs

Omer Angel and Tom Hutchcroft

ABSTRACT. We construct an example of a bounded degree, nonamenable, unimodular random rooted graph with $p_c = p_u$ for Bernoulli bond percolation, as well as an example of a bounded degree, unimodular random rooted graph with $p_c < 1$ but with an infinite cluster at criticality. These examples show that two well-known conjectures of Benjamini and Schramm are false when generalised from transitive graphs to unimodular random rooted graphs.

1. Introduction

In **Bernoulli bond percolation**, each edge of a connected, locally finite graph G is chosen to be deleted randomly with probability $1-p$, independently of all other edges, to obtain a random subgraph $G[p]$ of G. When G is infinite, the **critical parameter** is defined to be

$$p_c(G) = \inf\{p \in [0,1] : G[p] \text{ contains an infinite connected component almost surely}\}$$

and the **uniqueness threshold** is defined to be

$$p_u(G) = \inf\left\{p \in [0,1] : \begin{array}{l} G[p] \text{ contains a unique infinite} \\ \text{connected component almost surely} \end{array}\right\}.$$

Traditionally, percolation was studied primarily on the hypercubic lattice $\mathbb{Z}^d$ and other Euclidean lattices. In their seminal paper [**7**], Benjamini and Schramm proposed a systematic study of percolation on more general graphs, and posed many questions. They were particularly interested in **quasi-transitive** graphs, that is, graphs whose automorphism groups have only finitely many orbits. Two central questions concern the existence or non-existence of infinite clusters at p_c, and the equality or inequality of p_c and p_u. They made the following conjectures. Specific instances of these conjectures, such as those concerning $\mathbb{Z}^d$, are much older.

CONJECTURE 1.1. *Let G be a quasi-transitive graph, and suppose that $p_c(G) < 1$. Then $G[p_c]$ does not contain an infinite cluster almost surely.*

CONJECTURE 1.2. *Let G be a quasi-transitive graph. Then $p_c(G) < p_u(G)$ if and only if G is nonamenable.*

Key words and phrases. Unimodular random graphs, percolation, uniqueness, connectivity, nonamenable.

Given a set K of vertices in a graph G, we define $\partial_E K$ to be the set of edges of G that have exactly one endpoint in K. A graph is said to be **nonamenable** if

$$\inf\left\{\frac{|\partial_E K|}{\sum_{v\in K}\deg(v)} : K \subseteq V \text{ finite}\right\} > 0,$$

and **amenable** otherwise. It follows from the work of Burton and Keane [**9**] and Gandolfi, Keane and Newman [**11**] that $p_c(G) = p_u(G)$ for every amenable quasi-transitive graph, so that only the 'if' direction of Conjecture 1.2 remains open. It was also proven by Häggström, Peres, and Schonmann [**13**, **14**, **20**] that there is a unique infinite cluster for every $p > p_u$ when G is quasi-transitive. We refer the reader to [**12**] for an account of what is known in the Euclidean case $G = \mathbb{Z}^d$, and to [**17**] for percolation on more general graphs.

Substantial progress on Conjecture 1.1 was made in 1999 by Benjamini, Lyons, Peres, and Schramm [**5**], who proved that the conjecture is true for any *nonamenable, unimodular* quasi-transitive graph. Here, a graph is unimodular if it satisfies the *mass-transport principle*, see [**17**, Chapter 8]. (More recently, the conjecture has been verified for all quasi-transitive graphs of *exponential growth* [**15**, **21**], and in particular for all nonamenable quasi-transitive graphs, without the assumption of unimodularity.) In the mid 2000's, Aldous and Lyons [**1**] showed that this result, as well as several other important results such as those of [**6**, **9**, **11**, **13**, **14**, **18**–**20**] can be generalized, with minimal changes to the proofs, to *unimodular random rooted graphs*. These graphs appear naturally in many applications: For example, the connected component at the origin in percolation on a unimodular transitive graph is itself a unimodular random rooted graph. An important caveat is that when working with unimodular random rooted graphs one should consider a different, weaker notion of nonamenability than the classical one, which we call *invariant nonamenability* [**1**, §8].

In this note, we construct examples to show that, in contrast to the situation for the classical results mentioned in the previous paragraph, Conjectures 1.1 and 1.2 are in fact both *false* when generalized to unimodular random rooted graphs, even with the assumption of bounded degrees.

THEOREM 1.3. *There exists a bounded degree unimodular random rooted graph (G,ρ) such that $p_c(G) < 1$ but there is an infinite cluster $G[p_c]$ almost surely.*

THEOREM 1.4. *There exists a unimodular random rooted graph (G,ρ) such that G has bounded degrees, is nonamenable, and has $p_c(G) = p_u(G)$ for Bernoulli bond percolation almost surely.*

We stress that the example in Theorem 1.4 is nonamenable in the classical sense (which is a stronger property than being invariantly nonamenable). Thus, any successful approach to Conjectures 1 and 2 cannot rely solely on mass-transport arguments. See [**8**] for some further examples of unimodular random rooted graphs with unusual properties for percolation, and [**4**] for another related example.

2. Basic constructions

2.1. Unimodularity and normalizability of unrooted graphs. We assume that the reader is familiar with the basic notions of unimodular random rooted graphs, referring them to [**1**] otherwise. Since it will be important to us and is perhaps less widely known, we quickly recall the theory of unimodular random rooted graphs with fixed underlying graph from [**1**, Section 3].

Let G be a graph, let $\Gamma \subseteq \mathrm{Aut}(G)$ be a group of automorphisms of G, and for each $v \in V$ let $\mathrm{Stab}_v = \{\gamma \in \Gamma : \gamma v = v\}$ be the stabilizer of v in Γ. The group Γ is said to be **unimodular** if

$$|\,\mathrm{Stab}_v\, \gamma v| = |\,\mathrm{Stab}_{\gamma v}\, v|$$

for every $v \in V$ and $\gamma \in \Gamma$, where $\mathrm{Stab}_v\, u$ is the orbit of u under Stab_v. The graph G is said to be unimodular if $\mathrm{Aut}(G)$ is unimodular. Let G be a connected, locally finite, unimodular graph and let $\mathcal{O}$ be a set of **orbit representatives** of Γ. That is, $\mathcal{O} \subseteq V$ is such that for every vertex $v \in V$, there exists exactly one vertex $o \in \mathcal{O}$ such that $\gamma v = o$ for some $\gamma \in \Gamma$. We say that (G, Γ) is **normalizable** if there exists a measure μ_G on $\mathcal{O}$ such that if ρ is distributed according to μ_G then the random rooted graph (G, ρ) is unimodular. It is easily seen that the measure μ_G is unique when it exists.

It is proven in [**1**, Theorem 3.1] that a connected, locally finite, unimodular graph G is normalizable if and only if

$$Z_v(G) = \sum_{o \in \mathcal{O}} |\,\mathrm{Stab}_o(v)|^{-1} < \infty$$

for some (and hence every) vertex $v \in V$, and moreover the measure μ_G can be expressed as

$$\mu_G(\{o\}) = Z_v(G)^{-1} |\,\mathrm{Stab}_o(v)|^{-1} \qquad o \in \mathcal{O}.$$

2.2. Building new examples from old via replacement. We will frequently make use of the following construction, which allows us to construct one normalizable unimodular graph from another. Constructions of this form are well-known, see [**16**] and [**8**] for further background.

Let $G = (V, E)$ be a connected, locally finite graph, let $\Gamma \subseteq \mathrm{Aut}(G)$ be a unimodular subgroup of automorphisms, and let $G' = (V', E')$ be a connected, locally finite graph. Let $M_1(V')$ be the set of functions $m : V' \to [0, 1]$ with $|m| := \sum_{v \in V'} m(v) < \infty$, and suppose that there exists a function $m : V \to M(V')$, $m : v \mapsto m_v$ such that

(1) The functions $\{m_v : v \in V\}$ are a partition of unity on V' in the sense that $\sum_{v \in V} m_v(u) = 1$ for every $u \in V'$.

and

(2) m is automorphism-equivariant on V^2 in the following sense: If $u, v, w, x \in V$ are such that $(w, x) = (\gamma u, \gamma v)$ for some $\gamma \in \mathrm{Aut}(G)$, then there exists an automorphism γ' of G' such that $(m_w, m_x) = (\gamma m_u, \gamma m_v)$.

Then G' is also unimodular. If furthermore G is normalizable and

$$\sum_{o \in \mathcal{O}(G)} \mu_G(\{o\}) |m_o| < \infty, \tag{2.1}$$

then G' is normalizable with

$$\mu_{G'}(\{o'\}) = \sum_{v \in V'} \mathbb{1}\left[o' \in \mathrm{Aut}\,(G')\, v\right] \sum_{o \in \mathcal{O}(G)} \frac{\mu_G(\{o\}) m_o(v)}{\sum_{o \in \mathcal{O}(G)} \mu_G(\{o\}) |m_o|} \qquad o' \in \mathcal{O}(G').$$

Following [**8**], we call this method of constructing new normalizable unimodular graphs from old ones **replacement**. To give a simple example of replacement, suppose that G is a connected, locally finite, normalizable unimodular graph, and let G' be the graph in which each edge of G is replaced with a path of length two. Define $m : V \to M_1(v')$ by setting $m_v(u)$ to be 1 if u is equal to v, and to be 1/2 if u is the midpoint of a path of length 2 emanating from v in G' that was formerly an edge of G. It is easily verified that m satisfies conditions 1 and 2 above. If furthermore G has finite expected degree in the sense that $\sum_{o\in\mathcal{O}(G)} \mu_G(\{o\}) \deg(o) < \infty$, then (2.1) is satisfied and G' is normalizable.

One can also consider a variation of this procedure allowing for randomization: Let $G = (V, E)$ be a connected, locally finite, unimodular graph, let V' be a set, and let $G' = (V', E')$ be a random connected, locally finite graph with vertex set V', which we consider to be a random element of $\{0,1\}^{V^2}$. Suppose that there exists a function $m : V \to M(V')$, $m : v \mapsto m_v$ such that

(1) The functions $\{m_v : v \in V\}$ are a partition of unity on V' in the sense that $\sum_{v\in V} m_v(u) = 1$ for every $u \in V'$.

and

(2) m is automorphism-equivariant on V^2 in the following sense: If $u, v, w, x \in V$ are such that $(w, x) = (\gamma u, \gamma v)$ for some $\gamma \in \mathrm{Aut}(G)$, then there exists a bijection $\gamma' : V' \to V'$ such that $(m_w, m_x) = (\gamma m_u, \gamma m_v)$ and the *law* of G' is invariant under the action of γ' on V^2.

Let $\tilde{\rho}$ be a random element of $\mathcal{O}(G)$ drawn from biased measure $\tilde{\mu}_G$ defined by

$$\tilde{\mu}_G(\{o\}) = \frac{\mu_G(\{o\})|m(o)|}{\sum_{o\in\mathcal{O}(G)} \mu_G(\{o\})|m(o)|}, \qquad o \in \mathcal{O}(G)$$

and, conditional on $\tilde{\rho}$, let $\rho' \in V'$ be chosen according to the conditional distribution

$$\mathbb{P}(\rho' = u \mid \tilde{\rho}) = \frac{m_\rho(u)}{|m_\rho|}.$$

Then the random rooted graph (G', ρ') is unimodular.

Fixing the vertex set of G' in advance is of course rather unnecessary and restrictive, but it is sufficient for the examples we consider here.

3. A discontinuous phase transition

3.1. Trees of tori. Let $d \geq 2$. The **d-ary canopy tree** T_d is the tree with vertex set $\mathbb{Z} \times \mathbb{N}$ and edge set

$$\{\{(i,j),(k,j-1)\} : j \geq 1,\ di \leq k \leq d(i+1) - 1\}.$$

In other words, T_d is the tree that has infinitely many leaves (that have no children), and such that every vertex that is not a leaf has exactly d children, that is, neighbours that are closer to the leaves than it is. Note that the isomorphism class of (T_d, v) depends only on the distance between v and the leaves, called the **height** of v, and denoted $|v|$. We also say that vertices with height k for $k \geq 0$ are in **level** k. It is well known and easily verified that T_d is unimodular and normalizable, with $\mu_{T_d}(\{o\}) = d^{-|o|+1}/(d-1)$.

Let $n \geq 1$, and let $d, r \geq 2$. We define the **tree of tori** $\mathbb{T}^n(d,r)$ to be the connected, locally finite graph with vertex set

$$V\left(\mathbb{T}^n(d,r)\right) = \left\{(v,x) : v \in V(T_d),\ x \in \mathbb{Z}^n/r^{|v|}\mathbb{Z}^n\right\},$$

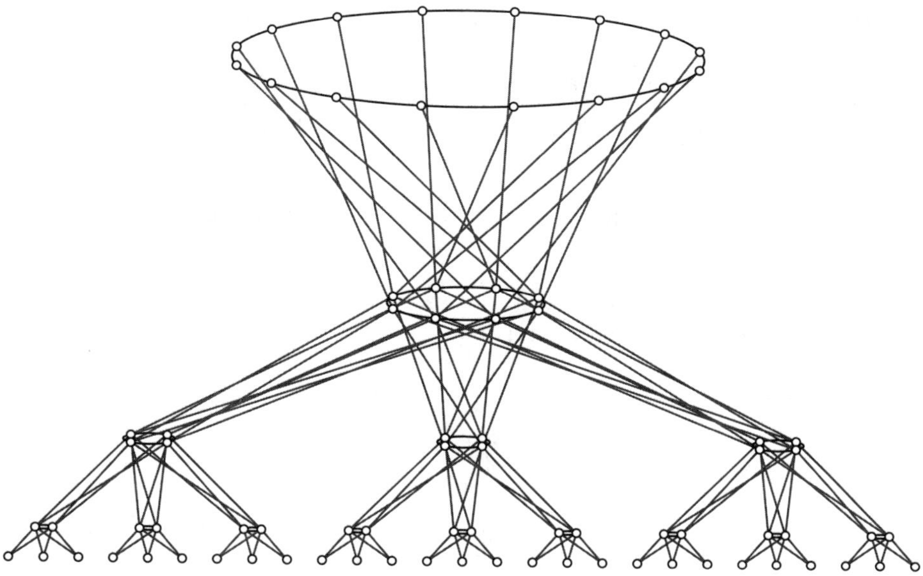

FIGURE 1. The canopy tree of one-dimensional tori $\mathbb{T}^1(3,2)$. The grey edges give a 3-to-2 correspondence between levels l and $l+1$ for each $l \geq 0$. For general $\mathbb{T}^n(d,r)$ the correspondence is d-to-r^n.

and where we connect two vertices (v,x) and (u,y) of $\mathbb{T}^n(d,r)$ by an edge if and only if either

(1) $v = u$ and x and y are adjacent in the torus, or else
(2) u is adjacent to v in T_d, and either $|v| \geq |u|$ and x is mapped to y by the quotient map $\mathbb{Z}^n/r^{|v|}\mathbb{Z}^n \to \mathbb{Z}^n/r^{|u|}\mathbb{Z}^n$ or, symmetrically, $|u| \geq |v|$ and y is mapped to x by the quotient map $\mathbb{Z}^n/r^{|u|}\mathbb{Z}^n \to \mathbb{Z}^n/r^{|v|}\mathbb{Z}^n$.

See Figure 1 for an illustration. (Note that removing the torus edges from this graph yields the horocyclic product of the d-ary canopy tree with the r^n-ary tree, which also arises as a half-space of the Diestel-Leader graph $DL(d, r^n)$ [**10**].)

The following is an easy consequence of replacement.

PROPOSITION 3.1. *If $r^n < d$ then $\mathbb{T}^n(d,r)$ is unimodular and normalizable, and*

$$\mu_{\mathbb{T}^n(d,r)}(\{o\}) = \frac{d^{-|o|+1}r^{n|o|}}{d - r^n}.$$

It will also be useful to consider a more general version of this construction, in which we let the sizes of the tori grow as a specified function of the height. Let $d \geq 2$, let $n \geq 1$, and let $r : \mathbb{N} \to \mathbb{N}$ be an increasing function. We define the tree of tori $\tilde{\mathbb{T}}^n(d,r)$ similarly to above, with vertex set

$$V\left(\tilde{\mathbb{T}}^n(d,r)\right) = \left\{(v,x) : v \in V(T_d),\, x \in \mathbb{Z}^n/2^{r(|v|)}\mathbb{Z}^n\right\},$$

and where we connect two vertices (v,x) and (u,y) of $\mathbb{T}^n(d,r)$ by an edge if and only if either

(1) $v = u$ and x and y are adjacent in the torus, or else

(2) u is adjacent to v in T_d, and either $|v| \geq |u|$ and x is mapped to y by the quotient map $\mathbb{Z}^n/2^{r(|v|)}\mathbb{Z}^n \to \mathbb{Z}^n/2^{r(|u|)}\mathbb{Z}^n$ or, symmetrically, $|u| \geq |v|$ and y is mapped to x by the quotient map $\mathbb{Z}^n/2^{r(|u|)}\mathbb{Z}^n \to \mathbb{Z}^n/2^{r(|v|)}\mathbb{Z}^n$.

We now have, by replacement, that $\tilde{\mathbb{T}}^n(d,r)$ is unimodular and is normalizable if and only if

$$\sum_{\ell \geq 0} d^{-\ell} 2^{nr(\ell)} < \infty, \qquad \text{in which case} \qquad \mu_{\tilde{\mathbb{T}}^n(d,r)}(\{o\}) = \frac{d^{-|o|+1}2^{nr(|o|)}}{\sum_{\ell \geq 0} d^{-\ell}2^{nr(\ell)}}.$$

3.2. Proof of Theorem 1.3. In this section we prove Theorem 1.3. We begin with unbounded degree example, and then show how it can be modified to obtain a bounded degree example.

Let $d \geq 2$ and let T_d be d-ary canopy tree. We write

$$\log^+ x = \begin{cases} 1 & x \leq e \\ \log x & x > e, \end{cases}$$

and write $\asymp$ for equalities that hold up to positive multiplicative constants. For each $\gamma \in \mathbb{R}$, let $G_{d,\gamma}$ be obtained from T_d by replacing each edge connecting a vertex at height n to a vertex at height $n+1$ with

$$m_\gamma(n) := \left\lceil \frac{\log^+ n + \gamma \log^+ \log^+ n}{\log 2} \right\rceil$$

parallel edges, which is chosen so that

$$\left(\frac{1}{2}\right)^{m_\gamma(n)} \asymp \frac{1}{n \log^\gamma n}.$$

It follows by replacement that $G_{d,\gamma}$ is unimodular.

The basic idea behind this construction is that the coefficient of $\log^+ n$ above determines the value of p_c (set here to be $1/2$), while the coefficient of $\log^+ \log^+ n$ determines the behaviour of percolation *at* p_c.

Proposition 3.2. *$p_c(G_{d,\gamma}) = 1/2$ for every $\gamma \in \mathbb{R}$. If $\gamma > 1$, then critical percolation on $G_{d,\gamma}$ contains an infinite cluster almost surely.*

Proof. If v is a vertex of $G_{d,\gamma}$, then the cluster of v in $G_{d,\gamma}[p]$ is infinite if and only if every ancestor of v has an open edge connecting it to its parent. This event occurs with probability

$$\theta_\gamma(v,p) = \prod_{n \geq |v|} \left(1 - (1-p)^{m_\gamma(n)}\right).$$

In particular, $\theta_\gamma(v,p) > 0$ if and only if

$$\sum_{n \geq 0} (1-p)^{m_\gamma(n)} < \infty.$$

Since

$$n^{\frac{\log p}{\log 2}} (\log n)^{\frac{\gamma \log p}{\log 2}} \leq (1-p)^{m_\gamma(n)} \leq (1-p) n^{\frac{\log p}{\log 2}} (\log n)^{\frac{\gamma \log p}{\log 2}},$$

it follows that $\theta_\gamma(v,p) > 0$ if and only if

$$\sum_{n \geq 0} n^{\frac{\log p}{\log 2}} (\log n)^{\frac{\gamma \log p}{\log 2}} < \infty.$$

Recall that the series

$$\sum_{n\geq 0} \frac{1}{n^\alpha(\log^+ n)^\beta}$$

converges if and only if either $\alpha > 1$ or $\alpha = 1$ and $\beta > 1$. Thus, $\theta_\gamma(v,p) > 0$ for some (and hence every) vertex v of G_m if and only if either $p > 1/2$, or $p = 1/2$ and $\gamma > 1$. In particular, $p_c(G_{d,\gamma}) = 1/2$ for every d and every value of γ, while if $\gamma > 1$ then $\theta_\gamma(v,1/2) > 0$ for every vertex v of G as desired. □

We now build a bounded degree variation on this example using trees of tori. Let $d, r \geq 2$, be such that $d > r^2$, and let $m : \mathbb{N} \to \mathbb{N} \setminus \{0\}$ be a function. Let $\mathbb{G}(d,r,m)$ be the graph obtained by replacing each edge connecting two vertices of height ℓ and $\ell+1$ in $\mathbb{T}^2(d,r)$ with a path of length $m(\ell)$. It follows by replacement that G is unimodular, and is normalizable if

$$\sum_{\ell\geq 0} d^{-\ell} r^{2\ell} m(\ell) < \infty. \tag{3.1}$$

Thus, Theorem 1.3 follows immediately from the following proposition.

PROPOSITION 3.3. *Let $0 < q < 1$ be sufficiently large that $\theta_q(\mathbb{Z}^2) > 3/4$, and let $m : \mathbb{N}\setminus\{0\} \to \mathbb{N}\setminus\{0\}$ be such that there exists a positive constant c such that*

$$c\,4^{-\ell}(\ell+1)^2 \leq q^{m(\ell)} \leq 4^{-\ell}(\ell+1)^2$$

for every $\ell \geq 1$. Then $\mathbb{G} = \mathbb{G}(5,2,m)$ is a normalizable, bounded degree, unimodular graph, $p_c(\mathbb{G}) = q$, and $\mathbb{G}[q]$ contains an infinite cluster almost surely.

For an example of a function m of the form required by Proposition 3.3, we can take

$$m(\ell) \equiv \left\lceil \frac{(\ell+2)\log 4 - 2\log^+ \ell}{\log(1/q)} \right\rceil.$$

PROOF. It is clear that $\mathbb{G} = \mathbb{G}(5,2,m)$ has bounded degrees, and we have already established that it is unimodular and normalizable. We now prove the statements concerning percolation on $\mathbb{G}$. Suppose that v is a vertex of the canopy tree T_d with $|v| = \ell$, and suppose that u is the parent of v in T_d. Thus, the torus $\{v\} \times (\mathbb{Z}^2/2^\ell\mathbb{Z}^2)$ is connected in $\mathbb{G}$ to the torus $\{u\} \times (\mathbb{Z}^2/2^{\ell+1}\mathbb{Z}^2)$ by $4^{\ell+1}$ paths of length $m(\ell)$. If $p < q$ then $p = q^{1+\delta}$ for some $\delta > 0$, and so the expected number of these paths that are open in $\mathbb{G}[p]$ is

$$4^{\ell+1}p^{m(\ell)} = 4^{\ell+1}q^{(1+\delta)m(\ell)} \asymp \ell^2 4^{-\delta\ell}.$$

Since this expectation converges to zero, it follows that $\mathbb{G}[p]$ does not contain an infinite cluster almost surely, and we conclude that $p_c(\mathbb{G}) \geq q$.

It remains to prove that $\mathbb{G}[q]$ contains an infinite cluster almost surely. Broadly speaking, the idea is that, since $\theta_q(\mathbb{Z}^2) > 3/4$, each torus in $\mathbb{G}$ has a high probability to contain a giant open component which contains at least three quarters of its vertices, which is necessarily unique. The logarithmic correction in the definition of m then ensures that the giant component in each torus is very likely to be connected by an open path to the giant component in its parent torus, which implies that an infinite open component exists as claimed.

To make this argument rigorous, we will apply the following rather crude estimate.

PROPOSITION 3.4. *Consider Bernoulli bond percolation on the $n \times n$ torus, $\mathbb{Z}^2/n\mathbb{Z}^2$, for $p > p_c(\mathbb{Z}^2)$ supercritical. There exist positive constants c_1 and c_2 depending on p such that for every $\varepsilon > 0$, the probability that $\mathbb{Z}^2/n\mathbb{Z}^2$ does* not *contain an open cluster C with $|C| \geq (\theta_p(\mathbb{Z}^2) - \varepsilon)n^2$ is at most*

$$\frac{c_1}{\varepsilon^2 n^2} + n^2 e^{-c_2 \varepsilon n}.$$

PROOF. It suffices to prove the analogous statement for the box $[1,n]^2$, which we consider as a subgraph of $\mathbb{Z}^2$. It follows from [**3**, Theorem 1.1] that if $p > p_c(\mathbb{Z}^2)$, $\delta > 0$, and $x, y \in [\delta n, (1-\delta)n]^2$, then there exists a positive constant c_p such that

$$\mathbb{P}\left(x \leftrightarrow \infty \text{ and } y \leftrightarrow \infty, \text{ but } x \nleftrightarrow y \text{ in } [0,n]^2 \cap \mathbb{Z}^2\right) \leq e^{-c_p \delta n}.$$

Thus, it follows by a union bound that the probability that the largest cluster in $[1,n]^2$ has size at most $(\theta_p(\mathbb{Z}^2) - \varepsilon)n^2$ is at most

$$\mathbb{P}_p\left(\sum_{x,y \in [\delta n, (1-\delta)n]^2} \mathbb{1}\left(x \leftrightarrow \infty, y \leftrightarrow \infty\right) \leq (\theta_p(\mathbb{Z}^2) - \varepsilon)n^2\right) + n^2 e^{-c_p \delta n}.$$

On the other hand, we have that [**12**, Section 11.6]

$$\operatorname{Var}\left[\sum_{x \in [\delta n, (1-\delta)n]^2} \mathbb{1}\left(x \leftrightarrow \infty\right)\right] \leq C n^2$$

for some constant $C = C_p$, and it follows by Chebyshev's inequality that

$$\mathbb{P}\left(\sum_{x \in [\varepsilon n/2, (1-\varepsilon/2)n]^2} \mathbb{1}\left(x \leftrightarrow \infty\right) \leq (\theta_p(\mathbb{Z}^2) - \varepsilon)n^2\right)$$
$$\leq C\left[(1-\delta)^2 \theta_p(\mathbb{Z}^2) - \theta_p(\mathbb{Z}^2) + \varepsilon\right]^{-2} n^{-2}$$

when the right hand side is positive. We conclude by taking $\delta > 0$ so that $(1-\delta)^2\theta_p(\mathbb{Z}^2) - \theta_p(\mathbb{Z}^2) + \varepsilon = \varepsilon/2$. □

We now apply Lemma 3.4 to complete the proof of Proposition 3.3. Let v_0 be a leaf of T_5, let $v_1, v_2, \ldots$ be its sequence of ancestors, and let Λ_i be the torus $\{v_i\} \times \left(\mathbb{Z}^2/2^i\mathbb{Z}^2\right)$ in $\mathbb{G}$. It follows from Lemma 3.4 and the Borel-Cantelli Lemma that $\Lambda_i[q]$ contains a (necessarily unique) giant open cluster of size at least $(3/4)4^i$ for every $i \geq i_0$ for some random, almost surely finite i_0. Thus, for each $i \geq i_0$, there exist at least $4^i/2$ vertices of Λ_i that are both contained in the giant open cluster of $\Lambda_i[q]$, and have a parent in $\Lambda_{i+1}[q]$ that is contained in the giant open cluster of $\Lambda_{i+1}[q]$. Thus, conditional on this event, for each i sufficiently large, the probability that the giant open cluster of $\Lambda_i[q]$ is *not* connected by an open path to the giant open cluster of $\Lambda_{i+1}[q]$ is at most

$$\left(1 - q^{m(i)}\right)^{4^i/2} \leq \left(1 - q i^2 4^{-i}\right)^{4^i/2} \leq e^{-q i^2/2},$$

where we have used the inequality $(1-x) \leq e^{-x}$, which holds for all $x \geq 1$, to obtain the second inequality. Since these probabilities are summable, it follows by Borel-Cantelli that there exists a random, almost surely finite $i_1 \geq i_0$ such that the giant open cluster of $\Lambda_i[q]$ is connected to the giant open cluster of $\Lambda_{i+1}[q]$ for every $i \geq i_1$. It follows that $\mathbb{G}[q]$ contains an infinite cluster almost surely. □

4. Nonamenability and uniqueness

In this section we prove Theorem 1.4 by constructing a nonamenable, unimodular, normalizable, bounded degree graph G for which $p_c(G) = p_u(G)$ for Bernoulli bond percolation. We begin by constructing a family of partitions of the four regular tree.

4.1. Isolated, invariantly defined partitions of the tree. Let S be a 4-regular tree. If we draw S in the plane, then for each vertex v of S we obtain a cyclic ordering of the edges emanating from v that encodes the clockwise order that the edges appear around v in the drawing. We fix one such family of cyclic orderings, and let Γ be the group of automorphisms of S that fix this family of cyclic orderings. (In other words, we consider S as a **plane tree**.) It is well known that Γ is unimodular.

We define the **isolation** of a subset W of $V(S)$ to be the minimal distance between distinct points of S. If A and B are partitions of a set, we say that A **refines** B if every set $W \in A$ is contained in a set of B. We say further that A is a k**-fold refinement** of B if every set in B is equal to the union of exactly k sets in A. Similarly, we say that the fold of the refinement is **bounded by** k if every set in B is equal to the union of at most k sets in A.

PROPOSITION 4.1. *There exists a random sequence of partitions $(\tilde{\mathbb{V}}_k)_{k\geq 0}$ of $V(S)$ with the following properties.*

(1) *The law of $(\tilde{\mathbb{V}}_k)_{k\geq 0}$ is invariant under Γ.*
(2) *$\mathbb{V}_0 = \{V(S)\}$, and for each $n \geq 0$, $\tilde{\mathbb{V}}_{n+1}$ is a refinement of $\tilde{\mathbb{V}}_n$ with fold bounded by 4.*
(3) *Each $W \in \tilde{\mathbb{V}}_k$ has isolation at least $2\lfloor \log_2 k - 1 \rfloor$.*

The statement 'the law of $(\tilde{\mathbb{V}}_k)_{k\geq 0}$ is invariant under Γ' should be interpreted as follows: Γ naturally acts pointwise on subsets of $V(S)$, and hence also on partitions of $V(S)$. Then for any $\gamma \in \Gamma$, the sequence of partitions $(\gamma\tilde{\mathbb{V}}_k)_{k\geq 0}$ has the same law as the sequence of partitions $(\tilde{\mathbb{V}}_k)_{k\geq 0}$. (A partition is described as a subset of $V(S)^2$, with the product σ-algebra.)

We define a tree $\mathbb{D}$, such that $V(\mathbb{D}) = \bigcup_k \tilde{\mathbb{V}}_k$. The root of $\mathbb{D}$ is the trivial partition $V(S)$, and the children of a vertex in $\tilde{\mathbb{V}}_k$ are the included parts of $\tilde{\mathbb{V}}_{k+1}$. Thus $\mathbb{D}$ has bounded degrees.

PROOF. We begin by constructing a deterministic sequence of partitions which have isolation growing linearly and fold growing exponentially. We then construct a random sequence of partitions that intermediate between these partitions, which will satisfy the conclusions of proposition 4.1.

Let $(F_n)_{n\geq 0}$ be the sequence defined recursively by $F_0 = F_1 = 4$ and

$$F_{n+1} = F_n(F_{n-1} - 1) \qquad n \geq 1.$$

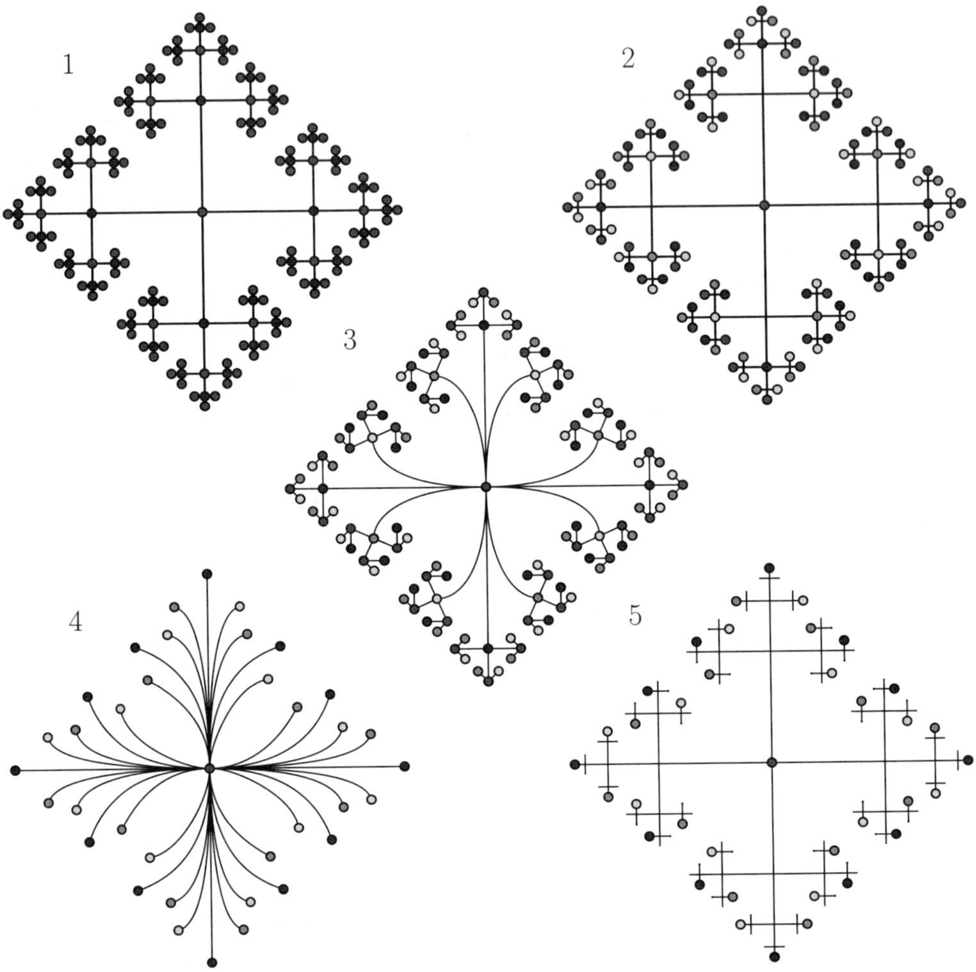

FIGURE 2. Recursively constructing the class containing the origin in the hierarchical partition of the 4-regular tree. (1) shows the decomposition of the tree into its two bipartite classes. (2) shows the subdivision of one of the two classes appearing in (1) into four classes as occurs in $\mathbb{V}_2(S)$. (3) shows the bipartite tree corresponding to the class of the origin in $\mathbb{V}_2(S)$, in which the two bipartite classes are 'red' and 'everything else' (where color is available). (4) shows the outcome of applying the same procedure another time, splitting the class of the origin in $\mathbb{V}_2(S)$ into four further subclasses and obtaining an associated bipartite tree for each of these classes. (5) shows the classes in (4) as they appear in the original 4-regular tree.

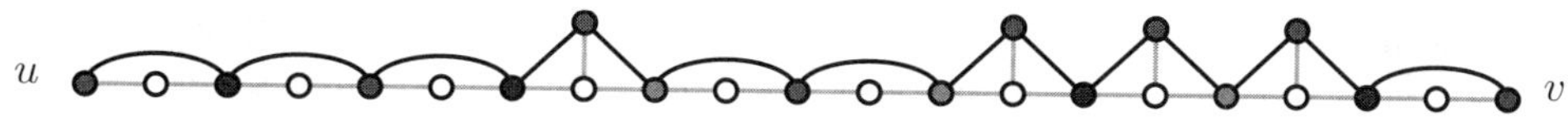

FIGURE 3. If $u, v \in V_1(S_{n,i})$ and the geodesic between u and v in S_n is given by the horizontal grey path, then the geodesic between u and v in $S_{n,i}$ is given by the black path. Red vertices represent vertices in $V_1(S_{n,i})$, blue and green vertices represent vertices in $V_1(S_n)\backslash V_1(S_{n,i})$, white vertices represent vertices in $V_2(S_n)$ (where color is available). The length of the black path is equal to the length of the grey path minus the number of red vertices. Grey curves are edges of S_n, black curves are edges of $S_{n,i}$.

Note that this sequence grows doubly-exponentially in n. In particular, $F_n \le 4^{2^n}$ for every $n \ge 0$. We construct a sequence of partitions $(\mathbb{V}_k)_{k\ge1}$ of $V(S)$, which we call the **hierarchical partition**, with the following properties:

(1) $(\mathbb{V}_k)_{k\ge1}$ is Γ-invariant in the sense that for any two vertices $u, v \in V(S)$, any $\gamma \in \Gamma$ and any $k \ge 1$, if $u, v \in V$ are in the same piece of the partition $\mathbb{V}_k$ (i.e., there exists $W \in \mathbb{V}_k$ such that $u, v \in W$) then $\gamma u, \gamma v$ are also in the same piece of the partition $\mathbb{V}_k$ (i.e., there exists $W' \in \mathbb{V}_k$ such that $\gamma u, \gamma v \in W$).
(2) $\mathbb{V}_1(S)$ is the partition of $V(S)$ into its two bipartite classes.
(3) For each $k \ge 1$, the partition $\mathbb{V}_{k+1}(S)$ is an F_{k-1}-fold refinement of the partition $\mathbb{V}_k$.
(4) Each $W \in \mathbb{V}_k$ has isolation at least $2k$.

The hierarchical partition may be constructed recursively as follows. Suppose that $n \ge 1$, and that S_n is the plane tree whose vertices are separated into bipartite classes V_1 and V_2 such that every vertex in V_1 has degree F_n and that every vertex in V_2 has degree F_{n-1}. We call vertices in V_1 **primary** and vertices in V_2 **secondary**. Consider a coloring of the primary vertices V_1 with the property that for every secondary vertex $v \in V_2$, the vertices $u_1, \ldots, u_{F_{n-1}}$ appearing in clockwise order adjacent to v in T have colors $1, \ldots, F_{n-1}$ up to a cyclic shift. Such a coloring is easily seen to exist and is unique up to a cyclic shifts of the colors.

For each color $1 \le i \le d_2$, let $S_{n,i}$ be the tree with vertex set V_1 in which two vertices are connected by an edge if and only if their distance in S_n is 2 and one of them has color i. This tree inherits a plane structure from S_n. Let $V_1(T_i)$ be the subset of $V(S_i) = V_1(S)$ containing the color i vertices and let $V_2(S_i)$ be the subset of $V(S_i) = V(S)$ containing the vertices with color other than i. It is easily verified that $V_1(S_i)$ and $V_2(S_i)$ are the two bipartite classes of $S_{n,i}$ and that vertices in these classes have degrees $F_n(F_{n-1}-1) = F_{n+1}$ and F_n respectively, so that $S_{n,i}$ is isomorphic to S_{n+1}. Moreover, the distance in $S_{n,i}$ between any two vertices in $V_1(S_{n,i})$ is at least two less than the distance of the corresponding vertices in S_n: Indeed, it is easily verified that the distance between $u, v \in V_1(S_{n,i})$ in $S_{n,i}$ is equal to their distance in S_n, minus the number of vertices in $V_1(S_{n,i})$ that are included in the geodesic between u and v in S_n (which is at least 2 due to the endpoints being in $V_1(S_{n,i})$). See Figure 3.

We apply this construction recursively, beginning with the 4-regular tree $S = S_1$ separated into its two bipartite classes. When we start step n of the recursion, we have constructed the sequence of partitions $(\mathbb{V}_k)_{k\le n}$ and have given each $W \in \mathbb{V}_n$

the structure of the bipartite plane tree S_n in such a way that the distance between any two vertices of W in the associated copy of S_n is at most their distance in S_1 minus $2(n-1)$. Given this data, we apply the above procedure to each of these copies of S_n to complete the next stage of the recursion, obtaining a F_{n-1}-fold refinement $\mathbb{V}_{k+1}$ of $\mathbb{V}_k$. This is the hierarchical partition: the above discussion implies inductively that it has the properties required above.

It remains to modify this hierarchical partition to have bounded fold. This is achieved by randomly adding further partitions that intermediate between $\mathbb{V}_n$ and $\mathbb{V}_{n+1}$. This is necessary, so that the construction in the next subsection gives a graph with unbounded degrees.

We define the **randomly intermediated hierarchical partition** $(\tilde{\mathbb{V}}_k)_{k\geq 0}$ of S as follows. For each $n \geq 0$, let $a_n = \lceil \log_4 F_n \rceil$, and let $b_n = \sum_{i=0}^{n} a_n$. Note that $a_n \leq 2^n$ and hence $b_n \leq 2^{n+1}$ for every $n \geq 0$.

(1) Let $\tilde{\mathbb{V}}_0 = \mathbb{V}_0 = \{V(S)\}$.
(2) For each $n \geq 1$, let $\tilde{\mathbb{V}}_{b_n} = \mathbb{V}_n$.
(3) We construct the partitions $(\tilde{\mathbb{V}}_{b_n - k})_{k=1}^{a_n - 1}$ recursively as follows: Given $\tilde{\mathbb{V}}_{b_n-k}$, for each set $W \in \tilde{\mathbb{V}}_{b_{n-1}} = \mathbb{V}_{n-1}$, choose uniformly at random a partition of the set $\{W' \in \tilde{\mathbb{V}}_{b_n-k} : W' \subseteq W\}$ into sets that all have size four except possibly for one of the sets. These random choices are made independently of each other, and independent of all other randomness used in the construction.

The definition of a_n and b_n ensure that $\tilde{\mathbb{V}}_{n+1}$ is a refinement of $\tilde{\mathbb{V}}_n$ with fold bounded by 4 for every $n \geq 1$. Moreover, the sequence of random variables $(\tilde{\mathbb{V}}_k)_{k\geq 0}$ is invariant *in distribution* under Γ. Finally, note that if we define c_k to be maximal such that $b_{c_k} \leq k$ for each $k \geq 1$, then every set in $\tilde{\mathbb{V}}_k$ is $2c_k$ isolated for every $k \geq 0$, since every such set is contained in a set in $\mathbb{V}_{c_k}$. Moreover, we have that

$$c_k \geq \lfloor \log_2 k - 1 \rfloor$$

for every $k \geq 1$. □

4.2. Proof of theorem 1.4. We now use the randomly intermediated hierarchical partition whose existence is stipulated by proposition 4.1 to construct the example required by Theorem 1.4. Let q be such that $\theta_q(\mathbb{Z}^2) > 3/4$, and let $m : \mathbb{N} \to \mathbb{N} \setminus \{0\}$ and $r : \mathbb{N} \to \mathbb{N} \setminus \{0\}$ be increasing functions such that

$$2^{r(\ell)} \asymp \ell + 1, \qquad q^{m(\ell)} \asymp (\ell+1)^{-2} \quad \text{and} \quad 2^{2r(\ell)} q^{m(\ell)} \leq 10^{-4}.$$

Suppose further that r and m have bounded increments. For example, we can take

$$r(\ell) = \left\lceil \frac{\log^+ \ell}{\log 2} \right\rceil \qquad \text{and} \qquad m(\ell) = \left\lceil \frac{2\log^+ \ell + 4 \log 10}{\log(1/q)} \right\rceil .$$

Let $\mathbb{T} = \tilde{\mathbb{T}}^2(100, r)$ be the canopy tree of tori, and let $\mathbb{G}$ be obtained from $\mathbb{T}$ by replacing edges between different levels of $\mathbb{T}$ with paths of length m, similarly to the construction in the previous section. For each vertex v of $\mathbb{G}$, we write $|v|$ for the unique $\ell \geq 0$ such that either v is a level-ℓ vertex of $\mathbb{T}$, or v lies on the interior of one of the paths of length $m(\ell)$ connecting level ℓ of $\mathbb{T}$ to level $\ell + 1$ of $\mathbb{T}$ that is added when constructing $\mathbb{G}$ from $\mathbb{T}$.

Let $(\tilde{\mathbb{V}}_k)_{k\geq 0}$ be a random sequence of partitions of $V(S)$ satisfying the conclusions of proposition 4.1, and $\mathbb{D}$ the associated tree. Conditional on $(\tilde{\mathbb{V}}_k)_{k\geq 0}$, we define the graph $\mathbb{H}$ as follows. The vertex set $V(\mathbb{H})$ is a subset of $\mathbb{G} \times S \times \mathbb{D}$ given by

$$\{(v_1, v_2, W) \in \mathbb{G} \times S \times \mathbb{D} \text{ such that } |W| = |v_1|\}.$$

Here, $|W| = k$ if $W \in \tilde{\mathbb{V}}_k$. This construction is somewhat similar to Diestel-Leader graphs (more precisely, half of the Diestel-Leader graph), since the tree structure of $\mathbb{G}$ branches towards level 0, and the tree $\mathbb{D}$ branches away from level 0.

We call a vertex (v_1, v_2, W) of $\mathbb{H}$ **type-1** if $v_2 \in W$ and **type-2** otherwise. We connect two vertices (v_1, v_2, W_1) and (v_1', v_2', W') of $\mathbb{H}$ by an edge if and only if one of the following hold:

(1) (v_1, v_2, W_1) and (v_1', v_2', W') are both type-1, $v_2 = v_2'$, and v_1, v_1' are adjacent in $\mathbb{G}$, and W, W' are adjacent in $\mathbb{D}$, or
(2) $v_1 = v_1'$, $W = W'$, and v_2 and v_2' are adjacent in S.

We call an edge of $\mathbb{H}$ a **$\mathbb{G}$-edge** if its endpoints have the same S-coordinate (in which case they must both be type-1 vertices), and an **S-edge** otherwise (in which case its endpoints have the same $\mathbb{G}$-coordinate, and at least one of the vertices must be type-2). Note that every connected component of the subgraph of $\mathbb{H}$ induced by the type-1 vertices (equivalently, spanned by the $\mathbb{G}$-edges) is isomorphic to $\mathbb{G}$. We call these **type-1 copies of $\mathbb{G}$ in $\mathbb{H}$.** Similarly, the type-2 edges span $\mathbb{H}$, and every connected component of the associated subgraph is isomorphic to the 4-regular tree S.

Let ρ_1 be a random root for the deterministic graph $\mathbb{G}$ chosen from the law $\mu_{\mathbb{G}}$, let ρ_2 be a fixed root vertex of the deterministic graph S, and let W_ρ be chosen uniformly from $\tilde{\mathbb{V}}_{|\rho_1|}$. Let $\rho = (\rho_1, \rho_2, W_\rho)$. It follows by replacement (applied to the product $\mathbb{G} \times S$, which is unimodular and normalizable by e.g. [**1**, Proposition 4.11]) that $(\mathbb{H}, \rho)$ is a unimodular random rooted graph.

We call a vertex (v_1, v_2, W) of $\mathbb{H}$ **type-1** if $v_2 \in W$ and **type-2** otherwise. We call an edge of $\mathbb{H}$ a **$\mathbb{G}$-edge** if its endpoints have the same S-coordinate (in which case they must both be type-1 vertices), and an **S-edge** otherwise (in which case its endpoints have the same $\mathbb{G}$-coordinate, and at least one of the vertices must be type-2). Note that every connected component of the subgraph of $\mathbb{H}$ induced by the type-1 vertices (equivalently, spanned by the $\mathbb{G}$-edges) is isomorphic to $\mathbb{G}$. We call these **type-1 copies of $\mathbb{G}$ in $\mathbb{H}$.**

Finally, given a constant $M \geq 1$, we define the graph $\tilde{\mathbb{H}}(M)$ by replacing each of the S-edges of $\mathbb{H}$ with a path of length M. It follows by replacement that $\tilde{\mathbb{H}}(M)$ can be rooted in such a way that it is a unimodular random rooted graph. Theorem 1.4 therefore follows immediately from the following proposition.

PROPOSITION 4.2. *The random graph $\tilde{\mathbb{H}}(M)$ described has bounded degrees and is nonamenable. If M is sufficiently large then $p_c(\tilde{\mathbb{H}}) = p_u(\tilde{\mathbb{H}}) = q$.*

The proof of Proposition 4.2 will apply the notion of disjoint occurrence and the BK inequality [**22**], see [**12**, Section 2.3] for background.

PROOF. Write $\tilde{\mathbb{H}} = \tilde{\mathbb{H}}(M)$. Moreover, it is immediate from the assumption that r and m have bounded increments that $\tilde{\mathbb{H}}$ has bounded degrees.

It is easily seen that stretching some edges by a bounded amount preserves nonamenability (indeed, nonamenability is stable under rough isometry), and so to prove that $\tilde{\mathbb{H}}$ is nonamenable it suffices to prove that $\mathbb{H}$ is nonamenable. Observe that we can partition the vertex set of $\mathbb{H}$ into sets $\{V_i : i \in I\}$ whose induced subgraphs are copies of the 4-regular tree S. Thus, given any finite set of vertices K in $\mathbb{H}$, we can write $K = \bigcup_{i\in I} K_i$ where $K_i = K \cap V_i$. Since the subgraph of $\mathbb{H}$ induced by V_i is a 4-regular tree for every $i \in I$, it follows that the external edge boundary of K_i in the subgraph induced by V_i has size at least $|K_i|$ for every $i \in I$, and so we have that

$$|\partial_E K| \geq \sum_{i\in I} |K_i| = |K|,$$

and hence that

$$\frac{|\partial_E K|}{\sum_{v\in K} \deg(v)} \geq \frac{1}{\max_{v\in\mathbb{H}} \deg(v)}$$

for every finite set of vertices K, so that $\mathbb{H}$ is nonamenable as claimed.

We now prove the statements concerning percolation on $\tilde{\mathbb{H}}$. It follows similarly to the proof of proposition 3.3 that $p_c(\mathbb{G}) = q$.

LEMMA 4.3. $p_u(\tilde{\mathbb{H}}) \leq p_c(\mathbb{G}) = q$.

PROOF. The proof is an an easy modification of the argument of Lyons and Schramm [**18**, Theorem 6.12], and applies the main theorem of that paper as generalised to unimodular random graphs by Aldous and Lyons [**1**, 6.15]; see also Theorem 6.17 of that paper. The sketch of the argument is as follows: It is easily verified that $\mathbb{G}$ is invariantly amenable (see e.g. [**1**, Section 8] and [**2**]), so that $p_u(\mathbb{G}) = p_c(\mathbb{G}) = q$ by [**1**, Corollary 6.11, 8.13]. Thus, for every $p > q$, every type-1 copy of $\mathbb{G}$ in $\tilde{\mathbb{H}}$ contains a unique infinite open cluster almost surely. It is easy to deduce using insertion tolerance and the mass-transport principle that these clusters must all be connected to each other by open edges in $\tilde{\mathbb{H}}$. Finally, indistinguishability implies that there cannot be any *other* infinite open cluster in $\tilde{\mathbb{H}}$, since the $\mathbb{G}$-edges within any such cluster would only have finite connected components. □

To complete the proof, it suffices to show that $p_c(\tilde{\mathbb{H}}) \geq q$ when M is sufficiently large. Let $V_1(\mathbb{H})$ be the set of type-1 vertices of $\mathbb{H}$. Define $\chi(\ell, k)$ and $\tilde{\chi}(\ell, k)$ for $\ell, k \geq 0$ by

$$\chi(\ell, k) = \sup_{v\in V(\mathbb{G}), |v|=\ell} \sum_{u\in V(\mathbb{G}), |u|=k} \mathbb{P}(v \leftrightarrow u \text{ in } \mathbb{G}[q]).$$

and

$$\tilde{\chi}(\ell, k) = \sup_{v\in V_1(\mathbb{H}), |v|=\ell} \sum_{u\in V_1(\mathbb{H}), |u|=k} \mathbb{P}(v \leftrightarrow u \text{ in } \tilde{\mathbb{H}}[q]).$$

It follows easily by mass transport and insertion tolerance that every infinite cluster of $\tilde{\mathbb{H}}[q]$ contains infinitely many type-1 vertices of $\mathbb{H}$, and so to prove that $p_c(\tilde{\mathbb{H}}) \geq q$ it suffices to prove that $\sum_{k\geq 0} \tilde{\chi}(\ell, k) < \infty$ for some (and hence every) $\ell \geq 0$. As a first step we bound the susceptibility in $\mathbb{G}$.

LEMMA 4.4. *There exists a constant C such that*

$$\chi(\ell, k) \leq C(k+1)^2$$

for every $\ell, k \geq 0$.

Note that the choice of M does not affect the definition of $\chi(\ell, k)$, and so the constant C here does not depend on the choice of M. A vertex of $\mathbb{G}$ at level at most k has a good chance of being connected in $\mathbb{G}[q]$ to the giant component in a torus at level k, and therefore the dependence on k cannot be improved here.

PROOF. For each vertex u in $\mathbb{G}$, let $t(u)$ be the associated vertex of the canopy tree T. Similarly, for each $x \in T$ let Λ_x be the associated torus in $\mathbb{T}$, and let $V_x = \{v \in V(\mathbb{G}) : x(v) = x\}$ be the associated set of vertices of $\mathbb{G}$. As in the proof of proposition 3.3, if y is the parent of x in T, then the probability that Λ_x is connected to Λ_y by an open path in $\mathbb{G}[q]$ is at most

$$2^{2r(|x|)}q^{m(|x|)} \leq 10^{-4}.$$

Thus, if x and y are vertices of T whose most recent common ancestor has height n, then the probability Λ_x is connected to Λ_y in $\mathbb{G}[q]$ is at most $10^{-8n+4|x|+4|y|}$.

Let $x \in V(T)$, let y be the parent of x in T, and let $u \in V_x$. Then

$$\begin{aligned}\sum_{u\in V(\mathbb{G}):|u|=k} &\mathbb{P}(v \leftrightarrow u \text{ in } \mathbb{G}[q])\\ &\leq \sum_{n\geq \ell\vee k}\sum_{|w|=k,|w\wedge x|=n} \mathbb{P}\left(\Lambda_x \cup \Lambda_{\sigma(x)} \leftrightarrow \Lambda_w \cup \Lambda_{\sigma(w)} \text{ in } \mathbb{G}[q]\right)|V_w|\\ &\leq C\sum_{n\geq \ell\vee k}\sum_{|w|=k,|w\wedge x|=n} 10^{-8n+4k+4\ell}|V_w|\\ &\leq C'\sum_{n\geq \ell\vee k} 10^{-8n+4\ell+4k}10^{2n-2k}(k+1)^2 \leq 2C'(k+1)^2\end{aligned}$$

as claimed, where C, C' are constants. □

We now apply lemma 4.4 to prove that $\sum_{k\geq 0}\tilde{\chi}(\ell,k) < \infty$. Observe that we may consider percolation on $\tilde{\mathbb{H}}$ as an inhomogeneous percolation on $\mathbb{H}$ in which every $\mathbb{G}$-edge of $\mathbb{H}$ is open with probability q, and every S-edge of $\mathbb{H}$ is open with probability q^M. We will work with this equivalent model for the rest of the proof.

We define a **traversal** in $\tilde{\mathbb{H}}$ to be a simple path in $\tilde{\mathbb{H}}$ that starts and ends at type-1 vertices, while every vertex in its interior is a type-2 vertex. Observe that every traversal uses only S-edges, and that every simple path in $\tilde{\mathbb{H}}$ that starts and ends at type-1 vertices can be written uniquely as a concatenation of traversals and $\mathbb{G}$-edges.

For each two type-1 vertices u, v in $\tilde{\mathbb{H}}$ let $\tau(u,v)$ be the probability that u and v are connected by an open path. Let $\mathscr{A}_i(u,v)$ be the event that u and v are connected by a simple open path containing exactly i traversals, let $\tau_i(u,v)$ be the probability of this event, and let

$$\tilde{\chi}_i(\ell,k) = \sup_{|u|=\ell}\sum_{|v|=k}\tau_i(u,v).$$

We have that $\tau(u,v) \leq \sum_{i\geq 0}\tau_i(u,v)$ and hence that

$$\tilde{\chi}(\ell,k) \leq \sum_{i\geq 0}\tilde{\chi}_i(\ell,k).$$

Furthermore, $\tau_0(u,v)$ is positive if and only if u and v are in the same type-1 copy of $\mathbb{G}$, and in this case it is equal to the probability that they are connected by an open path in this copy.

Let u, v be vertices of $\tilde{\mathbb{H}}$ with $|u| = \ell$, $|v| = k$, and let $i \geq 1$. For each type-1 vertex w, let Tr_w be the set of traversals starting at w. Given a traversal $t \in \mathrm{Tr}_w$, we write t^+ for the type-1 vertex at the other end of t.

Summing over possible choices of the ith traversal along a simple open path from u to v and applying the BK inequality, we obtain that

$$\begin{aligned} \tau_i(u,v) &\leq \sum_{j\geq 0} \sum_{w\in V_1(\mathbb{H}),|w|=j} \sum_{t\in \mathrm{Tr}_w} \mathbb{P}(\mathscr{A}_{i-1}(u,w)\circ \{t \text{ open}\}\circ \mathscr{A}_0(t^+,v)) \\ &\leq \sum_{j\geq 0} \sum_{w\in V_1(\mathbb{H}),|w|=j} \sum_{t\in \mathrm{Tr}_w} \tau_{i-1}(u,w)\tau_0(t^+,v)\mathbb{P}(t \text{ open}). \end{aligned} \tag{4.1}$$

In q^M-percolation on the 4-regular tree, the expected number of vertices that have distance at least k from the root and are connected to the root by an open path is equal to

$$\sum_{\ell\geq k} 3\cdot 4^{k-1}(q^M)^k \leq \frac{(4q^M)^k}{1-4q^M}.$$

Furthermore, by the isolation property of the hierarchical partition, for each type-1 vertex w of $\mathbb{H}$, every traversal in Tr_w has length at least $2\log^+|w|/\log 2$. Thus, we deduce that

$$\sum_{t\in\mathrm{Tr}_w} \mathbb{P}(t \text{ open}) \leq \frac{(4q^M)^{2\log^+|w|/\log 2}}{1-4q^M}. \tag{4.2}$$

Thus, substituting (4.2) into (4.1) and summing over v, we obtain that

$$\begin{aligned} \sum_{|v|=k} \tau_i(u,v) &\leq 4\sum_{j\geq 0} \tilde{\chi}_{i-1}(\ell,j)\chi(j,k)\frac{(4q^M)^{2\log^+ j/\log 2}}{1-4q^M} \\ &\leq \frac{C'}{1-4q^M}\sum_{j\geq 0}\tilde{\chi}_{i-1}(\ell,j)\chi(j,k)(j\vee e)^{-\alpha(M)} \end{aligned}$$

and hence that

$$\tilde{\chi}_i(\ell,k) \leq \frac{C'}{1-4q^M}\sum_{j\geq 0}(j\vee e)^{-\alpha(M)}\tilde{\chi}_{i-1}(\ell,j)\chi(j,k), \tag{4.3}$$

where

$$\alpha(M) = \frac{2M\log(1/q)}{\log 2} - 4$$

and C' is a constant.

Take M sufficiently large that

$$\frac{2C\cdot C'}{1-4q^M}\sum_{j\geq 0}(j\vee e)^{-\alpha(M)}(j+1)^2 \leq 1/2,$$

where C' is the constant above, C is the constant from lemma 4.4. We now prove by induction on i that, for this choice of M,

$$\tilde{\chi}_i(\ell,k) \leq C2^{-i}(k+1)^2 \tag{4.4}$$

for every $i \geq 0$ and $\ell, k \geq 0$. The case $i = 0$ follows from Lemma 4.4. If $i \geq 1$, then (4.3) and the induction hypothesis yield that

$$\tilde{\chi}_i(\ell,k) \leq \frac{2^{-i+1}C^2\cdot C'}{1-4q^M}\sum_{j\geq 0}(k+1)^2(j\vee e)^{-\alpha(M)}(j+1)^2,$$

and our choice of M yields that

$$\tilde{\chi}_i(\ell, k) \le C2^{-i}(k+1)^2$$

as claimed. This completes the proof of (4.4).

We conclude the proof by summing over i and k to deduce that $\tilde{\chi}(\ell, k) < \infty$ for every $\ell, k \ge 0$ as claimed. □

Acknowledgments. This was was carried out while TH was a PhD student at the University of British Columbia, during which time he was supported by a Microsoft Research PhD Fellowship.

References

[1] David Aldous and Russell Lyons, *Processes on unimodular random networks*, Electron. J. Probab. **12** (2007), no. 54, 1454–1508, DOI 10.1214/EJP.v12-463. MR2354165

[2] Omer Angel, Tom Hutchcroft, Asaf Nachmias, and Gourab Ray, *Unimodular hyperbolic triangulations: circle packing and random walk*, Invent. Math. **206** (2016), no. 1, 229–268, DOI 10.1007/s00222-016-0653-9. MR3556528

[3] Peter Antal and Agoston Pisztora, *On the chemical distance for supercritical Bernoulli percolation*, Ann. Probab. **24** (1996), no. 2, 1036–1048, DOI 10.1214/aop/1039639377. MR1404543

[4] Itai Benjamini and Gady Kozma, *Uniqueness of percolation on products with* $\mathbb{Z}$, ALEA Lat. Am. J. Probab. Math. Stat. **10** (2013), no. 1, 15–25. MR3083917

[5] Itai Benjamini, Russell Lyons, Yuval Peres, and Oded Schramm, *Critical percolation on any nonamenable group has no infinite clusters*, Ann. Probab. **27** (1999), no. 3, 1347–1356. MR1733151 (2000k:60197)

[6] Itai Benjamini, Russell Lyons, and Oded Schramm, *Percolation perturbations in potential theory and random walks*, Random walks and discrete potential theory (Cortona, 1997), Sympos. Math., XXXIX, Cambridge Univ. Press, Cambridge, 1999, pp. 56–84. MR1802426

[7] Itai Benjamini and Oded Schramm, *Percolation beyond* $\mathbb{Z}^d$*, many questions and a few answers [MR1423907]*, Selected works of Oded Schramm. Volume 1, 2, Sel. Works Probab. Stat., Springer, New York, 2011, pp. 679–690, DOI 10.1007/978-1-4419-9675-6_21. MR2883387

[8] Dorottya Beringer, Gábor Pete, and Ádám Timár, *On percolation critical probabilities and unimodular random graphs*, Electron. J. Probab. **22** (2017), Paper No. 106, 26, DOI 10.1214/17-EJP124. MR3742403

[9] R. M. Burton and M. Keane, *Density and uniqueness in percolation*, Comm. Math. Phys. **121** (1989), no. 3, 501–505. MR990777

[10] Reinhard Diestel and Imre Leader, *A conjecture concerning a limit of non-Cayley graphs*, J. Algebraic Combin. **14** (2001), no. 1, 17–25. MR1856226

[11] A. Gandolfi, M. S. Keane, and C. M. Newman, *Uniqueness of the infinite component in a random graph with applications to percolation and spin glasses*, Probab. Theory Related Fields **92** (1992), no. 4, 511–527, DOI 10.1007/BF01274266. MR1169017

[12] Geoffrey Grimmett, *Percolation*, 2nd ed., Grundlehren der Mathematischen Wissenschaften [Fundamental Principles of Mathematical Sciences], vol. 321, Springer-Verlag, Berlin, 1999. MR1707339

[13] Olle Häggström and Yuval Peres, *Monotonicity of uniqueness for percolation on Cayley graphs: all infinite clusters are born simultaneously*, Probab. Theory Related Fields **113** (1999), no. 2, 273–285, DOI 10.1007/s004400050208. MR1676835

[14] Olle Häggström, Yuval Peres, and Roberto H Schonmann, *Percolation on transitive graphs as a coalescent process: Relentless merging followed by simultaneous uniqueness*, Perplexing Problems in Probability: Festschrift in Honor of Harry Kesten, 69–90.

[15] Tom Hutchcroft, *Critical percolation on any quasi-transitive graph of exponential growth has no infinite clusters*, C. R. Math. Acad. Sci. Paris **354** (2016), no. 9, 944–947, DOI 10.1016/j.crma.2016.07.013. MR3535351

[16] Ali Khezeli, *Shift-coupling of random rooted graphs and networks* (2017).

[17] Russell Lyons and Yuval Peres, *Probability on trees and networks*, Cambridge Series in Statistical and Probabilistic Mathematics, vol. 42, Cambridge University Press, New York, 2016. MR3616205

[18] Russell Lyons and Oded Schramm, *Indistinguishability of percolation clusters*, Ann. Probab. **27** (1999), no. 4, 1809–1836. MR1742889 (2000m:60114)

[19] C. M. Newman and L. S. Schulman, *Infinite clusters in percolation models*, J. Statist. Phys. **26** (1981), no. 3, 613–628, DOI 10.1007/BF01011437. MR648202

[20] Roberto H. Schonmann, *Stability of infinite clusters in supercritical percolation*, Probab. Theory Related Fields **113** (1999), no. 2, 287–300, DOI 10.1007/s004400050209. MR1676831

[21] Ádám Timár, *Percolation on nonunimodular transitive graphs*, Ann. Probab. **34** (2006), no. 6, 2344–2364, DOI 10.1214/009117906000000494. MR2294985

[22] J. van den Berg and H. Kesten, *Inequalities with applications to percolation and reliability*, J. Appl. Probab. **22** (1985), no. 3, 556–569. MR799280

Department of Mathematics, University of British Columbia, Vancouver, British Columbia V6T 1Z2, Canada
Email address: angel@math.ubc.ca

Statslab, DPMMS, University of Cambridge, Cambridge CB3 0WB, United Kingdom
Email address: t.hutchcroft@maths.cam.ac.uk

Contemporary Mathematics
Volume **719**, 2018
https://doi.org/10.1090/conm/719/14466

Invariant ρ-percolation on regular trees

Itai Benjamini and Ori Gurel-Gurevich

ABSTRACT. Given an invariant bond percolation on the d-regular tree, where the probability of an edge to be open equals p, is it always possible to find, with positive probability, an infinite self-avoiding path along which the density of open edges is bigger than p? We give positive answer when $d \geq 4$ and explore related questions.

Let S be a bond percolation on the d-regular tree, which is invariant under all automorphism of the tree. Let p be the probability of an edge being open. We think of S as an invariant process with values in $\{0, 1\}$ (1 corresponds to open edges). For $\overline{x} = (x_0, x_1, x_2, \dots)$ an infinite self-avoiding path, let $D(\overline{x})$ be the density of S along $\overline{x}$, that is,

$$D(\overline{x}) = \limsup_{n\to\infty} \frac{1}{n} \sum_{k=1}^{n} S(x_{k-1}, x_k)$$

and let

$$D(S) = \sup_{\overline{x}} D(\overline{x}) .$$

This kind of concept was introduced by Menshikov and Zuev in [**6**] who called it ρ-percolation. The name ρ-percolation refers to the event that there exists a self-avoiding path with density at least ρ. Menshikov and Zuev explored properties of the critical probability for this event for Bernoulli percolation on regular trees and on $\mathbb{Z}^d$. See [**4**] and [**1**] for related results. In this paper, we are interested in the behaviour of $D(S)$ under all invariant percolation distributions, see [**2**] or [**5**] for background on invariant percolation.

$D(S)$ is a random variable which is measurable with respect to the tail σ-algebra. We may look at the essential supremum of this random variable and define

$$D_d(p) = \inf_{S} \operatorname{ess\,sup} D(S) ,$$

where the infimum is taken over all invariant percolation distributions with marginal p on the d-regular tree. Obviously, $D_d(p)$ is monotone in p and $D_d(p) \geq p$.

QUESTION 1. *Is $D_d(p) > p$ for all $d \geq 3$ and $0 < p < 1$?*

More generally we may ask

QUESTION 2. *What is $D_d(p)$?*

In his seminal paper [**3**], Olle Häggström proved that any invariant bond percolation on the d-regular tree, with marginal at least $\frac{2}{d}$, has an infinite cluster. In particular, we get that $D_d(\frac{2}{d}) = 1$ and specifically $D_3(\frac{2}{3}) = 1$. What about other values of D_3?

THEOREM 3. $D_3\left(1 - \frac{1}{\sqrt{3}}\right) \geq \frac{1}{2}$.

PROOF. Let S_1 be an invariant percolation with marginal probability p on the 3-regular tree and let S_2 be an i.i.d. percolation. Define S' to be the maximum of S_1 an S_2. If $p \geq 1 - \frac{1}{\sqrt{3}}$ then S' has marginal which is at least $\frac{2}{3}$, so by Häggström's Theorem there is almost surely an infinite cluster in S' and in particular there is $\overline{x}$ with all the edges open in S'. This means that in either S_1 or S_2 we have $D(\overline{x}) \geq \frac{1}{2}$, so $D(S_1) \geq \frac{1}{2}$ □

More generally, define

$$a(d, k) = 1 - \sqrt[k]{1 - \frac{2}{d}}.$$

THEOREM 4. *For any $d \geq 3$ and $k \geq 1$ we have $D_d(a(d, k)) \geq \frac{1}{k}$*

PROOF. The proof is the same as that of the previous theorem, except that you take k i.i.d. copies and work on the d-regular tree. □

Notice that for $d = 3$ and $2 \leq k \leq 5$ we have $a(d, k) < \frac{1}{k}$, so we get that $D_3(p) > p$ for any $p \in [a(3, k), \frac{1}{k})$, but if $k \geq 6$ then $a(3, k) > \frac{1}{k}$, so we obtain no new information.

However, for $d \geq 4$ we have $a(d, k) < \frac{1}{k}$ for all k, so we get that $D_d(p) > p$ for any $p \in \cup_{k=1}^{\infty}[a(d, k), \frac{1}{k})$.

In fact,

THEOREM 5. *For any $d \geq 4$ and any $0 < p < 1$ we have $D_d(p) > p$.*

PROOF. As mentioned before, $D_d(p) > p$ for any $p \in \cup_{k=1}^{\infty}[a(d, k), \frac{1}{k})$. We claim that for any $d \geq 4$ and any $k \geq 1$ we have $a(d, k) \leq \frac{1}{k+1}$ which implies that $\cup_{k=1}^{\infty}[a(d, k), \frac{1}{k}) = (0, 1)$, so that $D_d(p) > p$ for all $0 < p < 1$.

Indeed,

$$1 - \sqrt[k]{1 - \frac{2}{d}} \leq \frac{1}{k + 1}$$

is equivalent to

$$\left(1 - \frac{1}{k + 1}\right)^k \leq 1 - \frac{2}{d}$$

and the left hand side is decreasing (as a function of k) so the maximum is obtained for $k = 1$ and it is $\frac{1}{2} \leq 1 - \frac{2}{d}$. □

THEOREM 6. *For any d, the function D_d is continuous on $[0, 1]$.*

PROOF. Fix d. Let B be bernoulli percolation on the d-regular tree with marginal ε. For a path of length n the probability of getting at least an 1s is bounded above by

$$\binom{n}{\lceil an \rceil} \varepsilon^{\lceil an \rceil} \leq (2\varepsilon^a)^n.$$

There are $d(d-1)^{n-1}$ self avoiding paths of length n that begins at any fixed vertex x_0. Hence, for

$$a > f_d(\varepsilon) := \frac{\log(2(d-1))}{\log(1/\varepsilon)}$$

the probability of that one of these paths have at least an 1's decays exponentially in n. Using the first Borel-Cantelli lemma, we conclude that there are almost surely no infinite paths with upper density more than a. In other words,

$$D(B) \leq f_d(\varepsilon).$$

We now claim that if $0 \leq p < q \leq 1$ and $q-p \leq \varepsilon$ then $D_d(q) - D_d(p) \leq f_d(3\varepsilon)$. This implies continuity since $f_d(\varepsilon) \to 0$ when $\varepsilon \to 0$.

To show the claim, let S be an invariant percolation with marginal p and B bernoulli percolation with marginal 3ε. Let S' be their maximum. Then S' has marginal $p + (1-p)3\varepsilon \geq q$ (we may assume that $p \leq 2/3$, since for $p > 2/3$ we have $D_d(p) = D_d(q) = 1$). Therefore, for any $\delta > 0$ there is positive probability that there is an infinite path $\overline{x}$ such that the density of S' along $\overline{x}$ is at least $D_d(q) - \delta$. But the contribution of B to the density of S' along $\overline{x}$ is at most $f_d(3\varepsilon)$, so the density of S along $\overline{x}$ is at least $D_d(q) - \delta - f_d(3\varepsilon)$. Taking $\delta \to 0$ we get that $D(S) \geq D_d(q) - f_d(3\varepsilon)$ for any invariant percolation with marginal p, hence

$$D_d(q) - D_d(p) \leq f_d(3\varepsilon).$$

□

In particular, $D_d(p) \to 1$ as $p \to \frac{2}{3}$ so for some $a < \frac{2}{3}$ we have $D_3(p) > p$ for all $p \in [a, 1)$. However, we still don't know that $D_3(p) > p$ for all $0 < p < 1$ and specifically that $D_3(1/2) > (1/2)$.

Consider now what happens when $d \to \infty$. By Theorem 4 we know that $D_d(a(d,k)) \geq \frac{1}{k}$. For any fixed $k \geq 1$ we have

$$a(d,k) = \frac{2}{dk}(1 + o(1)),$$

where the o notation refers to the asymptotic behaviour when $d \to \infty$.

Hence, it make sense to define the limit

$$D_\infty(x) = \lim_{d\to\infty} D_d\left(\frac{2}{d}x\right)$$

and conclude that $D_\infty(x) \geq \frac{1}{k}$ for any $x > \frac{1}{k}$.

QUESTION 7. *Is it true that $D_\infty(x) = x$? Is D_∞ continuous?*

Consider now site percolation, instead of bond percolation. Let S be an invariant site percolation on the d-regular tree. Define S' to be a bond percolation, where you put an open edge in S' between any two adjacent open sites in S. The result is invariant and the marginal probability for an open edge in S' is at least $2p-1$, where p is the marginal probability of an open site in S. We can apply our theorems to S', but there is one big difference. As p goes to $\frac{1}{2}$, the marginal of S' goes to 0. This means that the analog of Theorem 5 fails for site percolation. Indeed, the tree is a bipartite graph and the partition into two sides is invariant under automorphisms of the tree, hence we can define a percolation that choose one of the sides with equal probabilities and then puts 1s on this side and 0s on the other. This gives a marginal of $\frac{1}{2}$ and also density of $\frac{1}{2}$ along any self-avoiding path. This percolation is ergodic, but has a nontrivial tail σ-algebra.

QUESTION 8. *What can be said about site percolation on regular trees if we require that the tail σ-algebra is trivial?*

We may also consider more general processes, i.e. not $\{0,1\}$-valued.

QUESTION 9. *Is it true that for any invariant, non-constant process S on the edges of a d-regular tree ($d \geq 3$), we have $\mathbb{E}[D(S)] > \mathbb{E}[S(e)]$, where e is some/any edge of the tree?*

An interesting side question is this:

QUESTION 10. *Is it true that when you replace the* $\limsup$ *in the definition of* $D(\overline{x})$ *by* $\liminf$ *you get the same function? If not, do our results still hold for the* $\liminf$ *version?*

Remark: Häggström's Theorem was extended to nonamenable Cayley graphs in [**2**]. The discussion above adapts to this set up.

Acknowledgements. Thanks to Tom Hutchcroft and an anonymous referee for useful comments.

References

[1] I. Benjamini and H. Kesten, *Percolation of arbitrary words in* $\{0,1\}^{\mathbf{N}}$, Ann. Probab. **23** (1995), no. 3, 1024–1060. MR1349161

[2] I. Benjamini, R. Lyons, Y. Peres, and O. Schramm, *Group-invariant percolation on graphs*, Geom. Funct. Anal. **9** (1999), no. 1, 29–66, DOI 10.1007/s000390050080. MR1675890

[3] O. Häggström, *Infinite clusters in dependent automorphism invariant percolation on trees*, Ann. Probab. **25** (1997), no. 3, 1423–1436, DOI 10.1214/aop/1024404518. MR1457624

[4] H. Kesten and Z.-G. Su, *Asymptotic behavior of the critical probability for ρ-percolation in high dimensions*, Probab. Theory Related Fields **117** (2000), no. 3, 419–447, DOI 10.1007/s004400050013. MR1774071

[5] R. Lyons and Y. Peres, *Probability on trees and networks*, Cambridge Series in Statistical and Probabilistic Mathematics, vol. 42, Cambridge University Press, New York, 2016. MR3616205

[6] M. V. Menshikov and S. A. Zuev, *Models of ρ-percolation*, Probabilistic methods in discrete mathematics (Petrozavodsk, 1992), Progr. Pure Appl. Discrete Math., vol. 1, VSP, Utrecht, 1993, pp. 337–347. MR1383148

DEPARTMENT OF MATHEMATICS, WEIZMANN INSTITUTE, 76100 REHOVOT, ISRAEL
Email address: itai.benjamini@weizmann.ac.il

EINSTEIN INSTITUTE OF MATHEMATICS, HEBREW UNIVERSITY OF JERUSALEM, 91904 JERUSALEM, ISRAEL
Email address: origurel@math.huji.ac.il

Contemporary Mathematics
Volume **719**, 2018
https://doi.org/10.1090/conm/719/14467

Sparse graph limits along balls

Itai Benjamini and Gabor Elek

ABSTRACT. Several problems and comments regarding sparse graph limits along balls are presented.

1. Hyperfiniteness

Let $\{G_n\}_{n=1}^\infty$ be a sequence of finite graphs. We say that the sequence $\{G_n\}_{n=1}^\infty$ is **hyperfinite** if for any $\varepsilon > 0$ there exists $K > 0$ such that for any $n \geq 1$ we can remove $\varepsilon|V(G_n)|$ vertices (with all the adjacent edges) from G_n such that the resulting graph has components of size at most K. A stronger condition is weighted hyperfiniteness. A sequence $\{G_n\}_{n=1}^\infty$ is **weighted hyperfinite** if for any $\varepsilon > 0$ there exists $K > 0$ such that for any $n \geq 1$ and any probability measure $w_n : V(G_n) \to \mathbb{R}$ we can remove a set of vertices from $V(G_n)$ with w_n-measure at most ε such that the resulting graph has components of size at most K. If we have a sequence of graphs $\{G_n\}_{n=1}^\infty$ with given probability measures $\kappa_n : V(G_n) \to \mathbb{R}$, we say that $\{G_n\}_{n=1}^\infty$ is **κ-hyperfinite** if for any $\varepsilon > 0$ there exists $K > 0$ such that for any $n \geq 1$ we can remove a set of vertices from $V(G_n)$ with κ_n-measure at most ε such that the resulting graph has components of size at most K. Thus, $\{G_n\}_{n=1}^\infty$ is hyperfinite if it is κ-hyperfinite with respect to the uniform measure and weighted hyperfinite if it is hyperfinite with respect to all probability measure.

DEFINITION 1.1. *Let $\{G_n\}_{n=1}^\infty$ be a sequence of finite graphs as above. We say that a sequence of probability measures $\kappa_n : V(G_n) \to \mathbb{R}$ is* **tempered** *if for any $\epsilon > 0$ there exists $Q_\varepsilon > 1$ such that for any $n \geq 1$ there exists a set $P_n \subset V(G_n)$, $\kappa_n(P_n) > 1 - \varepsilon$ such that if $p \in P_n$ and q is adjacent to p, then*

$$\frac{1}{Q_e} \leq \frac{\kappa_n(p)}{\kappa_n(q)} \leq Q_e\,.$$

QUESTION 1.1. *If G is the Cayley graph of a finitely generated group. Is it true that the sequence of balls $B_n(G, 1)$ is*

(i): *hyperfinite?*
(ii): *weighted hyperfinite?*
(iii): *hyperfinite with respect to some tempered probability measure?*

The answer to (iii) is Yes for arbitrary bounded degree graph.
Note: Sparse graph convergence can be defined for graph sequences with tempered probability measures. The limit object is a nonsingular measurable graph (or a nonsingular random graph). The limit measurable graph is amenable if and only

if the given sequence is hyperfinite with respect to the tempered measure (this is a generalization of Schramm's original theorem for uniform measures [**15**]).

Hyperfinite graph sequences were introduced in [**9**] as a combinatorial analogue of ergodic hyperfinite equivalence relations.

2. Yu's Property A

DEFINITION 2.1. *Let G be an infinite graph of bounded vertex degrees. We say that G has Property A, if there exists a sequence of maps $\{T_n : V(G) \to l^2(V(G))\}_{n=1}^{\infty}$ such that*

- *Each vector $T_n(x)$ has length one.*
- *If $d_G(x,y) \leq n$, then $\|T_n(x) - T_n(y)\| \leq \frac{1}{n}$.*
- *For any $n \geq 1$ we have some constant $W_n > 0$ such that the vector $T_n(x)$ is supported in the ball $B_{W_n}(G,x)$ for any $x \in V(G)$.*

It is not hard to see that if G and H are quasi-isometric then G has Property A if and only if H does. Thus, we can say that a finitely generated group has Property A.

PROPOSITION 2.1.

- *G has Property A if G is amenable.*
- *G has Property A if G is hyperbolic in the sense of Gromov.*
- *G has Property A if G has finite asymptotic dimension.*

PROPOSITION 2.2 (Hiroki Sako [**14**]). *G has Property A if and only if all the finite induced subgraph sequences of G are weighted hyperfinite. Hence, the answer to Question 2 (ii) is yes.*

CONJECTURE 2.1 (Brodzki et. al [**7**]). *G has Property A if and only if all the finite induced subgraph sequences of G are hyperfinite.*

Let G be a finitely generated group and $Cay(G)$ be its Cayley graph with respect to some finite generating system. Let us consider the simple random walk on $Cay(G)$. Then for any $n \geq 1$ we have a probability measure p_n on the ball $B_n(Cay(G), 1)$.

QUESTION 2.1. *Is it true that the measures p_n are tempered and the balls are hyperfinite with respect to p_n?*

All examples we know, including limits a la [**6**] along balls in hyperbolic Cayley graphs products of trees, are Liouville.

Consider the limit along balls of an hyperbolic group. It is a.s. a cone over polynomial growth Unimodular random graph (URG). The simple random walk (SRW) on the limit induced a polynomial tail random walk on the polynomial URG. After n steps with very high probability the SRW in the cone over a polynomial space is contained in a domain which is a product of a ball of polynomial radius in the polynomial base times height $Clogn$ in the cone. This set has polynomial volume, so the entropy of the SRW is sublinear. Use [**3**] to finish.

QUESTION 2.2. *Are there examples in which limits along balls are a.s non-Liouville?*

3. Further questions that arise

(1) Is it obvious that the sequence of balls in a Cayley graph are forming a convergent sequence?

(2) Is it completely trivial that the limit sequence of growing balls in an amenable unimodular random graphs is hyperfinite?

(3) It is easy to construct one single graph for which the conjecture does not hold, that is the sequence of balls are not forming a hyperfinite sequence (that is, the limit is non-amenable). If G is a graph, then H is in the closure of G, if any ball in H is isomorphic to a ball of G. Is it possible to construct a graph G, such that the conjecture does not hold for H if H is in the closure of G? That is, G is a sort of uniform counterexample.

(4) Is the conjecture that the sequence of balls in a Cayley graph is hyperfinite equivalent to the conjecture from **[2]** that there is no infinite vertex transitive graph, in which the sequence of balls is a uniform expander?

Gromov **[12]** constructed random groups that contain a uniform expander sequence, but these expanders are far from balls or any convex subset.

(5) Cayley graph could possibly be replaced by unimodular random graphs in the questions (see **[1, 6]** for unimodular random graphs).

(6) Construct a graph G as follows: add to each level of the binary tree edges to make the induced graph on the level a uniform large girth expander. It is shown in **[5]** that such a graph admits no non-constant bounded harmonic functions. Yet the limit along balls is a regular tree and is not Liouville. We conjecture that this can not happen if G is a unimodular random graph? In particular if G is unimodular, the speed of simple random walk on G is bigger or equal the speed on a limit along balls.

(7) Start with a unimodular random graph and take the limit along balls and repeat it again on the graph limit and so on, must it stabilize after one step?

For the regular tree, it stabilizes on the canopy tree after one step.

(8) Think the following procedure, suggested by Tom Hutchcroft, that recovers a transitive graph from the local limit of its balls (or any subsequential local limit). Suppose we have the limit of the balls in a transitive graph. We can tell which vertices are in the boundary and which are in the bulk by their degree. We can thus also tell how deep into the bulk a vertex is. By taking a sequences of vertices deeper and deeper in the bulk, we recover the original graph as a limit.

The limit provides an injection from vertex transitive graphs to the set of amenable (assuming positive answer to question 1.1) unimodular random graphs.

(9) Given a Cayley graph G, let H be the limit along balls of G. Assume G is quasi isometric to G', with limit along balls denoted H'. Does H quasi isometric to H' ?

(10) Given a Cayley graph G, let H be the limit along balls of G. Assume a Cayley graph G has infinitely many ends. Show that a.s. H has one end.

(11) Assume a Cayley graph G has no non constant harmonic Dirichlet functions? Is there exhaustion of G by finite subgraphs, (maybe balls), the limit along the subgraphs is a.s. transient for the simple random walk?

Assume a Cayley graph G admits non constant harmonic Dirichlet functions. Is the limit along balls of G, a.s. recurrent for the simple random walk?

(12) The limit along a Følner sequence of any amenable vertex transitive graph is a.s. the same graph.
(13) Let G be the Diestel Leader graph $DL(3,2)$, **[8]**, which is a vertex transitive graph but not unimodular and is not quasi isometric to a Cayley graph **[10]**. The limit along balls of $DL(3,2)$ is a horocyclic product of two Canopy trees. It will be cool if one can argue directly why such a limit can not be realized with a Cayley graph? Assume the limit along balls of G projects to two canopies. Show that G is not unimodular? One can show that the product of two canopies can not be a limit along balls.

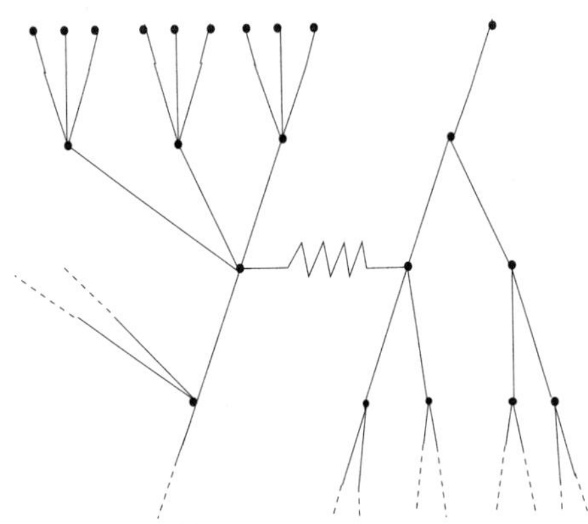

FIGURE 1. The graph limit along balls of $DL(3,2)$. (See **[4]** for first appearance of this graphical representation.)

For the grandfather graph one gets in the limit a graph which is quasi isometric to the canopy tree, see **[11]**. Describe that limit of balls of $T_3 \times T_3$, where T_3 is a 3-regular tree.

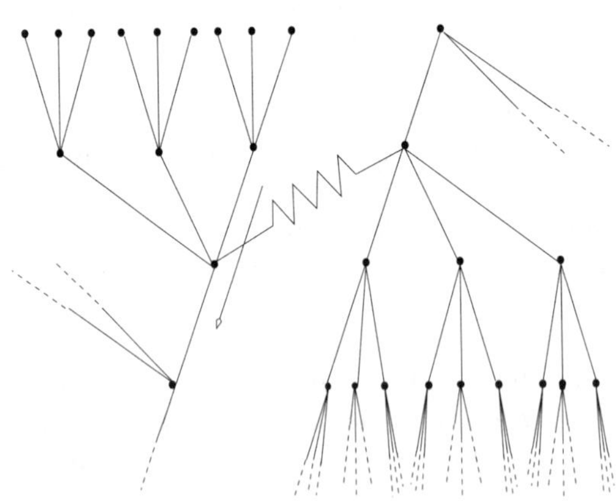

FIGURE 2. The graph limit along balls $T_3 \times T_3$.

4. An infinite model

Consider the set of all 4-regular connected vertex-transitive graphs.

By compactness, for every integer $R \geq R_0 \geq 0$, there is an integer N_R, so that for every 4-regular vertex-transitive graph of radius larger than N_R, there is an infinite vertex-transitive graph with identical R-balls.

QUESTION 4.1. *Do we know any bounds on N_R?*

Here is a comment by Mikael de la Salle: Note that, if one works with Cayley graphs with edges oriented and labeled by the corresponding generators $S = a, b$, then it is known that the analogous number N'_R grows at least exponentially fast with respect to R. (One can replace 4 by some fixed integer, or equivalently S by some larger fixed finite set.)

The basic observation is the following: if G is a group with finite generating set S, then its Cayley graph has the same labeled R-ball as some infinite Cayley graph if and only if the group $\{S, A\}$ if infinite, where A are the group words of length less than $2R$ that are trivial in G.

Indeed, $\{S, A\}$ is the largest group with the same R-balls as G, because every relation in G that one can see in a ball of radius R has length less than $2R$.

This observation leads to the following equivalent definition of N'_R . Consider all presentations $\{S, A\}$ for A a set of group words of length less than $2R$ with respect to S. We get finitely many group presentations, some of infinite groups and some of finite groups. Then N'_R is the maximal radius of those finite groups.

It is known that there are finite groups with relations of length less than R and radius bigger than e^{cR}. For example the finite nonabelian simple groups, see **[13]**. One concludes that $N'_R \geq e^{cR}$ for labeled Cayley graphs.

For unlabeled Cayley graphs, we wonder whether the finite simple groups can be used to show that N_R also grows at least exponentially. Our gut feeling is that N_R grows much faster.

Acknowledgements: Thanks to Lewis Bowen for useful discussions.

References

[1] D. Aldous and R. Lyons, *Processes on unimodular random networks*, Electron. J. Probab. **12** (2007), no. 54, 1454–1508, DOI 10.1214/EJP.v12-463. MR2354165

[2] I. Benjamini, *Expanders are not hyperbolic*, Israel J. Math. **108** (1998), 33–36, DOI 10.1007/BF02783040. MR1669333

[3] I. Benjamini and N. Curien, *Ergodic theory on stationary random graphs*, Electron. J. Probab. **17** (2012), no. 93, 20, DOI 10.1214/EJP.v17-2401. MR2994841

[4] D. Bertacchi, *Random walks on Diestel-Leader graphs*, Abh. Math. Sem. Univ. Hamburg **71** (2001), 205–224, DOI 10.1007/BF02941472. MR1873044

[5] I. Benjamini and G. Kozma, Nonamenable Liouville graphs. arXiv:1010.3365

[6] I. Benjamini and O. Schramm, *Recurrence of distributional limits of finite planar graphs*, Electron. J. Probab. **6** (2001), no. 23, 13, DOI 10.1214/EJP.v6-96. MR1873300

[7] J. Brodzki, G. A. Niblo, J. Špakula, R. Willett, and N. Wright, *Uniform local amenability*, J. Noncommut. Geom. **7** (2013), no. 2, 583–603, DOI 10.4171/JNCG/128. MR3054308

[8] R. Diestel and I. Leader, *A conjecture concerning a limit of non-Cayley graphs*, J. Algebraic Combin. **14** (2001), no. 1, 17–25, DOI 10.1023/A:1011257718029. MR1856226

[9] G. Elek, *The combinatorial cost*, Enseign. Math. (2) **53** (2007), no. 3-4, 225–235. MR2455943

[10] A. Eskin, D. Fisher, and K. Whyte, *Coarse differentiation of quasi-isometries I: Spaces not quasi-isometric to Cayley graphs*, Ann. of Math. (2) **176** (2012), no. 1, 221–260, DOI 10.4007/annals.2012.176.1.3. MR2925383

[11] J. Frisch and O. Tamuz, *Transitive graphs uniquely determined by their local structure*, Proc. Amer. Math. Soc. **144** (2016), no. 5, 1913–1918, DOI 10.1090/proc/12901. MR3460154

[12] M. Gromov, *Random walk in random groups*, Geom. Funct. Anal. **13** (2003), no. 1, 73–146, DOI 10.1007/s000390300002. MR1978492

[13] R. M. Guralnick, W. M. Kantor, M. Kassabov, and A. Lubotzky, *Presentations of finite simple groups: a computational approach*, J. Eur. Math. Soc. (JEMS) **13** (2011), no. 2, 391–458, DOI 10.4171/JEMS/257. MR2746771

[14] H. Sako, *Property A and the operator norm localization property for discrete metric spaces*, J. Reine Angew. Math. **690** (2014), 207–216, DOI 10.1515/crelle-2012-0065. MR3200343

[15] O. Schramm, *Hyperfinite graph limits*, Electron. Res. Announc. Math. Sci. **15** (2008), 17–23. MR2372897

Department of Mathematics, Weizmann Institute, 76100 Rehovot, Israel
Email address: itai.benjamini@weizmann.ac.il

Department of Mathematics, Lancaster University, Lancaster LA1 4YF, United Kingdom
Email address: elekgab@gmail.com

Contemporary Mathematics
Volume **719**, 2018
https://doi.org/10.1090/conm/719/14468

Percolation and coarse conformal uniformization

Itai Benjamini

Abstract. We formulate conjectures regarding percolation on planar triangulations suggested by assuming (quasi) invariance under coarse conformal uniformization.

1. Introduction

We start with a quick panoramic overview. A *conformal map*, between planar domains, is a function that infinitesimally preserves angles. The derivative of a conformal map is everywhere a scalar times a rotation.

Riemann's mapping theorem states that any open simply connected domain of the Euclidean plane admits a bijective conformal map to the open unit disk. In the 1940's Shizuo Kakutani observed that two dimensional Brownian motion is conformally invariant, up to a time reparametrization. Therefore the scaling limit of simple random walks on the Euclidean grid is conformal invariant.

In 2000 Stas Smirnov [**5**] proved that the scaling limit of critical Bernoulli site percolation on the triangular lattice is conformally invariant.

Poincaré (1907) proved that every simply connected Riemann surface is conformally equivalent to one of the following three surfaces: the open unit disk, the Euclidean plane, or the Riemann sphere. In particular it admits a Riemannian metric of constant curvature. This classifies Riemannian surfaces as elliptic (the shpere), parabolic (Euclidean), or hyperbolic (negatively curved).

The uniformization theorem is a generalization of the Riemann mapping theorem from proper simply connected open subsets of the plane to arbitrary simply connected Riemannian surfaces.

Conformal invariance of Brownian motion extends to the context of the uniformization. A simply connected Riemann surface is conformally equivalent to the hyperbolic plane iff the Brownian motion is transient.

How does surface uniformization manifest itself in the context of percolation?

Below we suggest that the discrete setup of planar triangulations is natural for this problem.

Recall that every planar graph admits a circle packing, Koebe (1936).

In 1995 He and Schramm [**4**] proved a discrete uniformization theorem for triangulations: Let G be the 1-skeleton of a triangulation of an open disk. If the random walk on G is *recurrent*, then G is circled packed in the Euclidean plane.

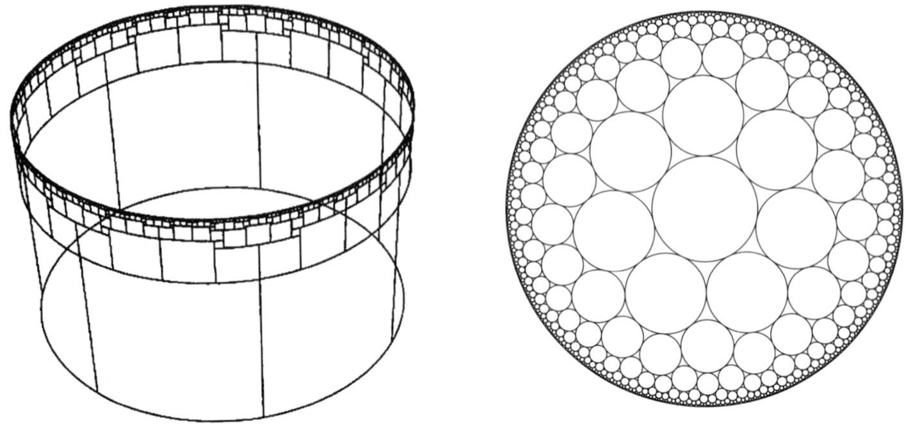

FIGURE 1. The square tiling and the circle packing of the 7-regular triangulation.

Conversely, if the degrees of the vertices in G are bounded and the random walk on G is *transient*, then G can be circle packed in the unit disc.

For an extended version of the Brooks, Smith, Stone and Tutte (1940) square tiling theorem, a related discrete uniformization theorem for graphs using squares, see [**1**].

In p-Bernoulli site percolation, each vertex is declared open independently with probability p, and clusters are connected components of open vertices. We will restrict to site percolation on the triangular lattice as it is self dual.

2. Two conjectures

Our general setup in the note is of G which is a 1-skeleton of a bounded degree triangulation of an open disk.

CONJECTURE 2.1. *Assume G is transient, then 1/2-Bernoulli site percolation on G admits an infinite cluster a.s.*

We don't know it even for any fixed $p > 1/2$.
What about a converse to conjecture 1.1?

Does recurrence imply no percolation at 1/2?

The motivation for the conjecture is outlined below and is based on conformal invariance of percolation. After more than two decades of thorough research, conformal invariance of critical Bernoulli percolation was established only for the triangular lattice [**5**].

One reason to be slightly skeptical about the conjecture is that for critical percolation on the triangular lattice, the probability the cluster of the origin reaches distance r decays polynomially in r [**6**], while there are transient triangulations of volume growth $r^2 \log^3 r$.

A heuristic

Tile the unit square with (possibly an infinite number) of squares of varying sizes so that at most three squares meet at corners, so the dual graph is a triangulation. Color each square black or white with equal probability independently.

CONJECTURE 2.2. *Show that there is a universal $c > 0$, so that the probability of a black left right crossing is larger than c.*

This conjecture is an extension of the RSW lemma for crossing in percolation, see [**7**] for the state of the art regarding the RSW lemma. If true, the same should hold for a tiling, or a packing of a triangulation, with a set of shapes that are of bounded Hausdorff distance to circles. At the moment we don't have a proof of the conjecture even when the squares are colored black with probability 2/3 or any $p < 1$.

Behind the second conjecture is a coarse version of conformal invariance. That is, the crossing probability is balanced if the tiles are of uniformly bounded distance to circles (rotation invariance), and the squares can be of different sizes, (dilation invariance).

Let G the 1-skeleton of bounded degree transient a triangulation of an open disk. By [**4**] it admits a circle packing with similar properties as the tiling in conjecture 2.2. And if the conformal invariance heuristic holds, we will a.s. see macroscopic crossings for 1/2-Bernoulli percolation.

By same reasoning we should see unboundely many macroscopic clusters for 1/2-Bernoulli percolation, suggesting that if G is a 1-skeleton of bounded degree transient triangulation of an open disk, then there are a.s. infinitely many infinite clusters for 1/2-Bernoulli site percolation?

On the critical probability of planar triangulations

We *believe* that $p_c \leq 1/2$ once the triangulation do not have very small (logarithmic) cut sets.

E.g. if there are $C > 0, \alpha > 0$, so that for every finite set of vertices S,

$$|\partial S| > C|S|^{\alpha}.$$

Then $p_c \leq 1/2$.

Maybe anchored version suffices.

We further believe that $p_c \geq 1/2$ for polynomial growth triangulations of the open disk. Note that if all degrees are at least 6, polynomial growth implies that vertices of higher degrees are polynomially sparse, this suggests that their critical probability for percolation is 1/2, as of the triangular lattice. For nonamenable transitive or sofic triangulations $p_c < 1/2$ [**3**], remove the transitivity assumption.

Since we believe that $p_c > 0$ for such G's, by planar duality we conjecture that $p_u < 1$ and *uniqueness monotonicity* holds as well. Where p_u is the threshold for uniqueness of the infinite cluster.

3. Conformal invariance and hyperbolicity

Here is another approach to conformal invariant of critical percolation.

Consider the disc model of the hyperbolic plane. Pick four points a, b, c, d on the circle at infinity, dividing the circle to four intervals, A, B, C, D.

What is the probability that when placing λ-intensity Poisson process in the disc, with respect to the hyperbolic metric, and coloring each Voronoi cell black or white independently with equal probability, there is a black crossing between intervals A and C on the boundary?

Since this process is invariant with respect to hyperbolic isometries, we get that this probability is a function of the crossratio of a, b, c, d and λ. There is no scale invariance for the Poisson process on the hyperbolic plane and increasing λ corresponds to the curvature approaching 0.

Fix the boundary intervals. It is reasonable to conjecture that the (annealed) crossing probabilities converge as λ increases to infinity. In particular they converge along a subsequence. We get that the subsequential limit is conformal invariant. The limit is Euclidean and should be given by Cardy's formula [**5**]. The argument above gives the conformal invariance of a subsequential limit.

The point is that conformal invariance of the subsequential limit follows from the hyperbolicity. Note that in [**3**], it is shown that $p_c(\lambda) < 1/2$ for any λ and suggested that this can be used to show that $p_c \leq 1/2$ in Euclidean Voronoi percolation. In [**2**], together with Oded Schramm we conjectured that changing the uniform measure in the disc (the measure used in sampling the Poisson points) in a uniformly absolutely continuous way, should not effect crossing probability, as the intensity grows and showed that conformal change of the metric do not effect crossing probabilities. Here we observe that when placing an infinite measure and unbounded metric, so that as the intensity grows the local tiling geometry also converges to that of the high intensity Euclidean, conformality follows via hyperbolicity.

As in conjecture 2.2 we want: show that the limiting of the crossing probabilities are bounded away from 0 and 1 for any non trivial intervals A and C?

Acknowledgements: Thanks to Nicolas Curien, Gady Kozma, Vincent Beffara and Vincent Tassion for useful sanity checks.

References

[1] I. Benjamini and O. Schramm, *Random walks and harmonic functions on infinite planar graphs using square tilings*, Ann. Probab. **24** (1996), no. 3, 1219–1238, DOI 10.1214/aop/1065725179. MR1411492

[2] I. Benjamini and O. Schramm, *Conformal invariance of Voronoi percolation*, Comm. Math. Phys. **197** (1998), no. 1, 75–107, DOI 10.1007/s002200050443. MR1646475

[3] I. Benjamini and O. Schramm, *Percolation in the hyperbolic plane*, J. Amer. Math. Soc. **14** (2001), no. 2, 487–507, DOI 10.1090/S0894-0347-00-00362-3. MR1815220

[4] Z.-X. He and O. Schramm, *Hyperbolic and parabolic packings*, Discrete Comput. Geom. **14** (1995), no. 2, 123–149, DOI 10.1007/BF02570699. MR1331923

[5] S. Smirnov, *Critical percolation in the plane: conformal invariance, Cardy's formula, scaling limits* (English, with English and French summaries), C. R. Acad. Sci. Paris Sér. I Math. **333** (2001), no. 3, 239–244, DOI 10.1016/S0764-4442(01)01991-7. MR1851632

[6] S. Smirnov and W. Werner, *Critical exponents for two-dimensional percolation*, Math. Res. Lett. **8** (2001), no. 5-6, 729–744, DOI 10.4310/MRL.2001.v8.n6.a4. MR1879816

[7] V. Tassion, *Crossing probabilities for Voronoi percolation*, Ann. Probab. **44** (2016), no. 5, 3385–3398, DOI 10.1214/15-AOP1052. MR3551200

DEPARTMENT OF MATHEMATICS, WEIZMANN INSTITUTE, 76100 REHOVOT, ISRAEL
Email address: itai.benjamini@weizmann.ac.il

Contemporary Mathematics
Volume **719**, 2018
https://doi.org/10.1090/conm/719/14469

Invariant tilings and unimodular decorations of Cayley graphs

Ádám Timár

ABSTRACT. Using the result that amenable graphs can be represented in the 3 dimensional Euclidean space as an isometry-invariant random tiling by bounded tiles, we prove that the 3-regular infinite tree can also be represented this way, if we allow unbounded tiles, and we can require them to be indistinguishable. Comparing this result to (deterministic) regular and periodic tilings, indistinguishability can be thought of as a stochastic version of the congruence of tiles, and isometry-invariance as a distributional analogue of periodicity. In proving indistinguishability of the constructed tiles, a key ingredient is a duality result on jointly unimodular decorations of Cayley diagrams by other Cayley diagrams, and the fact that factor of iid modifications preserve the indistinguishability of a family of random subgraphs.

Most results in this paper have been proved in [arXiv:1712.08210], sometimes in a stronger form. Some alternative proofs and corresponding generalizations are provided here.

1. Introduction

Suppose that G is an infinite transitive graph. When does it have a "nice" representation by domains in a Euclidean space? We will provide a precise definition for what we mean by a representation by domains (or domain representation, tiling) of G, but for the moment let us stay on the ground of intuition. Consider the standard representation of planar graphs by countries in 2 dimension, as in the 4-color theorem. In this representation, one considers some nice pathwise connected domains (e.g. polygons) of the plane, and says that two of them are adjacent if they share some 1-dimensional part of their boundary. For a higher dimensional version of this representation, we have a collection of pairwise disjoint "nice" subsets (e.g., connected polyhedra) in $\mathbb{R}^d$ such that the adjacency graph on these subsets is G. Here two such parts are adjacent if the intersection of their closures is $d-1$ dimensional. We further limit the set of such representations of G that interest us by requiring that the pairwise disjoint nice subsets (domains) are such that the union of their closures is equal to $\mathbb{R}^d$. That is, the countries cover the entire world. An example is the representation of $\mathbb{Z}^d$ in $\mathbb{R}^d$ by the tiling of the space by cubes. A nice, periodic representation that we have for $\mathbb{Z}^d$ in $\mathbb{R}^d$ does not exist for

This research was supported by a Marie Curie Intra European Fellowship within the 7th European Community Framework Programme and by grant LP 2016-5 of the Hungarian Academy of Sciences.

most other graphs G. The probabilistic relaxation of periodicity is to find random, isometry-invariant representations.

Extending further the notion of deteministic periodic tilings to the randomized setting, a natural requirement is that we want the domains to be indistinguishable. This can be thought of as the probabilistic version of using *congruent tiles* in periodic tilings. We say that the domains are indistinguishable, if for any property that "does not use the location of the origin" (that is, "internally defined"), either all the domains satisfy this property, or none. Such properties could be: congruence class of some prescribed tile, boundedness, lower/upper density, smoothness of the boundary, various topological properties... See Definition 4 for a more rigourous definition. We will be interested in domain representations in which the domains are indistinguishable. One arrives to the following question: when can a transitive graph G be represented by a *random, isometry-invariant* collection of domains, such that the domains are *indistinguishable*? A first candidate where this may not be possible could be the regular tree, which cannot be represented in $\mathbb{R}^2$ by indistinguishable domains, even though it is planar, [**8**].

QUESTION 1.1. (Itai Benjamini) Is there a domain representation of T^3 in $\mathbb{R}^3$, such that the domains are indistinguishable?

What are the possible obstacles for the existence of such a representation by domains? If G is not planar, then it cannot be represented in $\mathbb{R}^2$ by pathwise connected domains. This implies, for example, that there is no domain-representation of $\mathbb{Z}^3$ in $\mathbb{R}^2$. What can we say about the 3-regular infinite tree, T_3, which is planar? As we have mentioned, one can show that that there is no representation for T_3 in $\mathbb{R}^2$ with indistinguishable domains, see [**8**]. The requirement of indistinguishability is needed, as shown by the sketchy Figure 1.1. (The pattern on the figure has to continue both on the decreasing and on the increasing scale, and proper random translations of the resulting pattern have to be taken.) See [**8**] for a precise definition.

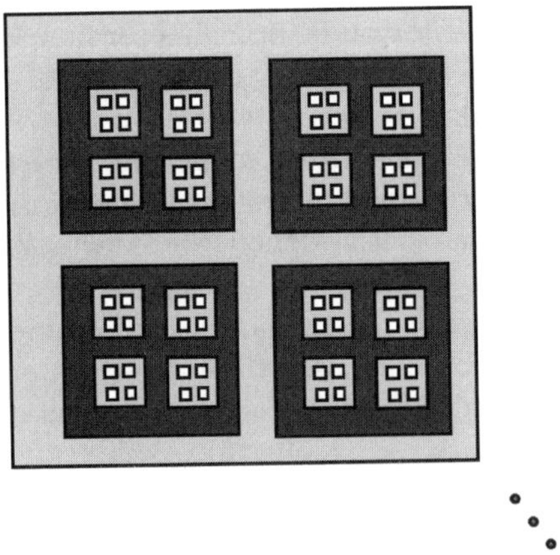

FIGURE 1.1. Domain representation of the 5-regular tree.

In [**8**], the present author proved a stronger form of the following theorem, which provides a positive answer to the question of Benjamini. We give a precise definition for *domain representations* right after the theorem.

THEOREM 1.2. There is an invariant full domain representation of T_3 in $\mathbb{R}^3$ with indistinguishable domains.

DEFINITION 1. *Domain representation; polyhedral representation* Let M be a collection of pairwise disjoint pathwise connected open subsets of $\mathbb{R}^3$. Suppose that any element of M has a boundary that is the countable union of hyperfaces (polyhedral subsets of a hyperplane). Say that two elements of M are adjacent, if they share one such hyperface. If the graph determined on M as vertex set is isomorphic to G, than we say that M is a domain representation for G. If M is such that the closure of the unions of the elements of M is $\mathbb{R}^3$, then we call it a *full* domain representation.

Call M a *polyhedral representation*, if every domain in M is a bounded polyhedron with finitely many faces. Define full polyhedral representations as full domain representations that are polyhedral representations.

In [**8**] a stronger form of Theorem 1.2 is proved, and not only the union of the domains contains $\mathbb{R}^3$ but also the union of the closures.

The same result is true for the r-regular tree ($r \geq 3$) and $\mathbb{R}^d$ ($d \geq 3$). All the arguments below would remain valid with such a setup. We stick to $d = r = 3$ for simplicity.

A major ingredient of that proof is the following:

THEOREM 1.3. Let G be an infinite transitive *amenable* graph with 1 end. Then there is an invariant full polyhedral representation of G in $\mathbb{R}^3$. Suppose further that there is some given invariant random embedding of $V(G)$ into $\mathbb{R}^3$. Then the polyhedral representation can be constructed from this embedded point set as a factor of iid.

See Definition 5 for the definition of a factor of iid in several contexts.

So, amenable graphs can be represented by (bounded) domains. E.g., $\mathbb{Z}^4$ can be invariantly represented by domains in $\mathbb{R}^3$. We will apply Theorem 1.3 to the group BS(1,2), which is known to be amenable. Then we will use the following observation, pointed out by Gaboriau in the context of Benjamini's question:

Observation (Damien Gaboriau) The usual Cayley graph of the Baumslag-Solitar group BS(1,2)$=< a, b | a^{-1}ba = b^2 >$ can be partitioned into connected pieces such that the adjecency graph between the pieces is T_3.

Instead of giving a formal proof, we refer the reader to Figure 1.2.

The proof of Theorem 1.2 is based on the following two steps:

- BS(1,2)$=: G$ has a full polyhedral representation in $\mathbb{R}^3$ by Theorem 1.3.
- By taking the unions of suitably chosen families of domains in this representation, we get a full domain representation of T_3. More precisely, we have to take the interior of the union of the closures in each such family. One can check that the result is a full domain representation of T_3.

But how can we guarantee that the resulting domains will be indistinguishable?

Showing indistinguishability is usually not easy. Even for the infinite components of Bernoulli percolation on Cayley graphs it is highly non-trivial; see [**5**]. In the remainder of the introduction, we will modify the above sketch (add more details), to guarantee that the resulting unions (the domains representing vertices of T_3) are indistinguishable.

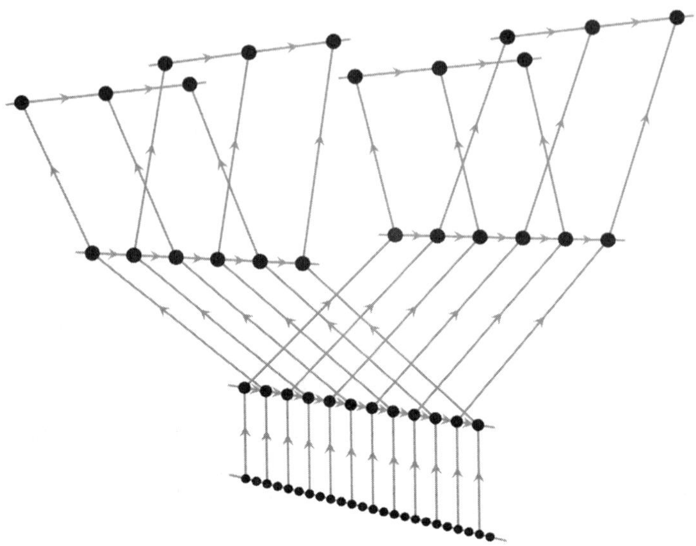

FIGURE 1.2. Part of the Cayley graph of BS(1,2). Red lines ("fibers") chosen as partition classes provide a partition whose adjacency graph is T_3 (where color is available). (Image by Jens Bossaert.)

The main ingredients for the proof of the indistinguishability part in Theorem 1.2 are the following:

(1) A new definition of indistinguishability is used. It is stronger than the usual one, but does not depend on the underlying graph. See Definitions 2 and 3.
(2) "Duality lemma": If we find an Aut(G)-invariant bijection from $V(G)$ to $V(H)$, that defines an Aut(H)-invariant bijection from $V(H)$ to $V(G)$. (Here $H = \mathbb{Z}^3$, and we need to take Cayley *diagrams* of H and G, to be precise.)
(3) "Decoration lemma": If a family of subgraphs is indistinguishable, then so is any factor of iid decoration of them.

The current paper is organized as follows. In the coming subsections of this Introduction, we provide the necessary definitions and additional higher level description for some of the proofs. In particular, we explain the above mentioned lemmas and deduce Theorem 1.2 from Theorem 1.3 in the coming subsections of this introduction. The claim and proof of Theorem 1.3 is in one way slightly more general than in [**8**]. However, we do not go into proving an extra geometric property that one can also obtain for such an embedding, namely, that the domains have no accumulation points. The proof of Theorem 1.3 that we present here is completely different from that of [**8**]. We also added a subsection that shows how the lamplighter group could be used instead of the BS(1,2), which might be of interest because of the technique to get connected families of domains out of unconnected ones. The Duality lemma (Lemma 2.1) is presented in Section 2.

1.1. Definitions and some tools. We will sketch the needed definitions and lemmas. The following definition for indistinguishability from [**8**] is the unimodular random graph version of the definition in [**5**]. (We mention that the results of [**5**] are valid with this stronger definition.) Its main advantage is that no underlying graph structure and corresponding group of automorphisms is needed. This will be important later, when we switch the roles of graph and its decoration in the Duality lemma. See also [**6**] for a related notion and futher references.

DEFINITION 2. (*Indistinguishability for graphs*) Let $\mathcal{G}$ be the set of all finite-degree connected rooted multigraphs up to rooted isomorphism. Define a distance on $\mathcal{G}$ by $d((G,o),(G',o')) := \min\{1/r : B_G(o,r) \text{ and } B_{G'}(o',r) \text{ are rooted isomorphic}\}$. Call a subset $\mathcal{A} \subset \mathcal{G}$ an *invariant property*, if it is Borel measurable and closed under the change of root (that is, if $(G,o) \in \mathcal{A}$ and $x \in V(G)$, then $(G,x) \in \mathcal{A}$).

The collection of random graphs is *indistinguishable*, if almost surely for every invariant property $\mathcal{A}$ either all of them belongs to $\mathcal{A}$, or none.

All the definitions extend to *decorated graphs* in an obvious way. But first we need to define what we mean by decorated graphs. This will be similar to the definition of *marked* graphs in [**1**], but more general. In a marked graph, labels from a space X are assigned to the vertices and/or the edges. In our definition, a further way of decorating is by adding extra edges and vertices to the underlying graph G. So a decoration would consist of a collection of extra edges and vertices, and labels from X on the vertices, edges of G and on the extra vertices, extra edges.

DEFINITION 3. (*Indistinguishability for decorated graphs*) Fix some complete separable metric space X. Let $\mathcal{G}^{\text{deco}}$ be the space of all finite-degree connected rooted multigraphs together with some decoration of (G,o), up to rooted isomorphisms that preserve the decoration. Here a decoration means a set E' of extra edges (possibly oriented) and a set V' of extra vertices on $V(G)$, and some coloring of $V(G) \cup E(G) \cup E'$ with elements of X. A distance on $\mathcal{G}^{\text{deco}}$ can be defined similarly to $\mathcal{G}$: (G,o) and (G,o') are at distance at most $1/r$ if the r-neighborhood of o in G is rooted isomorphic to the r neighborhood of o' in G', if such an isomorphism maps extra edges induced by the neighborhoods into each other bijectively, and finally, if for any pair of vertices or edges that are mapped to each other, their labels differ by at most $1/r$. Now, with $\mathcal{G}^{\text{deco}}$ and the metric on it, we can define indistinguishability of the infinite components of some percolation with decoration just the same way we did for the non-decorated case.

Denote the group of isometries of the Euclidean space $\mathbb{R}^d$ by $\text{Isom}(\mathbb{R}^d)$, and the group of automorphisms of G by $\text{Aut}(G)$.

DEFINITION 4. Let X be the set of all closed subsets of the Euclidean space $\mathbb{R}^3$, and consider the Hausdorff metric on X. Suppose that some $A \subset X$ is Borel measurable, and is closed under $\text{Isom}(\mathbb{R}^d)$ (i.e., if a set is in A then all its isometric copies are also in A). We say that the pieces (domains) of a domain representation M in $\mathbb{R}^3$ are *indistinguishable* if for any such A, either A contains the closure of every domain in M almost surely, or it contains none of them almost surely.

Let $G = (V(G), E(G))$ be an arbitrary transitive graph. Put iid Lebesgue[0,1] random variables on its vertices. Consider some decoration of $V(G)$. We call it a *factor of iid* or *fiid*, if the value of the decoration of every $v \in V(G)$ is determined by the labels in a bounded neighborhood of G up to an arbitrarily small error, and the rule that determines this value is independent of v. That is, the decoration is determined locally, using extra (local) randomness.

DEFINITION 5. Let X be some set $G = (V(G), E(G))$ be an arbitrary transitive graph with iid Lebesgue[0,1] labels $\omega \in [0,1]^{V(G)}$ on its vertices. Say that a function $f : (\omega, o) \mapsto x$ from $[0,1]^{V(G)} \times V(G)$ to X is a *factor of iid* or *fiid*, if it is a measurable function of ω, and for any isomorphism γ of G, $f(\gamma\omega, \gamma x) = f(\omega, x)$.

If π is a random discrete point set in $\mathbb{R}^3$ whose distribution is invariant with respect to the isometries, a factor of iid from π is defined similarly, with the iid labels $\omega \in [0,1]^\pi$ being on the points of π and the equation $f_{\gamma\omega}(\gamma x) = f_\omega(x)$ holding for almost every π, with respect to all isometries γ of $\mathbb{R}^3$.

See e.g. [**4**] for more details.

Suppose we have some random partition of a (decorated) graph G into pieces. When we speak about the indistinguishability of these pieces, we regard each piece as a decorated graphs, where the rest of G and the other pieces of the partition are viewed as the decoration. (We leave it to the reader to check that this information can in fact be encoded as a decoration of C.) This is essentially the same as the notion of indistinguishability used by Lyons and Schramm in [**5**], Remark 3.4. Of course, it makes sense to further decorate the pieces, as in the next lemma.

LEMMA 1.4 (Decoration lemma). *Let G be a unimodular transitive (decorated) graph. Consider some isomorphism-invariant partition of G into infinite pieces, and suppose that the pieces are indistinguishable. Then for any fiid decoration, the decorated pieces are also indistinguishable.*

In particular, no matter how we decorate the fibers (which are trivially indistinguishable subgraphs) of B(1,2), if the decoration is fiid, then the decorated fibers are indistinguishable. See [**8**] for a proof, which is omitted here.

REMARK 1.5. There exist factor of iid processes on transitive graphs whose infinite clusters do not satisfy indistinguishability. Consider some transitive graph G whose critical probability for site percolation is smaller than $1/3$. Let $\lambda(v)$ be a Lebesgue($[0,1]$) label on $v \in V(G)$, independently over all the v. Define a bond percolation: an edge $\{x, y\}$ is open in this percolation if $\lambda(x) < 1/2$ and $\lambda(y) < 1/2$, or if both $\lambda(x)$ and $\lambda(y)$ are between $1/2$ and $5/6$. It is easy to check that some infinite clusters look like those of Bernoulli($1/2$) site percolation, and some look like Bernoulli($1/3$) site percolation. The two are distinguishable (as can be seen, say, by the fact that their critical probabilities differ).

The Duality lemma is stated and proved in Section 2. Informally, it says the following. Let G and H be two Cayley diagrams. If G is decorated by a graph H on its vertices in an Aut(G)-invariant way, then we can think of it as an Aut(H)-invariant decoration of H with G.

The next theorem summarizes some results on invariant random copies of amenable Cayley graphs on other graphs or in Euclidean spaces. Here and in the rest of the paper, "invariant" means a random object whose *distribution* is invariant. Part 1 follows from an extension of the method in [**7**] or (as observed by Hutchcroft) from the fact that all countable amenable infinite groups are orbit equivalent. Part 2 is not needed for the current paper; it is proved in [**7**] for $G = \mathbb{Z}^d$ and can be easily generalized to any amenable group. Part 3 is a reformulation of Lemma 2.1, for completeness. By a Cayley diagram we mean a Cayley graph where the orientations of the edges and their labels by the generators are not ignored. In particular, the group of automorphisms of a Cayley diagram acts transitively and the only automorphism that stabilizes a fixed vertex is the identity.

THEOREM 1.6. (1) Let G and H be infinite amenable Cayley graphs or diagrams. Then one can decorate G by a fiid copy of H.

(2) If G is an infinite amenable Cayley graph or Cayley diagram then it arises as a factor of any point process, provided that the point process has no symmetry but the identity almost surely. I.e., there exists a mapping of the point configuration to a graph on it that is measurable and equivariant with the point process.

(3) If G and H are infinite amenable Cayley diagrams, then one can define a random bijective decoration of G by H in such a way that the result is invariant under the automorphisms of both the underlying graph ($\mathrm{Aut}(G)$) and of the decorating graph ($\mathrm{Aut}(H)$).

1.2. Sketch of proof: indistinguishable domains representing T_3 in $\mathbb{R}^3$. Theorem 1.3 says that we can represent any amenable Cayley graph or Cayley diagram G by domains. It goes through the following two steps:

(1) There is an $\mathrm{Aut}(G)$-invariant (random) bijection from G to $\mathbb{Z}^3$.

(2) Given the image set of vertices and edges of G on $\mathbb{Z}^3$ coming from (1), there is a factor of iid map from $\mathbb{Z}^3$ to a random polyhedral representation of G in $\mathbb{R}^3$, and this map preserves adjacencies of G. (Vertices adjacent by the image of G are mapped into adjacent polyhedra, and vica versa.) This fiid map can be taken as a fiid decoration of $\mathbb{Z}^3$, by elements of the suitable subspace of $X \subset 2^{\mathbb{R}^3}$, as in Definition 4. Applying a random uniform isometry from the compact quotient $\iota \in \mathrm{Isom}(\mathbb{R}^3)/\mathrm{Aut}(\mathbb{Z}^3)$ to this polyhedral representation, we obtain an isometry-invariant polyhedral representation of G in $\mathbb{R}^3$.

We mention that (2) works regardless of whether $\mathbb{Z}^3$ denotes a Cayley graph or a Cayley diagram of the cubic lattice (which can be potentially important in whether a map is a factor of iid or not). For simplicity, from now on $\mathbb{Z}^3$ *denotes the Cayley diagram* of the cubic lattice with respect to the usual generators. What we obtain, composing (1) and (2), is an $\mathrm{Isom}(\mathbb{R}^3)$-invariant polyhedral representation of G in $\mathbb{R}^3$. This representation is constructed as an $\mathrm{Aut}(G)$-invariant map from G to the polyhedra. So, *when we say that a representation is invariant, there is no ambiguity,* because invariance holds with respect to both $\mathrm{Aut}(G)$ and the isometry group of $\mathbb{R}^3$ (the first referring to the invariance of the mapping, the second to the invariance of the polyhedral representation). Because of this observation, we will sometimes use the term invariant without specifying the group action that is meant, but in such cases either possibility ($\mathrm{Isom}(\mathbb{R}^3)$ or the group of automorphisms of a Cayley diagram) works. The question of when the different senses of invariance coincides is investigated more thoroughly in [**8**].

After proving Theorem 1.3, we will turn to the main theorem: Theorem 1.2. Recall that one can obtain Theorem 1.2 from the polyhedral representation of G =BS(1,2) by taking the union of polyhedra (and some faces) for each of the fibers of BS(1,2). However, at this point it is not clear why the resulting pieces should be indistinguishable. This would be fixed if one could guarantee that the map (1) takes the (trivially indistinguishable) fibers of G to indistinguishable subgraphs of $\mathbb{Z}^3$, and also that the map (2) preserves indistinguishability.

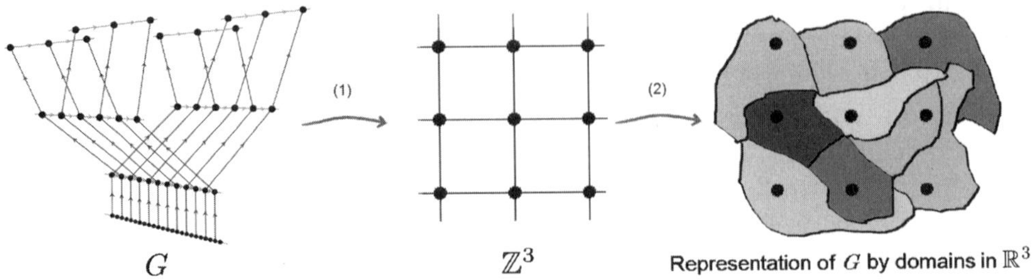

FIGURE 1.3. The two main steps in the proof of Theorem 1.2; we represent in dimension 2 a construction that is 3 dimensional in reality.

If the image sets of the fibers of G partition the vertex set of $\mathbb{Z}^3$ into indistinguishable pieces, then factor of iid decorations of these pieces are also indistinguishable by the "Decoration lemma" (Lemma 1.4). The mapping (2) is fiid from the decorated graph $\mathbb{Z}^3$ (its decoration provided as the image of G by the map (1)). Hence the corresponding pieces are indistinguishable. For (1), we need an $\text{Aut}(G)$-invariant random bijection from G to $\mathbb{Z}^3$. Such a (fiid) map exists between any two amenable Cayley diagrams, by Theorem 1.6. But here we also want the images of fibers to be indistinguishable subsets of $\mathbb{Z}^3$. Because in that case, the final image of the fibers as unions of domains will be indistinguishable (thinking of the domains as decorations), as we have just observed. It remains to show that:

(a) we can decorate $\mathbb{Z}^3$ by G in an $\text{Aut}(\mathbb{Z}^3)$-invariant way;
(b) we can make the fibers defined by the decoration indistinguishable.

(Here a fiber of $\mathbb{Z}^3$ is defined to be a subset of vertices in $\mathbb{Z}^3$ that got labelled by the vertices of one fiber by the decoration.)

Instead, we are able to show:

(a') we can decorate G by $\mathbb{Z}^3$, as a factor of iid;
(b') the decorated fibers of G are indistinguishable.

Part (a') follows from Theorem 1.6 (which could also be used to have (a), but (b) would be not clear), and (b') follows from the fact that the composition of two fiid maps as in (1) and (2) of Figure 1.3 is again fiid, hence the map from the fibers into corresponding sets of domains preserves indistinguishability.

By the Duality lemma (Lemma 2.1), the existence of an invariant decoration as in (a') implies (a). (For this we need to use the group of automorphisms of the Cayley *diagram* of $\mathbb{Z}^3$, that is, a subgroup of finite index in $\text{Aut}(\mathbb{Z}^3)$ such that the stabilizer of every vertex is trivial. This will not impact the validity of the proof, the only thing is that one has to take ι in (2) from a larger group.) The fibers in G are trivially indistinguishable, and the *decorated fibers* in (b') will be indistinguishable by the Decoration lemma, using that (a') provides a decoration which is fiid. Our definition of indistinguishability does not depend on the underlying graph, hence (b) follows from (b').

2. Duality

Suppose that there are two, possibly random graphs, G and H, defined on the same vertex set. We will use three, equivalent ways to jointly represent G and H. First, we will think about the pair (H, G) as one multigraph, with edges of two colors: let the vertex set of (H, G) be $V(H) = V(G)$, the edge set be $E(H) \cup^* E(G)$ (where $\cup^*$ denotes multiset union, in particular, we may have parallel edges), and color edges of $E(H)$ by one color, and edges of $E(G)$ by another color. Secondly, we can think about (H, G) as the graph H decorated by G. More precisely, in this setup every vertex $x \in V(H)$ is adjacent to its neighbors in H, and is labelled by the names of its neighbors in G. (The space of labels X in this case is just $2^{V(G)}$, with the discrete metric.) Thirdly, we can think of (H, G) as the graph G decorated by H. Denote the decoration of H by G as $\mathrm{Dec}(H, G)$ and the decoration of G by H as $\mathrm{Dec}(G, H)$. Although these three viewpoints correspond to the same random set, they are different from the point of view of group actions. On $\mathrm{Dec}(H, G)$ there is a natural action by the automorphisms of H. Similarly, $\mathrm{Aut}(G)$ is acting on $\mathrm{Dec}(G, H)$. Suppose now that there is a probability measure on $\mathrm{Dec}(H, G)$ that is preserved by $\mathrm{Aut}(H)$. We can think of this probability measure as one defined on $\mathrm{Dec}(G, H)$ (which we will call the *dual* of $\mathrm{Dec}(H, G)$). Under what condition is this measure on $\mathrm{Dec}(G, H)$ preserved by $\mathrm{Aut}(G)$? The next lemma says that invariance is preserved whenever G and H are Cayley diagrams. We call this phenomenon duality. In later sections we sketch an application of this result (from [**8**]), which was the original motivation of the question. In a draft under preparation, [**3**], generalizations and other questions of similar flavor are investigated.

LEMMA 2.1. *(Duality lemma)* Let G and H be Cayley diagrams. Let ϕ be a random map $\phi : V(G) \to V(H)$ that is bijective and satisfies $\phi(\mathrm{id}_G) = \mathrm{id}_H$. Suppose that the diagram $\phi^{-1}(H)$ is $\mathrm{Aut}(G)$-invariant. Then $\phi(G)$ is $\mathrm{Aut}(H)$-invariant.

In other words, for the decorations defined by ϕ, the $\mathrm{Aut}(G)$-invariance of $\mathrm{Dec}(G, H)$ implies the $\mathrm{Aut}(H)$-invariance of $\mathrm{Dec}(H, G)$.

By the $\mathrm{Aut}(G)$-invariance of $\phi^{-1}(H)$, the rooted graph $((G, \phi^{-1}(H)), \mathrm{id}_G)$ is unimodular. This random decorated rooted graph is naturally equivalent to $(\mathrm{Dec}(G, H), \mathrm{id}_G)$. Now, $((G, \phi^{-1}(H)), \mathrm{id}_G)$ is also equivalent to $((H, \phi(G)), \mathrm{id}_H)$, by the map ϕ (using the symmetry of colors); and finally, the latter is equivalent to $(\mathrm{Dec}(H, G), \mathrm{id}_H)$.

PROOF. Let's assume that H is a right-Cayley diagram, so its automorphisms are all of the form hH, where $h \in H$. For every vertex a in H there is a unique automorphism of H that takes id_H to a, namely, multiplication from the left by a. Consequently, the rooted graph (H, id_H) does not have any nontrivial rooted isomorphisms preserving the root, and therefore the same is true about the pair $((H, \phi(G)), o) = (\mathrm{Dec}(H, G), o)$. Therefore we may ignore the rooted equivalences when talking about $((H, \phi(G)), o)$ as a random element of $\mathcal{G}^{\mathrm{deco}}$ and just use the representative rooted graph. The pair $(\mathrm{Dec}(H, G), o)$, thought of as a rooted multigraph with colored edges, is the same as the pair $(\mathrm{Dec}(G, H), o) = ((G, \phi^{-1}(H)), o)$. This latter is unimodular, by the $\mathrm{Aut}(G)$-invariance of the distribution of $\phi^{-1}(H)$. To see this, check that the Mass Transport Principle (MTP) holds (see e.g. [**1**]). If one defines any nonnegative Borel function f on elements of the form $(\mathrm{Dec}(H, G), o, x)$, one defines a function f' from

$(\mathrm{Dec}(G,H),o,x)$ (here we are using the fact that the only stabilizer of o in G or in H is the identity, hence f' is well defined). Now, f' satisfies the equation for unimodularity (see [**1**]), hence f also satisfies it. This proves that $((H,\phi(G)),o)$ is unimodular.

The fact that a unimodular decoration of a unimodular transitive network is automorphism invariant is well known, but we prove it for completeness. Consider an automorphism γ of H. Pick an arbitrary event $\mathcal{A}$, with respect to the Borel σ-algebra on the equivalence classes of decorated rooted graphs with the usual metric. Define the following mass transport. Let every vertex x send mass 1 to vertex γx if the event $\mathcal{A}$ holds. Then the expected mass sent out is just $\mathbf{P}(\mathcal{A})$. By unimodularity and the MTP, this is the same as the expected mass received, which is $\mathbf{P}(\gamma^{-1}\mathcal{A})$. That is, the probability of event $\mathcal{A}$ is the same for o and for the vertex $\gamma^{-1}(o)$. Since this is true for any $\mathcal{A}$, we obtain that $(H,\phi(G))$ is invariant under the automorphism by γ^{-1}. As γ was an arbitrary automorphism, we obtain invariance as claimed. □

REMARK 2.2. The duality lemma does not hold in the present form if both G and H are Cayley *graphs*. First we show an example with finite Cayley graphs. Let H be a complete graph on 4 vertices, and let G be a cycle of length 4, with vertices v_1,v_2,v_3,v_4 along the cycle, in this order. Suppose that they are defined on the same vertex set. Then $\mathrm{Dec}(G,H)$ (with a uniformly chosen root) is invariant under $\mathrm{Aut}(G)$. But $\mathrm{Dec}(H,G)$ is not invariant under $\mathrm{Aut}(H)$: the automorphism of H that fixes v_1 and takes v_2 to v_3, does not preserve G. Taking some product of G and H with a fixed infinite Cayley graph (e.g., a direct product with $\mathbb{Z}$) will provide us an infinite example for the same phenomenon.

The duality lemma inspired more questions of similar flavor, which the author is investigating with Beringer in a paper under preparation, [**3**].

2.1. From embedded graphs to domain representations. In Section 1 we gave an overview of the proof of Theorem 1.2. In this and the next section we prove Theorem 1.3. The proofs of this section are different from the ones in [**8**]. We omit the technical part on accumulation points of domains here, while it is examined more thoroughly in [**8**].

Denote the closure of a subset A of $\mathbb{R}^3$ by $\bar{A}$, its boundary by ∂A, its interior by $\mathrm{int}(A)$, and its complement by A^c. For sets $A,B\subset\mathbb{R}^3$, let $\mathrm{dist}(A,B)$ be the Euclidean distance of $\bar{A}$ and $\bar{B}$. For $A\subset\mathbb{R}^3$, $c>0$, let $N(A,c)=\{x\in\mathbb{R}^3:\mathrm{dist}(x,A)<c\}$ be the c-neighborhood of A. If M is a collection of sets, let $\cup M$ be the short notation for $\cup_{A\in M}A$. Given a domain representation M, denote by $O(M)$ the interior of $\mathbb{R}^3\setminus\cup_{D\in M}D$, and call it the *ocean*. For $v\in V(G)$ denote by $M(v)$ the domain representing v; for $x\in\mathbb{R}^3$ let $M^{-1}(x)$ be the point $v\in V(G)\cup\{O(M)\}$ such that $x\in M(v)$, whenever there is such a point. If $D\in M$ is a domain and it shares a hyperface with the ocean, then we say that D is adjacent with the ocean. We think about domain representations as the 3-dimensional analogue of planar maps, and the ocean corresponds to the part of the space that is not covered by any region. Full domain representations are the ones where the ocean is empty.

In this section call polyhedral representations PR.

LEMMA 2.3. *Let G be an amenable transitive graph and an invariant random map $\lambda_0:V(G)\to\mathbb{R}^3$ of G into $\mathbb{R}^3$ such that $\lambda_0(V(G))$ has no accumulation points in $\mathbb{R}^3$. Then one can represent every edge by a broken line of finitely many segments.*

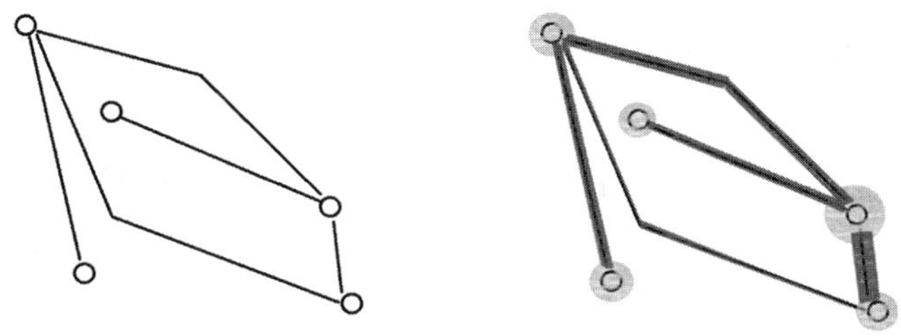

FIGURE 2.1. The embedded vertices and edges "fattened".

(In the future, when we say broken line, we always understand it to consist of finitely many straight line segments.) More precisely and furthermore, there is an embedding λ_1 of $E(G)$ such that every edge is mapped to a broken line segment with the property that for every vertex v there is some $\epsilon_v > 0$ and for every edge e there is some $\epsilon_e > 0$ such that:

- the collection of balls $\{B(\lambda_0(v), \epsilon_v) : v \in V(G)\}$ is pairwise disjoint;
- if $e = \{e^-, e^+\}$ is an arbitrary edge and $B(e)$ denotes the ϵ_e-neighborhood of $\lambda_1(e)$, then $B(e) \setminus (B(\lambda_0(e^-), \epsilon_{e^-}) \cup B(\lambda_0(e^+), \epsilon_{e^+}))$ is disjoint from every element of the collection $\{B(e') : e' \in E(G) \setminus \{e\}\} \cup \{B(\lambda_0(v), \epsilon_v) : v \in V(G)\}$. In words, every fattened embedded edge is disjoint from all other fattened edges or vertices outside of its fattened endpoints.

The above embedding of the edges can be done as a factor of iid from the embedded vertices.

In practice, the lemma aims for an embedding of the edges of G into $\mathbb{R}^3$ is such a way that if we thicken the edges and vertices properly (so that it becomes 3-dimensional, giving rise to a domain representation to be defined), then adjacencies are not distorted. By the latter we mean that if two thickened edges intersect, then they must share an endpoint; if a thickened edge intersects a thickened vertex, then the vertex is incident to the edge; and finally, thickened vertices do not intersect any other thickened vertices. See Figure 2.1 for an illustration.

PROOF. Define G_0 to be the graph on vertex set $\omega = \lambda_0(V(G))$. For each point $v \in V(G)$, let ϵ_v be $\min_{w \in V(G), w \neq v} \mathrm{dist}(\lambda_0(v), \lambda_0(w))/4$, and let $B(v) = B(\lambda_0(v), \epsilon_v)$. Let $G_n \supset G_0$ consist of every edge $\{u, v\} \in E(G)$ such that both $B(u)$ and $B(v)$ are in the same cell of $\mathbf{P}_n$. Then every component of G_n is finite (since the point process has no accumulation points in $\mathbb{R}^3$ almost surely). Let C be a cell of $\mathbf{P}_n$. Let $e_1, \ldots, e_{k(C)}$ be the edges of $(G_n \setminus G_{n-1})|_C$. We define the proper embeddings for the e_i as $i = 1, \ldots, k(C)$. Suppose that $\{e_1, \ldots, e_{i-1}\} \cup G_{n-1}|_C$ is already embedded in C, every embedded edge is represented by a line segment in C, and further that the ϵ_e for the embedded edges have been chosen so that $C \setminus (\bigcup_{e \in \{e_1, \ldots, e_{i-1}\} \cup G_{n-1} \cap C} B(e) \cup \bigcup_{v \in G_{n-1} \cap C} B(v)) =: C(n)$ is connected. Then we can choose a line segment between the endpoints of e_i in the interior of $C(n)$, and make this line segment be the embedded copy $\lambda_1(e_i)$ of e_i. Then we can choose ϵ_{e_i} to be small enough that $B(\lambda_1(e_i), 4\epsilon_{e_i})$ is within $C(n)$ (that is, disjoint from $\bigcup_{e \in \{e_1, \ldots, e_{i-1}\} \cup G_{n-1} \cap C} B(e) \cup \bigcup_{v \in G_{n-1} \cap C} B(v)$). Doing this repeatedly for every cell C (as $i = 1, \ldots, k(C)$), we define λ_1 and all the epsilons for the edges in G_n.

As $n \to \infty$, $G_n \to G$. Therefore, after repeating the above countably many times, all edges of G will be embedded in $\mathbb{R}^3$ and the ϵ_e's defined, as claimed by the lemma. $\square$

The next claim is almost trivial. Recall the definition of the ocean $O(M)$.

LEMMA 2.4. Let M be a random invariant PR representing the graph H with bounded domains, and $\epsilon > 0$ be arbitrary. Suppose that the ocean is connected. Let $S \subset V(H)$ be some invariant nonempty subset with the property that for every $x \in S$ the domain $M(x)$ is adjacent to the ocean. Then there is a PR M' that also represents H, and such that for every $v \in V(H) \setminus S$ we have $M(v) = M'(v)$, for every $v \in S$ we have $M(v) \subset M'(v)$, and finally, such that for every $x \in \mathbb{R}^3$ we have $\mathbf{P}(x \in O(M')) < \epsilon$. Furthermore, the changes from M to M' can be accomplished as a factor of iid from the vertices that the domains of M represent.

PROOF. Consider any $n > 0$ and the partition $\mathbf{P}_n$. As $n = 1, 2, \ldots$, repeat the following. For each $C \in \mathbf{P}_n$, and $T \in S$ that intersects C, do the following. List such T's somehow, and apply the following rule consecutively. If the union of components K of $C \cap O(M)$ that are neighbors of T has Lebesgue measure less than ϵ times the Lebesgue measure of T, then do nothing. Otherwise attach polyhedral pieces of all the above K to the T (in some arbitrary, but deterministically predefined way), in such a way that every augmented polyhedron after attachment is connected, and the complement of the domains in K is also connected. Do this so that the remaining part of the union of K's has Lebesgue measure less than ϵ times the Lebesgue measure of T. Doing this over every C, and as n goes to infinity, we obtain the desired M'. We leave it for the reader to show that the resulting domains are still polyhedra with finitely many faces, because every element of S is only modified finitely many times, by a MTP argument. $\square$

The next lemma is proved in a separate short paper, [**9**]. A weaker claim, with T having one or two ends (with no control on which one of these two), was shown by Benjamini, Lyons, Peres and Schramm [**2**]. (Their proof is about *invariant* spanning trees, but their method is straightforward to strengthen to obtain *factor of iid* spanning trees.)

LEMMA 2.5. [**9**] Let G be an amenable transitive graph with one end. Then there exists a factor of iid spanning tree T in G with one end.

In a one-ended tree, there is a unique infinite simple path from every vertex. Let us call its neighbor on this path its *parent*.

DEFINITION 6. Let G and T be as in the previous lemma. Define $T_0 := T$, and inductively, let T_n ($n \geq 1$) be the subgraph of T_{n-1} that results if we remove all the leaves (vertices of degree 1) from T_{n-1}. Define $G_n := G \setminus E(T_n)$.

The next lemma shows that (under proper conditions) one can delete one adjacency in a domain representation without changing the other adjacencies. Namely, the adjacency between X and O will disappear after a suitable modification of Y and O. Later we will use this to turn countries by the ocean into continental countries (i.e. countries nonadjacent to the ocean), with a relatively small modification of the domain representation, and without changing the adjacency structure in any other way.

FIGURE 2.2. Making the boundary shared with the ocean disappear. Illustration for Lemma 2.6, with X being the blue polyhedron, Y being the yellow one, and O being the ocean (where color is available).

Let $\epsilon > 0$ be arbitrary, $A, B \subset \mathbb{R}^3$. Define $\mathrm{Vor}(A, B, \epsilon) := \{x \in \mathbb{R}^3 : \mathrm{dist}(x, A) < \min(\mathrm{dist}(x, B), \epsilon)\}$.

LEMMA 2.6. Let M be a collection of pairwise disjoint polyhedra, $X, Y, O \in M$. Suppose that X is adjacent to both Y and O, that $\partial X \cap \partial O$ is connected, and that there is some ball that is intersected only by Y, X and O and that contains a point of $\partial Y \cap \partial X \cap \partial O$. Then there exists a collection of pairwise disjoint polyhedra M' that satisfies the following. First, $M' = M \setminus \{Y, O\} \cup \{Y', O'\}$, $Y \subset Y'$, $O' \subset O$. Furthermore, all adjacencies of M' are the same as in M (with Y' replacing Y and O' replacing O), with the only exception of X and O', which are not adjacent in M'.

The proof is best summarized on Figure 2.2, we only provide a sketch of it.

PROOF. Consider $\nu(X, c) := \mathrm{Vor}(\partial X \cap \partial O, \partial O \setminus (\partial X \cup \partial Y), c) \setminus X$, and define $\epsilon > 0$ to be a number such that $\mathrm{Vor}(\partial X \cap \partial O, \partial O \setminus (\partial X \cup \partial Y), \epsilon)$ does not intersect any element of M but O, X and Y. Now let $Y' = Y \cup \nu(X, \epsilon)$, $O' = \mathrm{int}(O \setminus Y')$. □

The next theorem will imply Theorem 1.3 right away.

THEOREM 2.7. Suppose that there is an invariant embedding λ_0 of G in $\mathbb{R}^3$, with $\lambda_0(V(G))$ having no accumulation point in $\mathbb{R}^3$ almost surely. Then there is a full polyhedral representation of G in $\mathbb{R}^3$, and such a representation can be obtained as a fiid from the (embedded) vertices of G.

PROOF. Define $\mathbf{P}_n$ to be a sequence of coarser and coarser partitions of $\mathbb{R}^3$, every class bounded, and such that any two points are in the same class of $\mathbf{P}_n$ if n is large enough. We may assume that every partition class of $\mathbf{P}_n$ is a polyhedron (with a boundary that may be partially closed); see [**7**] for such a construction. We can make $\mathbf{P}_n$ to be a factor of $\phi(V(G))$ in $\mathbb{R}^3$ if $\phi(V(G))$ has no nontrivial symmetries. If there are some nontrivial symmetries, we can break them by perturbing the locations of the points of $\phi(V(G))$ by tiny amounts, as factor of iid from this set of points. See [**7**] for the existence of such a sequence $\mathbf{P}_n$.

From Lemma 2.3 one can deduce that there is a representation M_0 of G in $\mathbb{R}^3$ by polyhedra, but with the extra property that every polyhedron is adjacent to the ocean $O(M_0)$. In other words, M_0 is a polyhedral representation of $G^* := G \cup \{\infty\}$, where ∞ is an extra vertex that is adjacent to every vertex of G. We explain this construction next.

Let ϵ_v, ϵ_e and λ_1 be as in Lemma 2.3, for $v \in V(G), e \in E(G)$. Simplify notation using $B(v) := B(\lambda_0(v), \epsilon_v)$ (where $v \in V(G)$) and $B(e) = B(\lambda_1(e), \epsilon_e)$ (where $e \in E(G)$). Let $B'(v) \subset B(v)$ be some polyhedron that is determined as a factor, from the neighborhood of v. (It is easy to check that one can fix such a factor map.) To each vertex v, we will assign a polyhedron D_v defined as follows. (See Figure 2.3 for a schematic example.) First, for every edge $e = \{x, y\}$ of G, divide $B(e)$ into two connected polyhedral pieces, one of them containing $B(e) \cap B(x)$, and the other one containing $B(e) \cap B(y)$. Call the interior of the former one $D(e, x)$, the interior of the latter one $D(e, y)$. Now let $D_v = B'(v) \cup \bigcup_{v \in e} D(e, v)$. It is easy to check that the collection $M_0 := \{D_v : v \in V(G)\}$ satisfies the claim for G, except for that the ocean for M_0 is nonempty, and M_0 represents G^*. We will repeatedly apply Lemma 2.6 to get rid of all the adjacencies to $O(M_0)$ by the end.

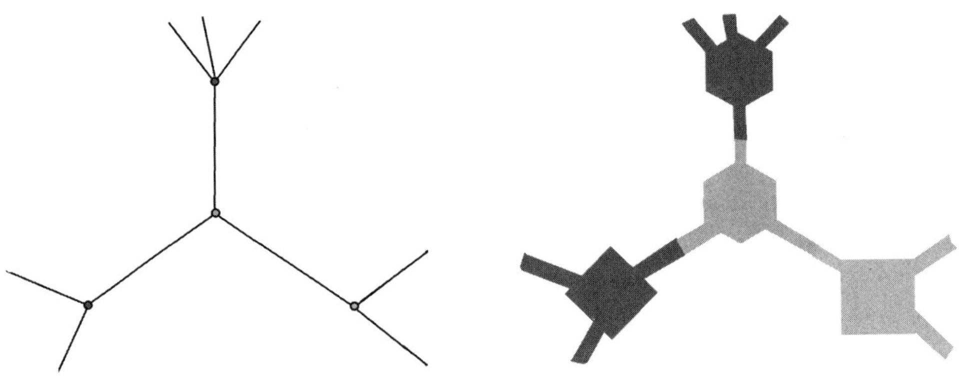

FIGURE 2.3. From graph embedding to polyhedral representation, allowing an ocean. Broken lines happen to be straight lines on the image, for simplicity. Note that the actual construction is in 3 dimensions, and therefore $\partial C \cap \partial O$ is connected for every polyhedron C.

Having defined M_0, we will construct a sequence $M_1, M_2, \ldots$ such that the following hold. (Recall the definition of G_n and T_n from Definition 6.)

(1) $M_n \setminus O(M_n)$ represents G_n.
(2) For every $C \in M_n$, $\partial C \cap \partial O(M_n)$ is connected (possibly empty).
(3) The domain $M_n(x)$ is not adjacent to the ocean if and only if x has full degree in G_n.
(4) $M_n(x) \neq M_{n-1}(x)$ if and only if x is the parent of some $y \in T_{n-1} \setminus T_n$.
(5) $\mathbf{P}(o \in O(M_n)) < 2^{-n}$.

So suppose that M_{n-1} has been defined. As $j = 1, 2, \ldots$, do the following. For every cell $C \in \mathbf{P}_j$, consider all the $y \in V(G)$ such that $M_{n-1}(y) \subset C$. Since there are finitely many such y for each C, we can list them in some order.

Going along this list of y one by one, do the following. Let $x_1(y), \ldots, x_{m(j,y)}(y)$ be some ordering of the neighbors of y with the property that $x_i \in T_{n-1} \setminus T_n$ (that is, x_i is a leaf of T_{n-1}), and $M_{n-1}(x_i) \subset C$. Apply Lemma 2.6 (with x_i in the role of X, y in the role of Y, and the ocean being O) to modify $M_{n-1}(y)$ so that the cell of x_i does not share a boundary with the ocean any longer. The conditions for Lemma 2.6 are satisfied by item 2 on the above list of assumptions (on M_{n-1}). Do this for $i = 1, \ldots, m(j, y)$. After having done all the finitely many modifications that correspond to a y as $j = 1, 2, \ldots$, we define the resulting cell for y to be $M_n(y)$. So for a fixed y we obtain that all the leaves of T_n whose parent was y cease to share a boundary with the ocean. All the other items above are satisfied too, except for the last one. To make that hold as well, apply Lemma 2.4 to M_n for $S = \{y \in T_n, \exists x \in T_{n-1} \setminus T_n \text{ adjacent to } y\}$. Redefine M_n according to this modification.

Taking the limit of the M_n (which exists by item 4), we get a domain representation M, which shows the theorem. Note that every domain gets modified only finitely many times along the process (that is, for any vertex x, $M_n(x) = M_{n-1}(x)$ for all but finitely many n). Therefore all the resulting domains remain in fact bounded polyhedra with finitely many faces, because they were polyhedra in M_0, and every modification preserves them being polyhedra. Furthermore, no boundary point of a domain is an accumulation point of the set of other domains. □

PROOF OF THEOREM 1.3. Apply Theorem 2.7 to a random embedding as given by Theorem 1.6. □

PROOF OF THEOREM 1.2. See Subsection 1.2. □

2.2. Connecting unconnected families of domains. By a slight abuse of terminology, a full domain representation of a graph G in $\mathbb{R}^d$ will be called a G-partition.

In Subsection 1.2 we explained how to use Theorem 1.3 to find a full domain representation of T_3 in $\mathbb{R}^3$ from the full polyhedral representation of BS(1,2). Furthermore, since every element of the representation could be defined as a fiid, we could also conclude that the pieces in the T_3-partition are indistinguishable.

A possible alternative way to complete the above line of proof is to use a Cayley graph G of the lamplighter group instead of BS(1,2). The existence of a T_3-partition for this G is very straightforward (see Lemma 2.8); on the other hand, the pieces that represent one vertex of T_3 are not connected, hence one has to work more to make the composed T_3-partition of $\mathbb{R}^3$ connected, as seen in the rest of this section. We decided to include this argument in this paper (while the one about BS(1,2) is more straightforward) in order to present the method of making disconnected "fibers" connected in the domain representation. While our proof does this only for the lamplighter, it is easy to generalize the technique to any amenable example. That is, if there is a domain representation in $\mathbb{R}^3$ and some invariant domain representation, then there is a domain representation that will make all the classes of the partitions connected, while leaving all other adjacencies and nonadjacencies unchanged. It also needs to be mentioned that the T_3-partition of $\mathbb{R}^3$ that we get from the lamplighter can be done to be a fiid only if we are using the group of automorphisms that preserve the fixed infinite rays of t_1 and t_2 (to be defined in the next paragraph). (That is, if in the definition of fiid we only require the mapping rule from iid-labelled vertices to be invariant under this narrower group of automorphisms, then we can make the T_3-partition of $\mathbb{R}^3$ fiid.) One more remark is about indistinguishability of subgraphs. In the current setup, we want to apply it to subsets of the lamplighter that are not connected, which is outside of the range of our definition. However, it is easy to generalize the notion to families of possible disconnected graphs; we omit the details.

Denote by L the Cayley graph of the lamplighter group represented as the horocyclic product of two 3-regular trees, or, in other words, the Diestel-Leader graph DL(2,2) of parameter (2,2). To recall that construction, consider two copies, t_1 and t_2 of the 3-regular tree T_3. Fix an infinite ray in t_1, and define a "level function" $\alpha_1 : V(t_1) \to \mathbb{Z}$ where $\alpha(x) = \alpha(x')+1$ if and only if x and x' are neighbors and x separates x' from the fixed infinite ray. Do similarly with t_2, to define the function α_2. Then $V(L) = \{(x,y) : x \in V(t_1), y \in V(t_2), \alpha_1(x) = -\alpha_2(y)\}$, and the edge set $E(L)$ consists of all edges of pairs (x,y), (x',y'), where x and x' are adjacent in t_1, and y and y' are adjacent in t_2. If x is a vertex of t_1, and it has a neighbor x^+ such that $\alpha(x)+1 = \alpha(x^+)$, then we say that x^+ is the parent of x and x is the child of x^+. If there is a sequence $x_0, \ldots, x_n$ such that x_i is the parent of x_{i-1} then we say that x_n is the n'th ancestor of x_0, and x_0 is the n'th descendant of x_n. Similar definitions will be used for t_2, with α_1 replaced by α_2. Call the group of those automorphisms that preserve the fixed infinite rays of t_1 and t_2 (of index 2 in the full automorphism group) as $\mathrm{Aut}^-(L)$.

LEMMA 2.8. There is an invariant T_3-partition $\mathcal{N}$ of L. If we only consider symmetries with respect to $\mathrm{Aut}^-(L)$, then $\mathcal{N}$ can be chosen to be deterministic.

PROOF. Suppose first that invariance is understood with respect to the full group of automorphisms. Choose the random variable i to be equal to 1 or 2, each with probability 1/2. If $i = 1$, define each class of $\mathcal{N}$ to be of the form $\{(x,y) \in V(L) : y \in V(t_2)\}$, one class corresponding to every $x \in V(t_1)$. If $i = 2$, define each class of $\mathcal{N}$ to be of the form $\{(x,y) \in V(L) : x \in V(t_1)\}$, one class corresponding to every $y \in V(t_2)$.

If we consider only $\mathrm{Aut}^-(L)$, then the extra randomness in the choice of i is not needed. We can set $i = 1$, and the partition as above is invariant. □

The next lemma says that the unions of domains that correspond to one fiber (partition class) of $\mathcal{N}$ can be made connected by small alterations of the domain representation, while preserving all invariance properties and applying local rules. Figure 2.4 provides a quick insight into how the proof works.

LEMMA 2.9. Consider a representation of L in $\mathbb{R}^3$ by bounded domains, which is invariant with respect to $\mathrm{Aut}(L)$ or with respect to $\mathrm{Isom}(\mathbb{R}^d)$. Let $\mathcal{N}$ be the partition as in the previous lemma, and $\mathcal{M}'$ be the partition of $\mathbb{R}^3$ that one obtains by composing the first representation by $\mathcal{N}$ (that is, by taking the unions of the representing domains over each partition class). Then there is a way to modify the domains of $\mathcal{M}'$ in such a way that every new domain is connected, and all other adjacencies remain unchanged. Furthermore, this modification can be done as a fiid, and therefore all the invariance properties remain valid.

PROOF. Let $\mathcal{M}'$ be the T_3-partition of $\mathbb{R}^3$ that we have just obtained. So there is a bijection $\beta : V(L) \to \mathcal{M}$ that preserves adjacency (i.e., adjacent vertices are mapped to adjacent domains, and vica versa). Similarly, there is a bijection $\beta : V(T_3) \to \mathcal{M}'$ that preserves adjacency. To complete the construction, what

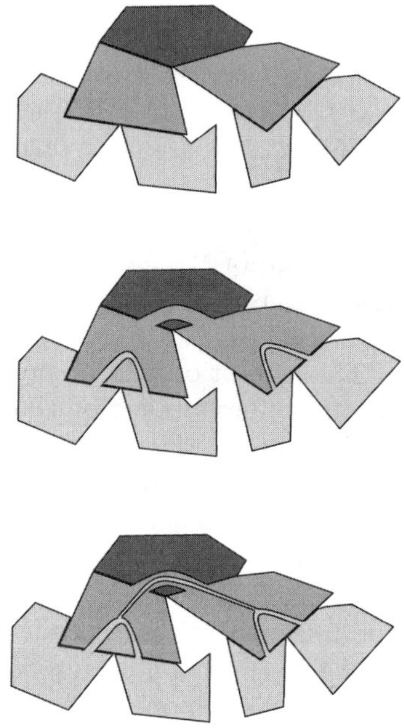

FIGURE 2.4. "Growing tunnels". Parts of $\mathcal{M}' = \mathcal{M}_0, \mathcal{M}_1, \mathcal{M}_2$, top to bottom. Components of the same colors belong to the same pieces. Note that in 3 dimensions, the "tunnels" do not disconnect the components that they are crossing.

remains is to show that one can modify $\mathcal{M}'$ to get a *connected* T_3-partition of $\mathbb{R}^3$. We finish this section by proving that. Figure 2.4 summarizes the key idea: we make the pieces of the T_3-partition connected by growing slim tunnels between some components of each piece.

Consider a domain $D \in \mathcal{M}$. There is some $(x, y) \in V(L) \subset V(t_1) \times V(t_2)$ (written in the horocyclic representation as before), such that $D = \beta((x, y))$. For simpler later reference, choose a point $p(x, y) \in \beta((x, y))$ uniformly at random, for every $(x, y) \in V(L)$. (This is doable because every domain $D \in \mathcal{M}$ has finite measure.) Let i be a random variable that is equal to 1 or 2, each with probability 1/2. Assume from now on that $i = 1$, by symmetry, there is no loss of generality in doing so. Now consider a piece $D' \in \mathcal{M}'$. Then D' can be identified with a vertex x of the 3-regular tree t_1, $D' = D'(x) = \cup_{y \in V(t_2), (x,y) \in V(L)} \beta((x, y))$. Note that the elements of the set $\{\beta(x, y) : y \in V(t_2), (x, y) \in V(L)\}$ are pairwise nonadjacent (because $\{(x, y) \in V(L) : y \in V(t_2)\}$ form an independent set in it L, and the PR $\mathcal{M}$ represents L). However, if y^+ is the parent of y_1 and y_2 in t_2, and x^- is a child of x in t_1, then $\beta(x, y_1)$ and $\beta(x, y_2)$ are both adjacent to $\beta(x^-, y^+)$. For every such triple, "grow a thin tunnel" connecting $\beta(x, y_1)$ and $\beta(x, y_2)$ inside $\beta(x^-, y^+)$, where x^- is one of the children of x chosen at random; make sure that these tunnels do not contain any of the $p(x, y)$. (More precisely, by growing a thin tunnel we mean that we choose a small enough neighborhood of some broken line from $\beta(x, y_1)$ and

$\beta(x, y_2)$, such that this neighborhood is completely contained in $\beta(x^-, y^+)$.) This can be done in parallel over all such triples, because the corresponding $\beta(x^-, y^+)$ (where the tunnel is grown) is distinct for any two such triples. After having grown the thin tunnels, we get a new T_3-partition of $\mathbb{R}^3$, where for every pair y_1, y_2 that have the same parent, $p(x, y_1)$ and $p(x, y_2)$ are now contained in the same connected domain. (We could say that $\beta(x, y_1)$ and $\beta(x, y_2)$ got united inside a new domain, except for that they may have lost some of their territories because of some tunnels crossing them.) Call this T_3-partition of $\mathbb{R}^3$ as $\mathcal{M}_1$, and use notation $\mathcal{M}_0 := \mathcal{M}'$.

Suppose we have defined $\mathcal{M}_{n-1}$, a T_3-partition of $\mathbb{R}^3$, and suppose that for every $x \in V(t_1)$, the union of the set $\{p(x, y) : y \in V(t_2), (x, y) \in V(L)\}$ is contained in one piece of the partition. This condition will be our inductive hypothesis. The validity of the condition naturally defines a bijection that assigns to every $x \in V(t_1)$ a piece $\mathcal{M}_{n-1}(x)$ of $\mathcal{M}_{n-1}$. Suppose further that $\mathcal{M}_{n-1}$ has the property that $p(x, y_1)$ and $p(x, y_2)$ are contained in the same *connected* component of $\mathcal{M}_{n-1}(x)$, whenever y_1, y_2 have the same $n-1$'th ancestor. Now modify $\mathcal{M}_{n-1}$ as follows. Suppose that $y_1, y_2 \in V(t_2)$ is a pair whose n'th ancestor y^+ is the closest one in common (i.e., they are at distance $2n$ from each other and $\alpha_2(y_1) = \alpha_2(y_2)$). Let $x \in V(t_1)$ and x^- be a randomly chosen n'th descendant of x in t_1. Then consider the connected components of $\mathcal{M}_{n-1}((x))$ containing $p(x, y_1)$ and $p(x, y_2)$. Both of them are adjacent to the connected component of $p(x^-, y^+)$ in $\mathcal{M}_{n-1}((x^-))$, by the assumption on $\mathcal{M}_{n-1}$. Grow a tunnel crossing the connected component of $p(x^-, y^+)$ in $\mathcal{M}_{n-1}((x^-))$, and connecting the two components. When choosing the tunnels, make sure that they are slim enough not to disconnect any of the domains, and that they do not contain any of the $p(x, y)$. Doing this for all pairs $y_1, y_2 \in V(t_2)$ that satisfy the above condition, we obtain a T_3-partition $\mathcal{M}_n$ that satisfies the inductive hypothesis.

Take the limit of $\mathcal{M}_n$ as $n \to \infty$, and call the resulting T_3-partition of $\mathbb{R}^3$ as $\mathcal{M}_\infty$. To summarize, the modified partition $\mathcal{M}_n$ had the property that $p(x, y)$ and $p(x, y')$ were in the same connected domain whenever $(x, y), (x, y') \in V(L)$, and y and y' are at distance at most $2i$ from each other in t_2. Therefore in $\mathcal{M}_\infty$ the entire piece of $\{(x, y) : y \in V(t_2), (x, y) \in V(L)\}$ is in the same component. □

REMARK 2.10. If one were to apply a similar procedure to some other amenable graph, one would have to use some other sequence of partitions into finite pieces that exhaust G (as in [**2**]), and grow the tunnels accordingly, to make every piece of some "fiber" within a partition connected.

Acknowledgements. I thank Dorottya Beringer for her comments on the manuscript. The valuable discussions with Russ Lyons, Damien Gaboriau and others who are acknowledged in [**8**] were of great help for the present paper as well.

References

[1] David Aldous and Russell Lyons, *Processes on unimodular random networks*, Electron. J. Probab. **12** (2007), no. 54, 1454–1508, DOI 10.1214/EJP.v12-463. MR2354165

[2] I. Benjamini, R. Lyons, Y. Peres, and O. Schramm, *Group-invariant percolation on graphs*, Geom. Funct. Anal. **9** (1999), no. 1, 29–66, DOI 10.1007/s000390050080. MR1675890

[3] Beringer, D., Timár, Á. (2017) Duality of unimodular random decorations of unimodular networks (work in progress).

[4] Russell Lyons, *Factors of IID on trees*, Combin. Probab. Comput. **26** (2017), no. 2, 285–300, DOI 10.1017/S096354831600033X. MR3603969

[5] Russell Lyons and Oded Schramm, *Indistinguishability of percolation clusters*, Ann. Probab. **27** (1999), no. 4, 1809–1836, DOI 10.1214/aop/1022677549. MR1742889
[6] Sébastien Martineau, *Ergodicity and indistinguishability in percolation theory*, Enseign. Math. **61** (2015), no. 3-4, 285–319, DOI 10.4171/LEM/61-3/4-2. MR3539840
[7] Ádám Timár, *Tree and grid factors for general point processes*, Electron. Comm. Probab. **9** (2004), 53–59, DOI 10.1214/ECP.v9-1073. MR2081459
[8] Timár, Á. (2016) A non-amenable "factor" of a Euclidean space (preprint), https://arxiv.org/abs/1712.08210.
[9] Timár, Á. (2017) Spanning trees of one-ended amenable graphs, preprint.

Alfréd Rényi Institute of Mathematics, Reáltanoda utca 13-15., H-1053 Budapest, Hungary

Email address: `madaramit@gmail.com`

Contemporary Mathematics
Volume **719**, 2018
https://doi.org/10.1090/conm/719/14470

Distributional lattices on Riemannian symmetric spaces

Elliot Paquette

Abstract. A Riemannian symmetric space is a Riemannian manifold in which it is possible to reflect all geodesics through a point by an isometry of the space. On such spaces, we introduce the notion of a distributional lattice, generalizing the notion of lattice. Distributional lattices exist in any Riemannian symmetric space: the Voronoi tessellation of a stationary Poisson point process is an example. We show that for an appropriate notion of amenability, the amenability of a distributional lattice is equivalent to the amenability of the ambient space. Using this equivalence, we show that the simple random walk on any nonamenable distributional lattice has positive embedded speed. For nonpositively curved, simply connected spaces, we show that the simple random walk on a Poisson–Voronoi tessellation has positive graph speed by developing some additional structure for Poisson–Voronoi tessellations.

1. Introduction

Riemannian symmetric spaces. A Riemannian symmetric space $\mathbb{M}$ is a connected Riemannian manifold where at each point p, there is an isometry σ_p of $\mathbb{M}$ that fixes p and whose differential at p is multiplication by -1.

Riemannian symmetric spaces provide many excellent examples of nonpositively and positively curved spaces, which include:

(1) the Euclidean spaces,
(2) the spheres $\mathbb{S}^n$ in Euclidean space,
(3) the real, complex, and quaternionic hyperbolic spaces ($\mathbb{H}^d, \mathbb{CH}^d, \mathbb{HH}^d$ respectively, see [**BH99**, Chapter 10] for a comprehensive treatment) of any dimension,
(4) $\mathrm{SL}_d(\mathbb{R})/\mathrm{SO}_d(\mathbb{R})$ for any $d \geq 1$, which can be identified with positive definite matrices modulo scalars,
(5) Riemannian products of any of the above examples, such as $\mathbb{H}^d \times \mathbb{R}^k$ or $\mathbb{H}^2 \times \mathbb{H}$.

All of the examples listed except for the spheres are nonpositively curved. Additionally, Riemannian symmetric spaces decompose nicely: any simply connected Riemannian symmetric space decomposes as a Riemannian direct product $\mathbb{M}_1 \times \mathbb{R}^d \times \mathbb{M}_2$ where $\mathbb{M}_1$ is nonpositively curved and $\mathbb{M}_2$ is compact (see [**Hel01**, V,Proposition 4.2]).

When a Riemannian symmetric space has no factors of $\mathbb{R}^k$ and no compact factors in its de Rham decomposition (its decomposition as a Riemannian product

2010 *Mathematics Subject Classification.* Primary 60G55, 22E40, 43A07.

into irreducible factors), it is called a Riemannian symmetric space of noncompact type. Such spaces can be identified as quotient spaces G/K where G is a semisimple Lie group with trivial center and K is a maximal compact subgroup.

Lattices and Voronoi tessellations. All Riemannian symmetric spaces $\mathbb{M}$ are diffeomorphic to quotient spaces G/K where G is the isometry group of $\mathbb{M}$ and K is a stabilizer of some point (see [**Hel01**, V, Theorem 3.3]). Going forward, we let $o \in \mathbb{M}$ be an arbitrary base point, and let K refer to the stabilizer of o in the isometry group of $\mathbb{M}$. The group G has a locally compact Lie group structure and so has a (left) Haar measure.

Generally, a discrete subgroup Γ of a Lie group is called a *lattice* if there is a measurable set of coset representatives of $\Gamma\backslash G$ so that $\Gamma\backslash G$ has finite Haar measure. When G is the isometry group of $\mathbb{M}$, one can identify an equivalent condition to Γ in terms of Voronoi tilings. Every simply connected Riemannian symmetric space has many lattices [**Bor63**]. In particular, each Riemannian symmetric space of noncompact type has many lattices.

The orbit $\Gamma \cdot o$ forms a closed discrete subset of $\mathbb{M}$. Hence, it is possible to define the *Voronoi tessellation* of $\mathbb{M}$ with *nuclei* $\Gamma \cdot o$. In general, for a closed discrete set $S \in \mathbb{M}$, the Voronoi tessellation with nuclei S is a decomposition of $\mathbb{M}$ into *cells*, which for a given point $x \in S$ is defined by

$$\mathcal{V}(x; S) = \left\{ y \in \mathbb{M} : d_{\mathbb{M}}(x, y) = \min_{s \in S} d_{\mathbb{M}}(s, y) \right\}.$$

Note that Γ permutes the voronoi cells $\{\mathcal{V}(\gamma \cdot o; \Gamma \cdot o) : \gamma \in \Gamma\}$, and hence all of them have equal volume. On account of the compactness of the stabilizer K, it is readily checked that the finiteness of the Riemannian volume of $\mathcal{V}(o)$ is equivalent to the finiteness of $\Gamma \setminus G$ under the Haar measure of G.

Stationary point processes. A (simple) point process $\mathcal{P}$ on a complete separable metric space $\mathcal{X}$ is informally a random countable subset of $\mathcal{X}$ which has finite intersection with any compact set. Formally, a *point process* is a random locally–finite, integer–valued measure. This is in correspondence with countable subsets by mapping a countable discrete set to a sum of point masses at each atom and conversely by mapping a random integer–valued measure to its atoms. When $\mathcal{X}$ is a homogeneous space, we say that a point process $\mathcal{P}$ is *stationary* if $\mathcal{P} \overset{\mathcal{L}}{=} \tau(\mathcal{P})$ for any isometry τ of $\mathcal{X}$.

We will only consider point processes that have finite intensity, which is to say that for any compact $A \subset \mathcal{X}$

$$\Lambda(A) := \mathbb{E}|A \cap \mathcal{P}| < \infty.$$

Note that for stationary point processes on a Riemannian homogeneous space $\mathbb{M}$, the *intensity* measure Λ will then be an invariant Radon measure, and hence it is a multiple of the Riemannian volume measure by the uniqueness of Haar measure. We will let λ denote this multiple.

The Palm process $\mathcal{P}_o$, which is again a point process, has distribution which can be considered as the distribution of $\mathcal{P}$ conditioned to have a point at o. For a general discussion of the theory, see [**Kal86**, Ch 6] or [**DVJ03**, II.13]. More formally, it is defined as a family of point processes $\{\mathcal{P}_o, o \in \mathcal{X}\}$ so that for any

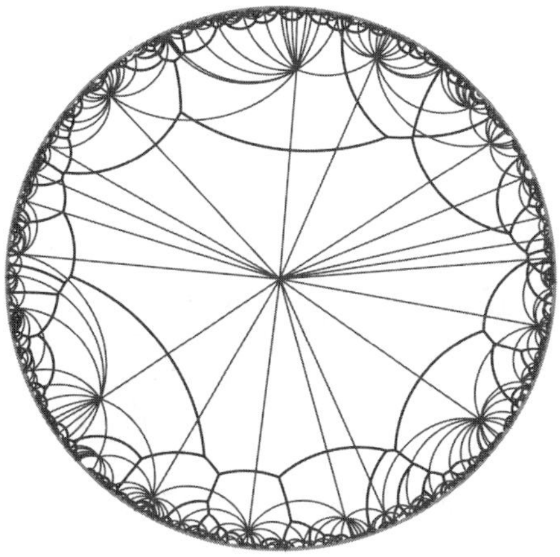

(A) The hyperbolic GAF (see [**HKPV09**, Ch. 5]), approximated by the roots inside the disk of a Kac polynomial of degree 1000, of which there are in expectation 500.

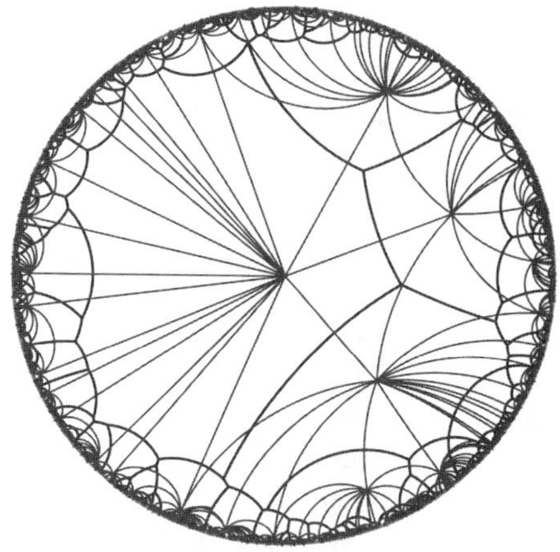

(B) A Poisson–Voronoi tiling with intensity matching the hyperbolic GAF, truncated in a disk of radius r where r is chosen to have 500 points in expectation (matching the Kac polynomial).

FIGURE 1. These are simulations depicting the hyperbolic tessellations for the palm processes of two different stationary point processes. They are chosen to have matching intensities. The GAF is statistically more rgular than the Poisson process. The points appear as the vertices of the triangulation (the blue graph, where color is available), and the dual tessellation is the Voronoi tessellation.

nonnegative measurable function $f : \mathcal{X} \times \mathcal{M}_{\mathcal{X}} \to \mathbb{R}$

$$\mathbb{E}\left[\int_{\mathcal{X}} f(o,\mathcal{P})\mathcal{P}(do)\right] = \mathbb{E}\left[\int_{\mathcal{X}} f(o,\mathcal{P}_o)\Lambda(do)\right].$$

Note that $\{\mathcal{P}_o, o \in \mathcal{X}\}$ is only well–defined up to Λ–null sets. For a stationary point process, one version of the Palm process is given by $\{\tau_{o,x}(\mathcal{P}_o), x \in \mathcal{X}\}$ where $\tau_{o,x}$ is an isometry of the space mapping o to x. In particular, for a stationary point process, the palm processes $\mathcal{P}_x$ are meaningful for all points $x \in \mathcal{X}$.

EXAMPLE 1.1 (Poisson–processes). While some of the results in this paper are applicable to general palm processes of stationary point processes, we are able to show much more for the case that $\mathcal{P}$ is a stationary Poisson process. This is a one–parameter family of point processes, which are determined by their intensities $\lambda \cdot dV$, where dV is the Riemannian volume measure and $\lambda > 0$ is a parameter. Such a process can be uniquely determined by the condition that for any disjoint collection of compact sets $A_1, A_2, \ldots, A_k$, the cardinalities $\{|\mathcal{P} \cap A_i|\}_1^k$ are independent Poisson–distributed variables with means $\lambda V(A_i)$. One can further see that conditional on the cardinality of $|\mathcal{P} \cap A|$, the points of $\mathcal{P} \cap A$ are independent and uniformly distributed over A.

The Palm process $\mathcal{P}_o$ has the attractive feature that $\mathcal{P}_o \overset{\mathcal{L}}{=} \{o\} \cup \mathcal{P}$, which is to say conditioning the Poisson process to have a point at the origin leaves the distribution unchanged on $\mathcal{X} \setminus \{o\}$.

EXAMPLE 1.2 (Translated–lattices). The lattice $\Gamma \cdot o$ forms another point process, albeit completely deterministic. If one randomizes the process by taking $g\Gamma \cdot o$ where g is distributed according to Haar measure on G/Γ then the resulting point process is invariant under all isometries by virtue of the G invariance of the Haar measure on G/Γ. The palm process that results from conditioning o to be at the origin has the distribution of $k\Gamma \cdot o$, where k is distributed according to Haar measure on K, the stabilizer of o.

As an example, in the case of $\mathbb{M} = \mathbb{R}^d$, the isometry group G is the 2–dimensional Euclidean group, generated by all translations, rotations and any one reflection of the plane. The stabilizer K of a point can be identified with the 2–dimensional orthogonal group. Hence the palm process $\mathcal{P}_o$ is simply a rotated version of $\Gamma \cdot o$, and the distributional lattice $\mathcal{P}$ is a random translate of $\mathcal{P}_o$ which takes o to a uniformly and independently chosen point in $\mathcal{V}(o; \Gamma \cdot o)$.

EXAMPLE 1.3 (Determinantal–processes). A more exotic class of examples are the determinantal processes. In two–dimensions there are relatively well–studied examples of stationary determinantal point processes for all of the constant curvature spaces. These examples have the additional interpretation as being zero sets of Gaussian random analytic functions. See [**HKPV09**, Chapter 2] for more details. Some properties of the Voronoi diagrams have been studied for Ginibre process [**G+10**].

These processes exist in much greater generality than suggested. Many stationary determinantal processes exist in any Riemannian symmetric space. In effect, it is possible to define a determinantal process for any isometry–invariant closed subspace S of $\mathrm{L}^2(dV)$ for which the evaluation maps $f \mapsto f(z)$ are bounded linear functionals (for example spaces of certain harmonic functions). On such spaces, one can find a reproducing kernel $\mathcal{K}(z,w) = \sum_{k=1}^{\infty} \psi_k(z)\psi_k(w)$, for a complete orthonormal system $\{\psi_k\}_1^{\infty}$ of S. The intensity of points will be given $\mathcal{K}(x,x)dV(x)$. By invariance, it follows that $\mathcal{K}(x,x)$ is constant. This construction leads to the canonical examples discussed in the prior paragraph. A particularly nice example is given by the Bergman kernel associated to any bounded symmetric domain in $\mathbb{C}^n$ (see [**Hel01**, VII.3,7] for details). The hyperbolic GAF depicted in Figure 1 is such an example, with the domain being the unit disk inside of $\mathbb{C}$.

Unimodular networks. We follow the notation and development of this material from [**AL07**], which should be consulted for more details. A network is a graph (V,E) in which vertices and edges are marked by some elements of a complete separable metric space, that is to say there is a complete separable metric space Ξ and maps $\psi_V : V \to \Xi$ and $\psi_E : E \to \Xi$. A network is rooted if there is a designated vertex $o \in V$, and birooted if there is an ordered pair (x,y) of distinguished vertices. Let $\mathcal{G}_*$ be the space of isomorphism classes of rooted networks, and let $\mathcal{G}_{**}$ be the isomorphism classes of birooted networks. A random rooted network is *unimodular* if it obeys the *mass-transport principle*, which is to say for all Borel $f : \mathcal{G}_{**} \to [0, \infty]$,

$$\mathbb{E}\left[\sum_{x \in V} f(G, o, x, \psi)\right] = \mathbb{E}\left[\sum_{x \in V} f(G, x, o, \psi)\right]. \tag{1.1}$$

The Voronoi tessellation on $\mathbb{M}$ with nuclei $\mathcal{P}_o$ naturally gives rise to a unimodular network. The graph G is the *Delaunay graph* on vertices $\mathcal{P}_o$. Two vertices x and y are connected by an edge if and only if $\mathcal{V}(x) \cap \mathcal{V}(y)$ has codimension-1. Equivalently, two vertices x and y are connected if and only if there is a ball $B_{\mathbb{M}}(z, r)$ having the property that $x, y \in \partial B_{\mathbb{M}}(z, r)$ and $B_{\mathbb{M}}(z, r) \cap \mathcal{P}_o$ is empty. Under some genericity conditions, which for example are satsified by Poisson points with probability 1, this graph is a 1–skeleton of a triangulation of $\mathbb{M}$.

The graph will be rooted at the vertex at o. The space of marks Ξ will be taken to be the manifold $\mathbb{M}$. The vertices will be marked by the locations of the points, and the edge marks will be the midpoints of the $\mathbb{M}$–geodesic between its endpoints. We let ψ denote this marking. We refer to this construction of random network as the *embedded Delaunay graph* with nuclei $\mathcal{P}_o$.

THEOREM 1.4. *Let $\mathcal{P}_o$ be the palm process of a stationary point process, and let (G, o, ψ) be the embedded Delaunay graph with these nuclei. Then (G, o, ψ) is a unimodular random network.*

The proof is given in Section 2

This leads us to the following generalization of lattices.

DEFINITION 1.5. A *distributional lattice* in a Riemannian symmetric space $\mathbb{M}$ is the palm process $\mathcal{P}_o$ of a stationary point process with the property that $\mathbb{E}[\mathrm{Vol}_{\mathbb{M}}(\mathcal{V}(o))] < \infty$ and $\mathbb{E}[\deg_G(o)] < \infty$.

When a Riemannian symmetric space has a lattice $\mathcal{L}$, it can be made into a distributional lattice by picking a Haar distributed coset (as in Example 1.2). Also, palm processes of stationary Poisson processes with positive finite intensity are always distributional lattices (see Theorem 3.3).

Amenability. A homogeneous Riemannian manifold G/H is amenable if for every compact subset $S \subset G$ there is a sequence of measurable sets V_n with finite volume so that

$$\lim_{n \to \infty} \max_{g \in S} \frac{|gV_n \Delta V_n|}{|V_n|} = 0,$$

with $|\cdot|$ denoting the volume measure. A locally compact topological group G is amenable if for every compact subset $S \subset G$, there is a sequence of measurable sets V_n with finite positive Haar measure so that

$$\lim_{n \to \infty} \max_{g \in S} \frac{\mu(gV_n \Delta V_n)}{\mu(V_n)} \to 0.$$

If G is amenable, then as H is necessarily closed (as G/H is a manifold), the amenability of H follows. On the other hand, if H is amenable (which is always the case for a Riemannian symmetric space, in which case we can take it to be compact), the amenability of G/H implies the amenability of G. In particular, from this we can conclude that all Riemannian symmetric spaces $\mathbb{M}$ of noncompact type are nonamenable, and hence, any Riemannian symmetric space with a de Rham factor of noncompact type is nonamenable as well.

A graph G is (edge) *non*-amenable if

$$i^E(G) := \inf_{\substack{V \subset G \\ |V| < \infty}} \frac{|\partial_E V|}{\mathrm{Vol}_G(V)} > 0. \tag{1.2}$$

where $\partial_E V$ is the set of edges crossing from V to its complement and $\mathrm{Vol}_G(V_n)$ is the sum of degrees of vertices in V. In a Riemannian symmetric space G/K having lattices, the amenability of the space is equivalent to the existence of a lattice $\mathcal{L} < G$ for which the Delaunay graph with nuclei $\mathcal{L} \cdot o$ is amenable. In particular the following theorem is classical (see **[Gel14]** for a discussion).

THEOREM 1.6. *In a Riemannian symmetric space* $\mathbb{M}$ *with distinguished point* o *for which lattices exist, the following are equivalent.*

(1) $\mathbb{M}$ *is amenable.*
(2) *For every lattice* $\mathcal{L}$ *in the isometry group of* $\mathbb{M}$, *the Delaunay graph with nuclei* $\mathcal{L} \cdot o$ *is amenable.*
(3) *There exists a lattice* $\mathcal{L}$ *in the isometry group of* $\mathbb{M}$ *so that the Delaunay graph with nuclei* $\mathcal{L} \cdot o$ *is amenable.*

Many random graphs, such as Galton Watson trees with offspring distribution having positive probability of fewer than 2 children and Poisson–Delaunay graphs are amenable due to random fluctuations, regardless of the amenability of the underlying space $\mathbb{M}$. For this reason, one needs a weaker notion of nonamenability (respectively a stronger notion of amenability) to characterize how nonamenability appears in these random graphs.

One such notion is *anchored amenability.* A rooted graph (G, o) is anchored amenable if there is a sequence of finite sets V_n so that $o \in V_n$, so that the induced subgraph on V_n is connected, and so that

$$\lim_{n \to \infty} \frac{|\partial_E V_n|}{\mathrm{Vol}_G(V_n)} = 0.$$

It is shown in **[BPP14]** that the Poisson–Delaunay graph in $\mathbb{H}^2$ is anchored nonamenable. Forthcoming work with Benjamini and Kraoz shows that a slightly weaker version of anchored nonamenability holds in $\mathbb{H}^d$.

Another, still weaker notion is *invariant amenability,* introduced in **[AL07]** and used to great effect in the study of unimodular random triangulations in **[AHNR16]**. For a unimodular random network (G, o, ψ) with marks Ξ, a *unimodular percolation* is another unimodular random network $(G, o, \psi \times \omega)$ with an augmented mark space $\Xi \times \{0, 1\}$ which is again unimodular. One may consider this as a random subnetwork of the original space by taking only those vertices v and edges e so that $\omega(v) = \omega(e) = 1$, and we assume that if an edge is present in the network, then so are both its endpoints. We let $K_\omega(v)$ denote the connected component of v in this subnetwork. If $\omega(v) = 0$, we take $K_\omega(v) = \{v\}$ by convention.

A percolation is *finitary* if all of its clusters are finite almost surely. A unimodular network is *invariantly amenable* if

$$\inf \left\{ \mathbb{E} \frac{|\partial_E K_\omega(v)|}{|K_\omega(v)|} : \omega \text{ a finitary unimodular percolation} \right\} = 0.$$

This notion allows us to give a precise generalization of Theorem 1.6 for stationary point processes.

THEOREM 1.7. *In a Riemannian symmetric space $\mathbb{M}$ with distinguished point o, the following are equivalent.*

(1) $\mathbb{M}$ *is amenable.*
(2) *For every distributional lattice $\mathcal{P}_o$ the embedded Delaunay graph with nuclei $\mathcal{P}_o$ is invariantly amenable.*
(3) *There exists a distributional lattice $\mathcal{P}_o$ so that the embedded Delaunay graph with nuclei $\mathcal{P}_o$ is invariantly amenable.*

In particular, if $\mathbb{M}$ is nonamenable, every palm process $\mathcal{P}_o$ of a stationary point process with finite volume Voronoi cells produces an invariantly nonamenable Delaunay graph.

Ergodic connections and applications. There are some immediate consequences that one can draw from the invariant nonamenability of a distributional lattice. The first conclusion we draw uses the following theorem.

THEOREM 1.8 (Theorem 3.2 of [**AHNR16**]). *Let (G, o) be an invariantly nonamenable unimodular random rooted graph with $\mathbb{E}[\deg(o) < \infty]$. Then G admits a unimodular percolation ω so that the open subgraph $G|_\omega$ has $i^E(G|_\omega) > 0$ (c.f. (1.2)) almost surely and so that every vertex in $G|_\omega$ has uniformly bounded degrees in G.*

In the cited theorem, the additional assumption of ergodicity is stated, but by the ergodic decomposition this version follows immediately from that one.

Let $(X_j : j = 0, 1, 2, \ldots)$ be simple random walk on (G, o), the Markov chain on the vertices of G where X_{j+1} is distributed uniformly over the neighbors of X_j for all $j = 0, 1, \ldots$. We take $X_0 = o$.

THEOREM 1.9 (Theorem 4.1 of [**AL07**] or Proposition 2.5 of [**BC12**]). *Define the probability measure $\mathbb{Q}$ absolutely continuous with respect to the law of $(G, o, \psi, (X_j)_1^\infty)$ with Radon-Nikodym derivative* $\deg(o)/\mathbb{E}\deg(o)$. *Fix a a representative set of isometries G/K, and let ψ_j be a translate of ψ by that representative isometry of $\mathbb{M}$ so that $\psi_j(X_j) = \psi(o)$. Under $\mathbb{Q}$ the random, rooted embedded Delaunay graphs $((G, X_j, \psi_j))_1^\infty$ are stationary. That is to say the shift map T defined by*

$$T(((G, X_j, \psi_j))_1^\infty) = ((G, X_j, \psi_j))_2^\infty$$

is $\mathbb{Q}$–measure preserving

Note that by mutual absolute continuity of $\mathbb{Q}$ and the original law, events that hold almost surely under one measure hold almost surely under the other.

COROLLARY 1.10. *Let $\mathbb{M}$ be nonamenable. For any distributional lattice $\mathcal{P}_o \subset \mathbb{M}$ the simple random walk on the embedded Delaunay graph (G, o, ψ) has positive speed in the sense that*

$$s = \lim_{n\to\infty} \frac{d_{\mathbb{M}}(X_j, o)}{j}$$

is positive almost surely. If in addition $\mathbb{M}$ is simply connected and nonpositively curved, then for the Poisson–Delaunay graph

$$s^G = \lim_{n\to\infty} \frac{d_G(X_j, o)}{j}$$

is positive almost surely.

This extends the conclusions of [**BPP14**] on random walk on the Poisson-Delaunay graph in the hyperbolic plane, where it was shown that $s > 0$ almost surely. See also [**CLP17**] where the low–intensity asymptotics of the speed are determined for the Poisson–Delaunay graph in $\mathbb{H}^d$.

PROOF. The proof is identical to [**BPP14**, Theorem 1.1], with the random graph $G|_\omega$ taking the place of $G \setminus \mathrm{IS}$. We outline the proof here.

By the subadditive ergodic theorem [**Dur10**, Theorem 6.1], both s and s^G exist almost surely. Hence it will suffice to show positive lim sup speed.

Using that the intensity of $\mathcal{P}$ is a multiple of Riemannian volume measure, by appealing to Markov's inequality and Borel-Cantelli, the number of points in an $\mathbb{M}$-ball of radius R has exponential growth in R, i.e.

$$|\{x \in \mathcal{P}_o : d_{\mathbb{M}}(x, o) \leq R\}| \leq Me^{\alpha R}, \quad \text{for all } R > 0$$

for some deterministic α and some almost surely finite random variable M.

Let $(T_k)_1^\infty$ be the times at which $X_j \in G|_\omega$. Then the induced random walk $(Y_k = X_{T_k} : k = 1, 2, \ldots)$ has a spectral gap, and so its transition probabilities decay exponentially in k. Hence applying Borel–Cantelli, we have that

$$d_{\mathbb{M}}(Y_k, o) \geq \beta k$$

for some $\alpha > 0$ and all k sufficiently large.

We then transfer the result to X_k by invoking the ergodic theorem (under $\mathbb{Q}$) due to which

$$\lim_{n \to \infty} \frac{1}{n} \sum_{j=1}^{n} \mathbf{1}[X_k \in G|_\omega] > 0,$$

which shows that there is at least a subsequence k_n along which

$$d_{\mathbb{M}}(X_{k_n}, o) > \gamma k_n.$$

As the limit s exists almost surely, it follows it is positive almost surely.

As for the graph speed, we have by Proposition 4.1 that there is a constant $t > 0$ so that for all r sufficiently large

$$\psi(B_G(o, r)) \subset B_{\mathbb{M}}(o, tr),$$

and hence the graph speed must also be positive. □

Outline of the paper. In Section 2, we prove Theorem 1.6. In Section 3, we show that the Poisson–Voronoi tessellation is always a distributional lattice. Some of the facts developed here are also applied in Section 2. In Section 4, we prove some additional properties of Poisson–Voronoi tessellations in nonpositively curved spaces. Specifically, we show that the graph ball $B_G(o, R)$ for large R fits with high probability inside the manifold ball $B_{\mathbb{M}}(o, tR)$ for some $t > 0$ (Proposition 4.1). We also show that Poisson–Voronoi tessellations in simply connected, nonpositively curved spaces are *sofic* in the sense that they are local limits of uniformly rooted finite random networks (Proposition 4.3).

Acknowledgements. The author would like to thank Matías Carrasco Piaggio and Pablo Lessa for many helpful conversations; the author would especially like to thank them for comments on Proposition 4.1, which was substantially improved following suggestions of Piaggio and Lessa. Thanks also to Itai Benjamini for helpful comments and inspiring conversations.

2. Distributional lattices

The main goal of this section is the proofs of Theorems 1.4 and 1.6. We begin with the proof of Theorem 1.4.

PROOF OF THEOREM 1.4. By [**AL07**, Proposition 2.2], it suffices to show that for nonnegative f supported on (G, x, y, ψ) with $x \sim y$ that the mass transport principle holds. For any two nonequal points $y, z \in \mathbb{M}$, let $\mathbb{B}_{y,z}$ be the bisector of y and z, i.e. the (not necessarily totally geodesic) submanifold of $\mathbb{M}$ consisting of points that are equidistant from y and z, and let $\mathcal{C}_{y,z}$ be the event that $\{y, z\} \subset \mathcal{X}$ and y connects to z in G. This event

$$\mathcal{C}_{y,z}(\mathcal{P}_o) = \{\exists\ u \in \mathbb{B}_{y,z}\ :\ B_{\mathbb{M}}(u, d_{\mathbb{M}}(u, y)) \cap \mathcal{P}_o = \emptyset\}.$$

For any bijection $\tau : \mathbb{M} \to \mathbb{M}$, let $\tau^* : \sigma(\mathcal{X}) \to \sigma(\mathcal{X})$ be the induced map on events, i.e. for measurable $V \subset \mathbb{M}^{\mathbb{N}}$

$$\tau^*(\{\mathcal{X} \in V\}) = \{\tau(\mathcal{X}) \in V\}.$$

Let $\tau_{y,z}$ denote the isometric involution of $\mathbb{M}$ that reverses the geodesics through the midpoint m between y and z and hence interchanges y and z. As $\mathbb{B}_{y,z}$ is stabilized as a set under $\tau_{y,z}$ it follows that for any $y \in \mathbb{M}$

$$\tau_{o,y}^*(\mathcal{C}_{o,y}(\mathcal{P}_o)) = \mathcal{C}_{o,y}(\mathcal{P}_y).$$

This is because if $B_{\mathbb{M}}(u, r)$ is a ball centered at a point on the bisector $\mathbb{B}_{o,y}$ not intersecting $\mathcal{P}_o$ then $B_{\mathbb{M}}(\tau_{o,y}(u), r)$ is a ball centered at a point on the bisector $\mathbb{B}_{o,y}$ not intersecting $\mathcal{P}_y = \tau_{o,y}(\mathcal{P}_o)$, and vice versa.

Using Palm theory,

$$\mathbb{E}\left[\sum_{x \in \mathcal{P}_o} f(G, o, x, \psi)\right] = \int_{\mathbb{M}} \mathbb{E}_{\mathcal{P}_{o,x}}\left[f(G, o, x, \psi)\mathbf{1}\left[\mathcal{C}_{o,x}(\mathcal{P}_{o,x})\right]\right] \Lambda_o(dx),$$

where Λ_o is the intensity of the reduced palm process $\mathcal{P}_o \setminus \{o\}$ and $\mathcal{P}_{o,x}$ is the point process conditioned to have points at both o and x. By invariance of $\mathcal{P}$, the palm process $\mathcal{P}_{o,x}$ is invariant under any isometry of $\mathbb{M}$ fixing the set $\{o, x\}$, in particular $\tau_{o,x}$, and hence

$$\begin{aligned}
\mathbb{E}_{\mathcal{P}_{o,x}}\left[f(G, o, x, \psi)\mathbf{1}\left[\mathcal{C}_{o,x}(\mathcal{P}_{o,x})\right]\right] &= \mathbb{E}_{\tau_{o,x}^2(\mathcal{P}_{o,x})}\left[f(G, o, x, \psi)\mathbf{1}\left[\mathcal{C}_{o,x}(\mathcal{P}_{o,x})\right]\right] \\
&= \mathbb{E}_{\tau_{o,x}(\mathcal{P}_{o,x})}\left[f(G, x, o, \tau_{o,x} \circ \psi)\mathbf{1}\left[\mathcal{C}_{o,x}(\tau_{o,x}(\mathcal{P}_{o,x}))\right]\right] \\
&= \mathbb{E}_{\mathcal{P}_{o,x}}\left[f(G, x, o, \psi)\mathbf{1}\left[\mathcal{C}_{o,x}(\mathcal{P}_{o,x})\right]\right].
\end{aligned}$$

In the second equation, we have changed the integration to be over the reflected point process. The graph topology does not change on doing this, however the embeddings change, as they correspond to the locations of the points. In the final line, we use the distributional invariance of $\mathcal{P}_{o,x}$. On changing the integration to be against this random variable, we no longer need to reflect the embeddings ψ. Integrating this against $\Lambda_o(dx)$, we conclude that

$$\mathbb{E}\left[\sum_{x \in \mathcal{P}_o} f(G, o, x, \psi)\right] = \mathbb{E}\left[\sum_{x \in \mathcal{P}_o} f(G, x, o, \psi)\right].$$

□

We turn to the proof of Theorem 1.6. This will depend greatly on the unimodularity, which we will use frequently in the following way.

LEMMA 2.1. *Suppose f is a Borel map from embedded rooted Delaunay graphs considered up to isomorphism to the nonnegative real numbers. Let (G, o, ψ) be the embedded Delaunay graph of an invariant Poisson point process $\mathcal{P}_o$. Let $(G, o, \psi \times \omega)$ be a finitary unimodular percolation, and let $K_\omega(v)$ be the connected component in this percolation of v.*

$$\mathbb{E}\left[f(G, o, \psi \times \omega)\right] = \mathbb{E}\left[\sum_{v \in K_\omega(o)} \frac{f(G, v, \psi \times \omega)}{|K_\omega(o)|}\right].$$

PROOF. We define a new nonnegative Borel function g of birooted networks by the rule

$$g(G, x, y, \psi \times \omega) = \frac{f(G, x, \psi \times \omega)\mathbf{1}\left[K_\omega(x) = K_\omega(y)\right]}{|K_\omega(x)|}.$$

Then, on the one hand

$$\mathbb{E}\left[\sum_{y \in \mathcal{P}_o} g(G, o, y, \psi \times \omega)\right] = \mathbb{E}\left[f(G, o, \psi \times \omega)\right].$$

On the other hand,

$$\mathbb{E}\left[\sum_{y \in \mathcal{P}_o} g(G, y, o, \psi \times \omega)\right] = \mathbb{E}\left[\sum_{v \in K_\omega(o)} \frac{f(G, v, \psi \times \omega)}{|K_\omega(o)|}\right],$$

and the equality of these two follows from unimodularity. □

We also need the following simple geometric observation about Voronoi tessellations.

LEMMA 2.2. *Let $S \subset \mathbb{M}$ be a closed discrete set, and let $r > 0$ be arbitrary. Suppose $o, x \in S$, have that $B_{\mathbb{M}}(o, r) \cap \mathcal{V}(x; S) \neq \emptyset$. Then $d_{\mathbb{M}}(x, o) \leq 2r$.*

PROOF. Suppose there is a point $y \in B_{\mathbb{M}}(o, r) \cap \mathcal{V}(x; S)$. Then $d_{\mathbb{M}}(o, y) \leq r$, and $d_{\mathbb{M}}(x, y) \leq d_{\mathbb{M}}(y, o)$ as $y \in \mathcal{V}(x; S)$. Hence $d_{\mathbb{M}}(o, x) \leq 2r$ by the triangle inequality. □

With these observations in place, we turn to proving Theorem 1.6.

PROOF. The implication that (2) $\implies$ (3) is trivial. We show that (1) $\implies$ (2) and that (3) $\implies$ (1), beginning with the former. Let $\mathcal{P}_o$ be a distributional lattice, and let (G, o, ψ) be its embedded Delaunay graph. We wish to show that for any $\epsilon > 0$, there is a finitary unimodular percolation ω for which $(G, o, \psi \times \omega)$ has

$$\mathbb{E}\left[\frac{|\partial_E K_\omega(o)|}{|K_\omega(o)|}\right] \leq \epsilon,$$

with $K_\omega(o)$ the connected component of o. By Lemma 2.1,

$$\mathbb{E}\left[\frac{|\partial_E K_\omega(o)|}{|K_\omega(o)|}\right] = \mathbb{E}\left[\sum_{x : d_G(o,x)=1} \mathbf{1}\left\{\omega(\{x, o\}) = 1\right\}\right]. \tag{2.1}$$

Let $\delta > 0$ and let $\mathcal{Q}$ be a stationary Poisson process on $\mathbb{M}$ with intensity $\delta \cdot \mathrm{Vol}_{\mathbb{M}}(\cdot)$. Define a percolation on (G, o, ψ) by letting $\omega(v) = 1$ for all $v \in \mathcal{P}_o$ and letting

$$\omega(\{x, y\}) = \mathbf{1}\left\{\exists\, q \in \mathcal{Q} \;:\; x, y \in \mathcal{V}(q; \mathcal{Q})\right\},$$

that is to say that x and y are connected if and only if they are in the same Voronoi cell formed by $\mathcal{Q}$. From the almost sure finiteness of the Voronoi cells of $\mathcal{Q}$, this percolation is therefore finite.

Let d_1 be the distance of the closest point in $\mathcal{Q}$ to o, and let d_2 be the distance of the second closest point in $\mathcal{Q}$. Then

$$\mathbb{P}[d_1 \geq t] = \exp(-\delta \operatorname{Vol}_{\mathbb{M}}(B_{\mathbb{M}}(o,t))),$$

and conditional on d_1,

$$\mathbb{P}[d_2 - d_1 \geq t \,|d_1] = \exp\Big(-\delta\Big(\operatorname{Vol}_{\mathbb{M}}(B_{\mathbb{M}}(o,d_1+t)) - (\operatorname{Vol}_{\mathbb{M}}(B_{\mathbb{M}}(o,d_1)))\Big)\Big).$$

Hence, integrating out the dependence on d_1, we get that

$$\mathbb{P}[d_2 - d_1 \geq t] = \int_0^\infty \exp\Big(-\delta\Big(\operatorname{Vol}_{\mathbb{M}}(B_{\mathbb{M}}(o,u+t))\Big) \cdot \delta\partial_u(\operatorname{Vol}_{\mathbb{M}}(B_{\mathbb{M}}(o,u)))\,du.$$

Observe that

$$\operatorname{Vol}_{\mathbb{M}}(B_{\mathbb{M}}(o,u+t)) \geq \operatorname{Vol}_{\mathbb{M}}(B_{\mathbb{M}}(o,u)) + \operatorname{Vol}_{\mathbb{M}}(B_{\mathbb{M}}(o,t/2)),$$

as it is possible to fit disjoint balls of radius u and $t/2$ inside one of radius $u+t$, and hence

$$\mathbb{P}[d_2 - d_1 \geq t] \leq e^{-\delta \operatorname{Vol}_{\mathbb{M}}(B_{\mathbb{M}}(o,t/2))} \cdot \int_0^\infty -\partial_u \exp\Big(-\delta\Big(\operatorname{Vol}_{\mathbb{M}}(B_{\mathbb{M}}(o,u))\Big)\,du.$$

In particular, this probability goes to 1 for any fixed $t > 0$ as $\delta \to 0$.

The relevance of this calculation is that if $o \in \mathcal{V}(q;\mathcal{Q})$, then the distance $d_{\mathbb{M}}(o,q) = d_1$. If $x \in \mathbb{M}$ is any point such that $d_{\mathbb{M}}(o,x) < (d_2-d_1)/2$, then

$$d_{\mathbb{M}}(x,q) < d_2/2 + d_1/2.$$

On the other hand, if $q' \in \mathcal{Q}$ is any other point

$$d_{\mathbb{M}}(x,q') \geq d_{\mathbb{M}}(o,q') - (d_2-d_1)/2 \geq d_2/2 + d_1/2.$$

Hence, we have shown that $B_{\mathbb{M}}(o,(d_2-d_1)/2) \subset \mathcal{V}(q;\mathcal{Q})$. Therefore, we have that

$$\mathbb{E}\left[\sum_{x:d_G(o,x)=1} \mathbf{1}\left\{\omega(\{x,o\})=0\right\}\right] \leq \mathbb{E}\left[\sum_{x:d_G(o,x)=1} \mathbf{1}\left\{d_{\mathbb{M}}(x,o) \geq (d_2-d_1)/2\right\}\right].$$

Since $(d_2-d_1)/2 \to \infty$ in probability as $\delta \to 0$, we get by dominated convergence and (2.1)

$$\lim_{\delta\to 0} \mathbb{E}\left[\sum_{x:d_G(o,x)=1} \mathbf{1}\left\{d_{\mathbb{M}}(x,o) \geq (d_2-d_1)/2\right\}\right] = 0,$$

and so it can be chosen smaller than ϵ, which completes the proof by (2.1).

We turn to showing the implication (3) $\implies$ (1). Let (G,o,ψ) be an invariantly amenable embedded Delaunay graph of a distributional lattice $\mathcal{P}_o$. Let S be a compact subset of the isometry group of $\mathbb{M}$. By Lemma 2.2, for any $r > 0$

$$\mathbb{P}[B_{\mathbb{M}}(o,r) \not\subset \mathcal{V}(o)] \leq \mathbb{P}[B_{\mathbb{M}}(o,2r) \cap \mathcal{P}_o \neq \{o\}] \to 0,$$

as $r \to 0$ as $\mathcal{P}_o$ is a simple point process. Hence for some $\delta > 0$,

$$\mathbb{P}[\operatorname{Vol}_{\mathbb{M}}(\mathcal{V}(o)) \leq \delta] \leq \frac{1}{100}.$$

Let $\epsilon > 0$ be arbitrary. In what follows, we let $(G, o, \psi \times \omega)$ be any finitary percolation, and let $K_\omega(o)$ be its component containing o. We will use this percolation to construct a finite volume measurable $V \subset \mathbb{M}$ which satisfies the condition

$$\max_{g \in S} \frac{\mathrm{Vol}_{\mathbb{M}}(gV \triangle V)}{\mathrm{Vol}_{\mathbb{M}}(V)} \leq C(\delta)\epsilon \tag{2.2}$$

for some $C(\delta) > 0$. Hence, on taking $\epsilon \to 0$ along some sequence, this gives the desired conclusion. The set in question will be given by

$$\mathscr{V} = \bigcup_{v \in K_\omega(o)} \mathcal{V}(v),$$

and we will show there is positive probability that this set satisfies the desired condition (2.2) for some constant $C(\delta)$. As this occurs with positive probability, the desired set must exist.

We begin by observing that the average volume of a Voronoi cell in $K_\omega(o)$ can not be too small. Using Lemma 2.1,

$$\mathbb{E}\left[\sum_{v \in K_\omega(o)} \frac{\mathbf{1}\left[\mathrm{Vol}_{\mathbb{M}}(\mathcal{V}(v)) \leq \delta\right]}{|K_\omega(o)|}\right] = \mathbb{P}\left[\mathrm{Vol}_{\mathbb{M}}(\mathcal{V}(o)) \leq \delta\right] \leq \frac{1}{100}.$$

Hence on applying Markov's inequality, we conlude

$$\mathbb{P}\left[\sum_{v \in K_\omega(o)} \frac{\mathbf{1}\left[\mathrm{Vol}_{\mathbb{M}}(\mathcal{V}(v)) \leq \delta\right]}{|K_\omega(o)|} \geq \frac{1}{2}\right] \leq \frac{1}{50}. \tag{2.3}$$

Hence with probability at least $49/50$ we have that $\mathrm{Vol}_{\mathbb{M}}(\mathscr{V}) \geq \frac{1}{2}\delta|K_\omega(o)|$.

Conversely, the contribution of large Voronoi cells can not be too big. From dominated convergence,

$$\lim_{M \to \infty} \mathbb{E}\left[\mathrm{Vol}_{\mathbb{M}}(\mathcal{V}(o))\mathbf{1}\left\{\mathrm{Vol}_{\mathbb{M}}(\mathcal{V}(o)) \geq M\right\}\right] \to 0.$$

Applying Lemma 2.1, we can find an M sufficiently large that

$$\mathbb{E}\left[\sum_{v \in K_\omega(o)} \frac{\mathrm{Vol}_{\mathbb{M}}(\mathcal{V}(v))\mathbf{1}\left\{\mathrm{Vol}_{\mathbb{M}}(\mathcal{V}(v)) \geq M\right\}}{|K_\omega(o)|}\right] \leq \epsilon.$$

Applying Markov's inequality,

$$\mathbb{P}\left[\sum_{v \in K_\omega(o)} \frac{\mathrm{Vol}_{\mathbb{M}}(\mathcal{V}(v))\mathbf{1}\left\{\mathrm{Vol}_{\mathbb{M}}(\mathcal{V}(v)) \geq M\right\}}{|K_\omega(o)|} \geq 50\epsilon\right] \leq \frac{1}{50}. \tag{2.4}$$

We turn to arguing that while Voronoi cells may be large (in fact they could be unbounded for a nonuniform lattice in a negatively curved space such as $\mathrm{SL}_2(\mathbb{Z}) < \mathrm{SL}_2(\mathbb{R})$), their volume is necessarily mostly contained in a bounded region. Specifically just by dominated convergence and the assumption that $\mathbb{E}\,\mathrm{Vol}_{\mathbb{M}}(\mathcal{V}(o)) < \infty$ almost surely,

$$\lim_{R \to \infty} \mathbb{E}[\mathrm{Vol}_{\mathbb{M}}(\mathcal{V}(o) \setminus B_{\mathbb{M}}(o, R))] = 0.$$

Hence, by Lemma 2.1 there is an R sufficiently large that

$$\mathbb{E}\left[\sum_{v \in K_\omega(o)} \frac{\mathrm{Vol}_{\mathbb{M}}(\mathcal{V}(v) \setminus B_{\mathbb{M}}(v, R))}{|K_\omega(v)|}\right] < \epsilon.$$

Applying Markov's inequality,

$$\mathbb{P}\left[\sum_{v\in K_\omega(o)} \frac{\mathrm{Vol}_{\mathbb{M}}(\mathcal{V}(v)\setminus B_{\mathbb{M}}(v,R))}{|K_\omega(v)|} \geq 50\epsilon\right] \leq \frac{1}{50}. \tag{2.5}$$

By compactness of S, there is an $N>0$ so that $d_{\mathbb{M}}(x,s(x))\leq N$ for all $s\in S$ and all $x\in B_{\mathbb{M}}(o,R)$.

We next show that most Voronoi cells have the property that shifting them by some $s\in S$ leaves them in the interior of $\mathscr{V}$. From the finiteness of the intensity of $\mathcal{P}_o$, we have that

$$\lim_{T\to\infty}\mathbb{P}\left[|\mathcal{P}_o\cap B_{\mathbb{M}}(o,2(R+N))|\geq T\right]\to 0.$$

Hence we may find a T sufficiently large that this probability is strictly less than ϵ. By Lemma 2.2, this T also bounds the number of Voronoi cells intersecting $B_{\mathbb{M}}(o,R+N)$. Using Lemma 2.1 and Markov's inequality in the same way as in (2.3) and (2.5),

$$\mathbb{P}\left[\sum_{v\in K_\omega(o)} \frac{\mathbf{1}\left\{|\mathcal{P}_o\cap B_{\mathbb{M}}(v,2(R+N))|\geq T\right\}}{|K_\omega(v)|} \geq 50\epsilon\right] \leq \frac{1}{50}. \tag{2.6}$$

By assumption there is a percolation ω so that

$$\mathbb{E}\left[\frac{|\partial_E K_\omega(v)|}{|K_\omega(v)|}\right] < \frac{\epsilon}{1+T}.$$

And hence,

$$\mathbb{P}\left[\frac{|\partial_E K_\omega(v)|}{|K_\omega(v)|}\geq\frac{50\epsilon}{1+T}\right]\leq\frac{1}{50}. \tag{2.7}$$

From here on, we will work under the event that the complements of the union of events whose probability is estimated in (2.3),(2.4),(2.5),(2.6),and (2.7). This complementary event has probability at least 45/50.

Call a vertex $v\in K_\omega(o)$ *good* if every nucleus of every Voronoi cell intersecting $B_{\mathbb{M}}(v,R+N)$ is not adjacent to a vertex of $K_\omega(o)^c$, and call it bad otherwise. Let B denote the set of bad vertices, and let Q be the set of bad vertices $v\in K_\omega(o)$ that have

$$|\mathcal{P}_o\cap B_{\mathbb{M}}(v,2(R+N))|\leq T,$$

then $|Q|\leq T\cdot|\partial_E K_\omega(o)|\leq\epsilon\cdot|K_\omega(o)|$. as $|Q|$ is at most a T-fold overcounting of the number of vertices in $K_\omega(o)$ adjacent to something outside $K_\omega(o)$, which is itself a lower bound for $|\partial_E K_\omega(o)|$. Hence, the total number of bad vertices is at most $51\epsilon|K_\omega(o)|$.

Now for any good vertex v and any $s\in S$ we have

$$s(\mathcal{V}(v)\cap B_{\mathbb{M}}(v,R))\subset B_{\mathbb{M}}(v,R+N)\subset\mathscr{V}.$$

Therefore, we have that

$$\begin{aligned}\mathrm{Vol}_{\mathbb{M}}(s\mathscr{V} \setminus \mathscr{V}) &\leq \sum_{v \in K_\omega(o)} \mathrm{Vol}_{\mathbb{M}}(\mathcal{V}(v) \setminus B_{\mathbb{M}}(v,R)) + \sum_{v \in B} \mathrm{Vol}_{\mathbb{M}}(\mathcal{V}(v)) \\ &\leq 50\epsilon|K_\omega(o)| + M|B| + \sum_{v \in K_\omega(o)} \mathrm{Vol}_{\mathbb{M}}(\mathcal{V}(v)) \geq M\mathbf{1}\left\{\mathrm{Vol}_{\mathbb{M}}(\mathcal{V}(v))\right\} \\ &\leq 50\epsilon|K_\omega(o)| + 51M\epsilon|K_\omega(o)| + 50\epsilon|K_\omega(o)|.\end{aligned}$$

Now, as $|K_\omega(o)| \leq 2\delta^{-1}\,\mathrm{Vol}_{\mathbb{M}}(\mathscr{V})$, we have shown that

$$\mathrm{Vol}_{\mathbb{M}}(s\mathscr{V} \setminus \mathscr{V}) \leq (200 + 102M)\epsilon\delta^{-1}\,\mathrm{Vol}_{\mathbb{M}}(\mathscr{V}).$$

The same bound holds for

$$\mathrm{Vol}_{\mathbb{M}}(\mathscr{V} \setminus s(\mathscr{V})) = \mathrm{Vol}_{\mathbb{M}}(s^{-1}(\mathscr{V}) \setminus \mathscr{V}),$$

on account of s^{-1} satisfying the same displacement bound as s, and so we have shown

$$\mathrm{Vol}_{\mathbb{M}}(s\mathscr{V} \triangle \mathscr{V}) \leq 2(200 + 102M)\epsilon\delta^{-1}\,\mathrm{Vol}_{\mathbb{M}}(\mathscr{V}) := \epsilon C(\delta)\,\mathrm{Vol}_{\mathbb{M}}(\mathscr{V}),$$

as desired. □

3. General Properties of Poisson–Voronoi tilings in Symmetric Spaces

In this section, we study in some more detail the Poisson–Voronoi tessellations. We will let Π^λ denote a Poisson point process with intensity $\lambda \cdot \mathrm{Vol}_{\mathbb{M}}$. The main purpose of this section is to show that these tessellations are always distributional lattices, i.e. they have finite expected volume and degree. In fact, more is true.

We will let $f(r) = \mathrm{Vol}_{\mathbb{M}}(B_{\mathbb{M}}(o,r))$, the volume growth function. We will need some simple estimates on this function, which we summarize as follows.

LEMMA 3.1. *Suppose $\mathbb{M}$ is a noncompact Riemannian symmetric space. Then there are constants c_1 and $c_2 > 0$ so that for all $r \geq 1$,*

$$e^{c_1 r} \geq f(r) \geq c_2 r.$$

Further $\lim_{r\to\infty} f(r)^{1/r}$ exists.

PROOF. By homogeneity, the scalar curvature is bounded below by some $-\alpha < 0$. From this lower bound, we get an exponential upper bound on the volume growth by comparing with the corresponding constant curvature space of the corresponding dimension (see for example [**Lee09**, Theorem 11.1/2]).

When $\mathbb{M}$ is noncompact, there is an infinite geodesic γ connecting $o \to \infty$, and hence we can bound the volume growth below by comparing with the portion of the volume of a unit–distance tubular neighborhood of γ that is contained within $B_{\mathbb{M}}(o,r)$, which gives rise to the lower bound.

The existence of the limit follows from a subadditivity argument. □

We use this to find a tail bound for the diameter of $\mathcal{V}(o;\Pi^\lambda)$.

LEMMA 3.2. *Suppose $\mathbb{M}$ is a noncompact Riemannian symmetric space. There is a constant $C > 0$ so that for all $R > 1$,*

$$\mathbb{P}\left[\mathcal{V}(o;\Pi^\lambda) \not\subset B_{\mathbb{M}}(o,R)\right] \leq Cf(R)\exp(-\lambda f(R-1)).$$

PROOF. Let $R > 0$ be fixed. Let $\{\Delta_j\}_1^M$ be a minimal 1–net of the sphere $\partial B_{\mathbb{M}}(o, R)$. By a sphere packing argument, $M \leq C \cdot f(R)$ for some $C > 0$ depending only on $\mathbb{M}$. Hence,

$$\mathbb{P}\left[\exists\, 1 \leq j \leq M : B_{\mathbb{M}}(\Delta_j, R-1) \cap \Pi^\lambda = \emptyset\right] \leq Cf(R)\exp(-\lambda f(R-1)).$$

If $q \in \Pi^\lambda$ is a point whose cell is adjacent to $\mathcal{V}(o;\Pi^\lambda)$, then there is a point u with $r = d_{\mathbb{M}}(u,q) = d_{\mathbb{M}}(u,o)$ and so that $B_{\mathbb{M}}(u,r) \cap \Pi^\lambda = \emptyset$. If $r \geq R$, then if we let u' be the point on the geodesic from u to o with $d_{\mathbb{M}}(u',o) = R$, the ball $B_{\mathbb{M}}(u',R) \subset B_{\mathbb{M}}(u,r)$. Further, there is a closest point Δ_j to u' which is necessarily at distance less than 1 from it. Hence $B_{\mathbb{M}}(\Delta_j, R-1) \subset B_{\mathbb{M}}(u',R)$, and so is again empty. Thus,

$$\mathbb{P}\left[\mathcal{V}(o;\Pi^\lambda) \not\subset B_{\mathbb{M}}(o,R)\right] \leq \mathbb{P}\left[\exists\, 1 \leq j \leq M : B_{\mathbb{M}}(\Delta_j, R-1) \cap \Pi^\lambda = \emptyset\right],$$

completing the proof. □

THEOREM 3.3. *Let $\mathbb{M}$ be a Riemannian symmetric space. Let $\lambda > 0$ be arbitrary, and let $\mathcal{P}^\lambda$ be a stationary Poisson process on $\mathbb{M}$. Then for any $k \geq 0$,*

$$\mathbb{E}\left[\mathrm{Vol}_{\mathbb{M}}(\mathcal{V}(o;\Pi^\lambda))^k\right] < \infty \qquad \textit{and} \qquad \mathbb{E}\left[\deg(o)^k\right] < \infty.$$

PROOF. In the case that $\mathbb{M}$ is compact, the result is trivial, as the volume of the whole manifold is finite and the degree is bounded by the total number of points, which is Poisson($\lambda\,\mathrm{Vol}_{\mathbb{M}}(\mathbb{M})$).

In the case that $\mathbb{M}$ is noncompact, we use the bound in Lemma 3.2. Conditional on the smallest R so that

$$\mathcal{V}(o;\Pi^\lambda) \subset B_{\mathbb{M}}(o,R),$$

we can estimate $\mathrm{Vol}_{\mathbb{M}}(\mathcal{V}(o;\Pi^\lambda)) \leq \mathrm{Vol}_{\mathbb{M}}(B_{\mathbb{M}}(o,R)) = f(R)$. Hence,

$$\mathbb{E}[\mathrm{Vol}_{\mathbb{M}}(\mathcal{V}(o;\Pi^\lambda))^k] \leq \int_0^\infty kf'(R)f(R)^k C\exp(-\lambda f(R-1))\,dR. \tag{3.1}$$

Since $\lim_{R\to\infty} f(R)^{1/R}$ exists, for any $\delta > 0$ we can estimate $f(R)^k \leq C_{\delta,k}e^{\delta f(R-1)}$. Likewise, we can estimate $f(R-1) \geq cf(R)$ for some $c \in (0,1)$ for all $R > 2$. In particular, we can estimate the tail of the integral by

$$\begin{aligned}&\int_2^\infty f'(R)f(R)^k \exp(-\lambda f(R-1))\,dR\\ &\leq C\int_2^\infty f'(R)\exp(-\eta f(R))\,dR < \infty,\end{aligned}$$

for some constant $\eta > 0$.

Turning to the degree bound, with the same random R as used to control the volume, we have that all neighbors of o are contained in $B_{\mathbb{M}}(o, 2R)$. Hence

$$\begin{aligned}
\mathbb{E}[\deg(o)^k] &\leq \sum_{j=0}^{\infty} \mathbb{E}\left[|\Pi^\lambda \cap B_{\mathbb{M}}(o, 2j+2)|^k \mathbf{1}\{j+1 \geq R \geq j\}\right] \\
&\leq \sum_{j=0}^{\infty} \mathbb{E}\left[|\Pi^\lambda \cap B_{\mathbb{M}}(o, 2j+2)|^k \mathbf{1}\{R \geq j\}\right] \\
&\leq \sum_{j=0}^{\infty} \left(\mathbb{E}\left[|\Pi^\lambda \cap B_{\mathbb{M}}(o, 2j+2)|^{2k}\right] \mathbb{P}[R \geq j] \right)^{1/2} \\
&\leq C_0 + C(k,\lambda) \sum_{j=2}^{\infty} f(2j+2)^k f(j)^{1/2} e^{-\lambda f(j-1)/2} \\
&\leq C_0 + C(k,\lambda)' \sum_{j=2}^{\infty} f(j-1)^{2k+1} e^{-\lambda f(j-1)/2} < \infty,
\end{aligned}$$

where we have used that the volume grows at most exponentially to compare $f(2j+2) \leq Cf(j-1)^2$ for some $C > 0$ and all $j \geq 2$. □

4. Additional structure for Poisson–Voronoi tessellations in nonpositively curved spaces

Non-positive curvature is beneficial for many reasons, one of which is that the notion of convexity translates well to nonpositively curved space (see [**Ebe96**, Section 1.6]). We will also deal exclusively with the simply connected case, for which we recall that $\mathbb{M} = \mathbb{R}^d \times \prod_{i=1}^r \mathbb{M}_i$ for some Riemannian symmetric spaces of noncompact type $\mathbb{M}_i$. In particular, these spaces are amenable if and only if the space is some Euclidean space. As a consequence, when there is a non-Euclidean factor, the volume growth is necessarily exponential, i.e.

$$\lim_{r\to\infty} f(r)^{1/r} = h > 0. \tag{4.1}$$

For Riemannian symmetric spaces of noncompact type, more precise estimates are available [**Kni97**].

We also let $\hbar$ denote the isoperimetric constant of $\mathbb{M}$, that is the largest real number so that for all piecewise smooth compact submanifolds A with boundary,

$$|\partial A| \geq \hbar \operatorname{Vol}_{\mathbb{M}}(A),$$

where $|\partial A|$ denotes the surface measure of ∂A. The constant $\hbar = 0$ if and only if $\mathbb{M}$ is amenable. For the particular case that $\mathbb{M} = \mathbb{H}^d$, it is a Theorem of [**Yau75**] (and also a relatively straightforward consequence of the divergence theorem) that $\hbar = d - 1$.

By a theorem of Borel [**Bor63**], in any simply connected Riemannian symmetric space, there is a co-compact lattice Γ. This means there is a countable collection of points Λ and a group of isometries of $\mathbb{M}$ that act transitively on Λ, with the further property that the Voronoi cells with nuclei Λ are bounded. Let $\mathscr{L}$ denote the dual graph of these cells, which by virtue of the transitive action on Λ becomes a transitive graph. To a point $y \in \Lambda$, we define $\mathcal{L}(y)$, the Voronoi cell in the Λ-nucleated tessellation that is centered at y. The large-scale geometry of $\mathbb{M}$ is captured by the large-scale geometry of $\mathscr{L}$: any map $\pi : \mathbb{M} \to \Lambda$ with the property that if $\pi(x) = y$ then $x \in \mathcal{L}(y)$ is a quasi-isometry of the two spaces.

PROPOSITION 4.1. *Let $\mathbb{M}$ be a nonpositively curved, nonamenable, simply connected Riemannian symmetric space. Let (G, o, ψ) be the embedded Delaunay graph with nuclei Π^λ for some $\lambda > 0$. For any $\alpha < \hbar$, there are constants $C, t_0 > 0$ depending only on α so that for all $\lambda > 0$, $t > t_0$ and all $R > 1$,*

$$\mathbb{P}\left[B_G(o, R) \not\subseteq B_{\mathbb{M}}(x_0, tR)\right] \leq e^{R(C(1+t) - \lambda e^{\alpha t/2})}.$$

As a consequence, there are constants $\kappa, \beta, R_0 > 0$ depending only on $\mathbb{M}$ and λ so that for all $R > R_0$,

$$\mathbb{P}\left[|B_G(o, R)| \geq e^{\kappa R}\right] \leq e^{-\beta R}.$$

In particular, we have that

$$\limsup_{r \to \infty} |B_G(o, R)|^{1/R} < \infty,$$

almost surely.

To prove this, we begin with the following Lemma on the volume of a union of balls in a nonamenable space.

LEMMA 4.2. *Let $\mathbb{M}$ be a nonamenable Riemannian symmetric space. For any $\Delta > 0$ and any $0 < \alpha < \hbar$ there is a constant $C > 0$ so that for any finite collection of balls $\{B_{\mathbb{M}}(x_i, r_i)\}_{i=1}^t$ with $d_{\mathbb{M}}(x_i, x_j) \geq \Delta$ whenever $i \neq j$, we have*

$$\mathrm{Vol}_{\mathbb{M}}\left(\cup_{i=1}^t B_{\mathbb{M}}(x_i, r_i)\right) \geq \sum_{i=1}^t \frac{e^{\alpha r_i}}{C} \mathbf{1}\left\{r_i \geq \tfrac{\Delta}{2}\right\}.$$

PROOF. Set

$$W = \bigcup_{\substack{i \in \{1, \dots, t\} \\ r_i \geq \Delta/2}} B_{\mathbb{M}}(x_i, r_i).$$

We have that for any continuous $\phi : \mathbb{R} \to \mathbb{R}^+$, setting $\Phi(x) = \int_0^x \phi(y)\, dy$, that

$$\int_W \phi(d_{\mathbb{M}}(x, W^c))\, dV(x) \geq h \int_W \Phi(d_{\mathbb{M}}(x, W^c))\, dV(x), \tag{4.2}$$

by decomposing the integral into level sets of $d_{\mathbb{M}}(x, W^c)$ and applying integration by parts.

Using (4.2) with $\phi(x) = e^{\alpha x}$, we therefore get that

$$\int_W e^{\alpha d_{\mathbb{M}}(x, W^c)}\, dV(x) \geq \frac{h}{\alpha}\left(\int_W e^{\alpha d_{\mathbb{M}}(x, W^c)}\, dV(x) - \mathrm{Vol}_{\mathbb{M}}(W)\right).$$

Rearranging, we get that

$$\mathrm{Vol}_{\mathbb{M}}(W) \geq \frac{h - \alpha}{h} \int_W e^{\alpha d_{\mathbb{M}}(x, W^c)}\, dV(x). \tag{4.3}$$

We note that there is a constant $C = C(\alpha, \Delta) > 0$ so that for any $1 \leq i \leq t$ with $r_i \geq \frac{\Delta}{2}$,

$$e^{\alpha r_i} \leq C \int_{B_{\mathbb{M}}(x_i, \Delta/2)} e^{\alpha d_{\mathbb{M}}(x, W^c)}\, dV(x).$$

Hence

$$\sum_{i=1}^{t} e^{\alpha r_i} \mathbf{1}\left\{r_i \geq \tfrac{\Delta}{2}\right\} \leq C \sum_{i=1}^{t} \mathbf{1}\left\{r_i \geq \tfrac{\Delta}{2}\right\} \int_{B_{\mathbb{M}}(x_i,\Delta/2)} e^{\alpha d_{\mathbb{M}}(x,W^c)}\, dV(x).$$
$$\leq C \int_W e^{\alpha d_{\mathbb{M}}(x,W^c)}\, dV(x).$$

By (4.3), the proof is complete. □

Proof of Proposition 4.1. Recall that Λ is a collection of points on whose Voronoi cells $\{\mathcal{L}(y)\}_{y\in\Lambda}$ there is a subgroup of the isometry group of $\mathbb{M}$ that acts transitively. In particular this implies that all $\{\mathcal{L}(y)\}_{y\in\Lambda}$ have the same diameter and volume. Let Δ denote this diameter.

We let G denote the dual graph of the Poisson Voronoi tessellation. Let G' be the graph formed by adding to G an edge between every pair of Poisson Voronoi cells that are at distance at most 2Δ. Note that this implies that $B_G(o,R) \subseteq B_{G'}(o,R)$ for all $R > 0$, and hence it suffices to show the claim for G'.

Suppose that $x_0 x_1 x_2 \cdots x_R$ is a collection of Π^λ whose Poisson Voronoi cells form a geodesic path in G'. Let $J \subseteq \{0,1,2,\ldots,R-1\}$ be those indices for which x_i and x_{i+1} are adjacent in G. For every $i \in J$, there is an open ball $B_{\mathbb{M}}(z_i,r_i)$ with $B_{\mathbb{M}}(z_i,r_i) \cap \Pi^\lambda = \emptyset$ and $\{x_i,x_{i+1}\} \subset \partial B_{\mathbb{M}}(z_i,r_i)$. Let $y_i \in \Lambda$ be a closest point to z_i (that is a point $y_i \in \Lambda$ so that $z_i \in \mathcal{L}(y_i)$), and let ρ_i be the largest integer (possibly 0) so that $V_i = B_{\mathbb{M}}(y_i,\rho_i) \subset B_{\mathbb{M}}(z_i,r_i)$.

For $i \notin J$, there are points u,v with u on the boundary of $\mathcal{V}(x_i)$ and v on the boundary of $\mathcal{V}(x_{i+1})$ so that $d_{\mathbb{M}}(u,v) \leq 2\Delta$. Let $r = d_{\mathbb{M}}(u,x_i)$ and $s = d_{\mathbb{M}}(v,x_{i+1})$. Both of $B_{\mathbb{M}}(u,r)$ and $B_{\mathbb{M}}(v,s)$ do not intersect Π^λ. Let z_i be the midpoint of uv. Set $r_i = \max(r,s) - \Delta$. Then we have $B_{\mathbb{M}}(z_i,r_i)$ is contained in one of $B_{\mathbb{M}}(u,r)$ or $B_{\mathbb{M}}(v,s)$. As before, let $y_i \in \Lambda$ be a closest point to z_i, and let ρ_i be the largest integer so that $V_i = B_{\mathbb{M}}(y_i,\rho_i) \subset B_{\mathbb{M}}(z_i,r_i)$.

Note that for all $0 \leq i \leq R-1$, $r_i \leq 1 + \Delta + \rho_i$, as y_i and z_i are at most Δ apart. Note that for $i \notin J$, we have that

$$d_{\mathbb{M}}(x_i,x_{i+1}) \leq d_{\mathbb{M}}(x_i,u) + d_{\mathbb{M}}(u,v) + d_{\mathbb{M}}(v,x_{i+1}) \leq 2r_i + 4\Delta \leq 2\rho_i + 2 + 6\Delta.$$

For $i \in J$ we have by the triangle inequality, $d_{\mathbb{M}}(x_i,x_{i+1}) \leq 2r_i \leq 2\rho_i + 2 + 2\Delta$

$$d_{\mathbb{M}}(x_0,x_R) \leq \sum_{i=0}^{R-1} d_{\mathbb{M}}(x_i,x_{i+1})$$
$$\leq \sum_{i\notin J} d_{\mathbb{M}}(x_i,x_{i+1}) + \sum_{i\in J} d_{\mathbb{M}}(x_i,x_{i+1})$$
$$\leq (6\Delta + 2)\cdot R + \sum_{i=0}^{R-1} 2\rho_i. \tag{4.4}$$

Hence provided that $d_{\mathbb{M}}(x_0,x_R)/R$ is large, so too will be the sum of radii of these balls.

For $0 \leq i \leq R-1$, we can estimate the lattice distance,

$$d_{\mathscr{L}}(y_i,y_{i+1}) \leq C d_{\mathbb{M}}(z_i,z_{i+1}) + C$$
$$\leq C(d_{\mathbb{M}}(z_i,x_i) + d_{\mathbb{M}}(z_{i+1},x_{i+1}) + d_{\mathbb{M}}(x_i,x_{i+1})) + C$$
$$\leq C(\rho_i + \rho_{i+1} + 2\rho_i) + C'$$

for a sufficiently large constant C' depending on Δ. Also, let $y_{-1} \in \Lambda$ be a point so that $x_0 \in \mathcal{L}(y_{-1})$. Then we have that $d_{\mathbb{M}}(x_0, z_0) \le r_0 + 2\Delta$, and so we get a similar bound for $d_{\mathscr{L}}(y_{-1}, y_0)$ in terms of ρ_0 and Δ. We conclude that

$$\sum_{i=-1}^{R-1} d_{\mathscr{L}}(y_i, y_{i+1}) \le CR + C\sum_{i=0}^{R-1} \rho_i \tag{4.5}$$

for some constant C depending on Δ.

We now show that the centers $\{y_i\}$ of these balls are mostly distinct (and hence separated from each other as they are elements of Λ). In fact, by this construction, it could be that $y_i = y_{i+1}$. However, it will transpire that if $i+1 < \ell$, then $y_i \ne y_\ell$.

If $i \in J$, the point z_i is contained in $\mathcal{V}(x_i) \cap \mathcal{V}(x_{i+1})$. The point z_ℓ is within distance Δ of $\mathcal{V}(x_{\ell+1})$. Hence if $d_{\mathbb{M}}(z_i, z_\ell) \le \Delta$, the distance between $\mathcal{V}(x_i)$ and $\mathcal{V}(x_{\ell+1})$ is at most 2Δ. In this case the path $x_0x_1x_2\cdots x_ix_{\ell+1}x_{\ell+2}\cdots x_R$ is a shorter path from x_0 to x_R, contradicting that the path was a geodesic. The same proof works if $\ell \in J$, now using that z_ℓ is in $\mathcal{V}(x_{\ell+1})$. Therefore, the separation of z_i and z_ℓ is strictly greater than Δ, and hence they can not lie in the same Voronoi cell $\mathcal{L}(y)$ for any $y \in \Lambda$. In particular, $y_i \ne y_\ell$.

Suppose that both of i and ℓ are not in J. Then by construction z_i and z_ℓ are at most distance Δ from $\mathcal{V}(x_i)$ and $\mathcal{V}(x_{\ell+1})$ respectively. The midpoint of z_i and z_ℓ is contained in some $\mathcal{V}(w)$ for some $w \in \Pi^\lambda$. If $d_{\mathbb{M}}(z_i, z_\ell) \le 2\Delta$, then the distance between $\mathcal{V}(w)$ and $\mathcal{V}(x_i)$ is at most 2Δ, as is the distance between $\mathcal{V}(y)$ and $\mathcal{V}(x_{\ell+1})$. Hence the path $x_0x_1x_2\cdots x_iwx_{\ell+1}x_{\ell+2}\cdots x_R$ is a shorter path from x_0 to x_R, contradicting that the path was a geodesic. Hence, we once again conclude that z_i and z_ℓ can not be contained in the same Voronoi cell $\mathcal{L}(y)$ for any $y \in \Lambda$, and so $y_i \ne y_\ell$.

Because the centers are mostly distinct, we can bound below the volume of the union of the balls $\{V_i\}_0^{R-1}$. Specifically, we have shown that $(y_i : i \text{ odd})$ and $(y_i : i \text{ even})$ are distinct. Hence, applying Lemma 4.2, for any $\alpha < h$ there is a constant $C(\alpha)$ so that

$$\begin{aligned} \mathrm{Vol}_{\mathbb{M}}(\cup_{i \text{ odd}} V_i) &\ge C \sum_{i \text{ odd}} e^{\alpha\rho_i} \mathbf{1}\{\rho_i > 0\} \text{ and} \\ \mathrm{Vol}_{\mathbb{M}}(\cup_{i \text{ even}} V_i) &\ge C \sum_{i \text{ even}} e^{\alpha\rho_i} \mathbf{1}\{\rho_i > 0\}. \end{aligned}$$

As one of these lower bounds is larger, we conclude that

$$\mathrm{Vol}_{\mathbb{M}}(\cup_i V_i) \ge \frac{C}{2} \sum_{i=0}^{R-1} e^{\alpha\rho_i} \mathbf{1}\{\rho_i > 0\},$$

for the same constant C. By convexity, we conclude that

$$\mathrm{Vol}_{\mathbb{M}}(\cup_i V_i) \ge \frac{RC}{2} \exp\left(\frac{\alpha}{R}\sum_{i=0}^{R-1} \rho_i\right). \tag{4.6}$$

Hence we have a lattice path $\gamma \in \mathscr{L}$ connecting y_{-1} to y_{R-1}, and at each y_i, $0 \le i \le R-1$ we have an integer radius ρ_i so that the ball $B_{\mathbb{M}}(y_i, \rho_i)$ is empty. The union of these balls will have a large volume when the sum of ρ_i is a large multiple of R. Let $s = \sum_{i=0}^{R-1} \rho_i$, and let ℓ denote the length of γ. Recall that by (4.5) there is a $D > 0$ so that $\ell \leq D(R+s)$. In conclusion, if we define the events

$$\begin{aligned} &A((\rho_i, y_i)_{i=0}^{R-1}) = \{V_i = B_{\mathbb{M}}(y_i, \rho_i) \cap \Pi^\lambda = \emptyset, \forall\, 0 \leq i \leq R-1\}, \\ &A_\gamma((\rho_i)_{i=0}^{R-1}) = \cup_{(y_i) \subset \gamma} A((\rho_i, y_i)_{i=0}^{R-1}), \\ &A_*(s) = \cup_{\substack{\gamma : \ell \leq D(R+s) \\ (\rho_i) : \sum \rho_i = s}} A_\gamma((\rho_i)_{i=0}^{R-1}), \end{aligned} \tag{4.7}$$

then on the complement of $A_*(s)$, we have by (4.4) that every length-R geodesic started at x_0 in G' satisfies

$$d_{\mathbb{M}}(x_0, x_R) \leq (6\Delta + 2) \cdot R + 2s.$$

Hence, we need to estimate the probability of $A_*(s)$.

For the first event in (4.7), using (4.6), we have that for any non-negative integral choices of ρ_i for $(y_i)_0^{R-1}$ satisfying the assumptions of (4.6)

$$\mathbb{P}[A((\rho_i, y_i)_{i=0}^{R-1})] \leq e^{-\lambda \operatorname{Vol}_{\mathbb{M}}(\cup_i V_i)} \leq e^{-\frac{\lambda RC}{2} \cdot \exp\left(\frac{\alpha}{R} \sum_{i=0}^{R-1} \rho_i\right)}.$$

For the second event in (4.7), the number of possible ways to choose which vertices will be centers of V_i is at most $\binom{\ell}{\lceil R/2 \rceil}\binom{\ell}{\lfloor R/2 \rfloor}$, as the even vertices are pairwise distinct and the odd vertices are pairwise distinct. Hence applying the union bound over all such choices and increasing C as needed,

$$\begin{aligned} \mathbb{P}[A_\gamma((\rho_i)_{i=0}^{R-1})] &\leq \binom{\ell}{\lceil R/2 \rceil}\binom{\ell}{\lfloor R/2 \rfloor} e^{-\frac{\lambda RC}{2} \cdot \exp\left(\frac{\alpha}{R} \sum_{i=0}^{R-1} \rho_i\right)} \\ &\leq e^{C(R+s) - \frac{\lambda RC}{2} \cdot \exp\left(\frac{\alpha}{R} \sum_{i=0}^{R-1} \rho_i\right)}, \end{aligned}$$

where we have used that ℓ is at most $D(R+s)$.

For the final even in (4.7), we sum over the possible choices of γ and the possible choices of $(\rho_i)_0^{R-1}$ summing to s. The graph $\mathcal{L}$ is regular, and so the number of paths γ is at most $C^\ell \leq C^{D(R+s)}$ with C the degree of a vertex in the Delaunay graph on $\mathcal{L}$. As for the choices of non-negative integers $(\rho_i)_{i=0}^{R-1}$ summing to s, there are exactly $\binom{s+R-1}{R-1} \leq 2^{s+R}$ choices. Increasing C as needed,

$$\mathbb{P}[A_*(s)] \leq e^{C(R+s) - \frac{\lambda RC}{2} \cdot \exp\left(\frac{\alpha s}{R}\right)},$$

which concludes the proof of the first part of the proposition.

To prove the second part of the proposition, note that by the first part, we can find some t, β, and R_0 so that for all $R > R_0$,

$$\mathbb{P}\left[B_G(o, R) \not\subseteq B_{\mathbb{M}}(x, tR)\right] \leq e^{-\beta R}.$$

On the event that $B_G(o,R) \subseteq B_{\mathbb{M}}(x,tR)$, we have that $|B_G(o,R)| \leq |\Pi^\lambda \cap B_{\mathbb{M}}(x,tR)|$. And by Poisson tails, we have that

$$\mathbb{P}[|\Pi^\lambda \cap B_{\mathbb{M}}(x, tR)| \geq 2\lambda f(tR)] \leq e^{-\lambda f(tR)/C}$$

for some constant. Hence, combining this with the previous bound and adjusting constants gives the second consequence. The third part of the proposition follows immediately by Borel-Cantelli. □

PROPOSITION 4.3. *When $\mathbb{M}$ is a Riemannian symmetric space of noncompact type, the embedded Delaunay network (G, o, ψ) with nuclei Π^λ is a random weak limit of finite random networks.*

REMARK 4.4. This was observed earlier in [**BS01**, "Hyperbolic Surfaces" proof of Theorem 6.2].

PROOF. The core of the proof is the existence of the following family of spaces (see [**DW78**, Theorem 2.1]). There is a family of Riemannian manifolds $\{S_r\}_{r=1}^{\infty}$ so that S_r has the property that any ball of radius r in S_r is isometric to $B_{\mathbb{M}}(0,r) \subseteq \mathbb{M}$. Hence, on S_r we can define a Poisson point process $\Pi_{S_r}^{\lambda}$ whose intensity measure on any ball of radius r is the pullback of the intensity of Π^{λ} on $B_{\mathbb{M}}(0,r)$. We can also associate to $\Pi_{S_r}^{\lambda}$ its associated Voronoi tessellation, and we define (G_r, ρ_r, ψ_r) to be the embedded Delaunay network, where ρ_r is a uniformly chosen vertex of G_r. We claim that (G, o, ψ) is the local limit of (G_r, ρ_r, ψ_r), i.e. (G, o, ψ) is the random weak limit of G_r. From the almost sure finiteness of the Voronoi cells with nuclei Π^{λ} on $\mathbb{M}$, we have that for each $r > 0$,

$$\lim_{q\to\infty} \mathbb{P}\left[\max_{x\in B_G(\rho,r)} \operatorname{diam}_{\mathbb{M}}(\mathcal{V}(x)) > q\right] = 0.$$

Hence by diagonalization, we can find some sequence $\{q_r\}_{r=1}^{\infty} \subset \mathbb{N}$ with $q_r \to \infty$ so that

$$\lim_{r\to\infty} \mathbb{P}\left[\max_{x\in B_G(\rho,r)} \operatorname{diam}_{\mathbb{M}}(\mathcal{V}(x)) > q_r\right] = 0.$$

Let $\mathcal{E}_r$ be the event $\left\{\max_{x\in B_G(\rho,r)} \operatorname{diam}_{\mathbb{M}}(\mathcal{V}(x)) \le q_r\right\}$. On $\mathcal{E}_r$, every nucleus $x \in B_G(\rho,r)$ is contained in $B_{\mathbb{M}}(x_0, q_r \cdot (r+1))$. Further, if $y \in \Pi^{\lambda}$ is a neighbor of some $x \in B_G(\rho,r)$, then there is a point $z \in \mathcal{V}(x) \cap \mathcal{V}(y)$ that is equidistant to x and to y. Hence $d_{\mathbb{M}}(x,y) \le 2d_{\mathbb{M}}(x,z) \le 2q_r$. In particular, the event $\mathcal{E}_r$ is measurable with respect to $\Pi^{\lambda} \cap B_{\mathbb{M}}(x_0, q_r \cdot (r+2))$.

For every $r \in \mathbb{N}$, let

$$j(r) = \max\{s \in \mathbb{N} : q_s \cdot (s+2) \le r\}.$$

Then we have that

$$\lim_{r\to\infty} \mathbb{P}\left[\max_{x\in B_{G_r}(\rho_r,j(r))} \operatorname{diam}_{S_r}(\mathcal{V}(x)) > q_{j(r)}\right] = 0.$$

Moreover, the law of $B_{G_r}(\rho_r, j(r))$ on the event $\max_{x\in B_{G_r}(\rho_r,j(r))} \operatorname{diam}_{S_r}(\mathcal{V}(x)) \le q_{j(r)}$ coincides with the law of $B_G(\rho, r)$ on the event $\mathcal{E}_{j(r)}$, from which the local weak convergence follows. □

References

[AHNR16] Omer Angel, Tom Hutchcroft, Asaf Nachmias, and Gourab Ray, *Unimodular hyperbolic triangulations: circle packing and random walk*, Invent. Math. **206** (2016), no. 1, 229–268, DOI 10.1007/s00222-016-0653-9. MR3556528

[AL07] David Aldous and Russell Lyons, *Processes on unimodular random networks*, Electron. J. Probab. **12** (2007), no. 54, 1454–1508, DOI 10.1214/EJP.v12-463. MR2354165

[BC12] Itai Benjamini and Nicolas Curien, *Ergodic theory on stationary random graphs*, Electron. J. Probab. **17** (2012), no. 93, 20, DOI 10.1214/EJP.v17-2401. MR2994841

[BH99] Martin R. Bridson and André Haefliger, *Metric spaces of non-positive curvature*, Grundlehren der Mathematischen Wissenschaften [Fundamental Principles of Mathematical Sciences], vol. 319, Springer-Verlag, Berlin, 1999. MR1744486

[Bor63] Armand Borel, *Compact Clifford-Klein forms of symmetric spaces*, Topology **2** (1963), 111–122, DOI 10.1016/0040-9383(63)90026-0. MR0146301

[BPP14] I. Benjamini, E. Paquette, and J. Pfeffer, *Anchored expansion, speed, and the hyperbolic Poisson Voronoi tessellation*, submitted (2014).

[BS01] Itai Benjamini and Oded Schramm, *Percolation in the hyperbolic plane*, J. Amer. Math. Soc. **14** (2001), no. 2, 487–507, DOI 10.1090/S0894-0347-00-00362-3. MR1815220

[CLP17] Matìas Carrasco, Pablo Lessa, and Elliot Paquette, *A Furstenberg type formula for the speed of distance stationary sequences*, ArXiv e-prints (2017).

[Dur10] Rick Durrett, *Probability: theory and examples*, 4th ed., Cambridge Series in Statistical and Probabilistic Mathematics, vol. 31, Cambridge University Press, Cambridge, 2010. MR2722836

[DVJ03] D. J. Daley and D. Vere-Jones, *An introduction to the theory of point processes. Vol. I*, 2nd ed., Probability and its Applications (New York), Springer-Verlag, New York, 2003. Elementary theory and methods. MR1950431

[DW78] David L. de George and Nolan R. Wallach, *Limit formulas for multiplicities in* $L^2(\Gamma\backslash G)$, Ann. of Math. (2) **107** (1978), no. 1, 133–150, DOI 10.2307/1971140. MR0492077

[Ebe96] Patrick B. Eberlein, *Geometry of nonpositively curved manifolds*, Chicago Lectures in Mathematics, University of Chicago Press, Chicago, IL, 1996. MR1441541

[G+10] André Goldman, *The Palm measure and the Voronoi tessellation for the Ginibre process*, Ann. Appl. Probab. **20** (2010), no. 1, 90–128, DOI 10.1214/09-AAP620. MR2582643

[Gel14] Tsachik Gelander, *Lectures on lattices and locally symmetric spaces*, Geometric group theory, IAS/Park City Math. Ser., vol. 21, Amer. Math. Soc., Providence, RI, 2014, pp. 249–282. MR3329730

[Hel01] Sigurdur Helgason, *Differential geometry, Lie groups, and symmetric spaces*, Graduate Studies in Mathematics, vol. 34, American Mathematical Society, Providence, RI, 2001. Corrected reprint of the 1978 original. MR1834454

[HKPV09] J. Ben Hough, Manjunath Krishnapur, Yuval Peres, and Bálint Virág, *Zeros of Gaussian analytic functions and determinantal point processes*, University Lecture Series, vol. 51, American Mathematical Society, Providence, RI, 2009. MR2552864

[Kal86] Olav Kallenberg, *Random measures*, 4th ed., Akademie-Verlag, Berlin; Academic Press, Inc., London, 1986. MR854102

[Kni97] G. Knieper, *On the asymptotic geometry of nonpositively curved manifolds*, Geom. Funct. Anal. **7** (1997), no. 4, 755–782, DOI 10.1007/s000390050025. MR1465601

[Lee09] Jeffrey M. Lee, *Manifolds and differential geometry*, Graduate Studies in Mathematics, vol. 107, American Mathematical Society, Providence, RI, 2009. MR2572292

[Yau75] Shing Tung Yau, *Isoperimetric constants and the first eigenvalue of a compact Riemannian manifold*, Ann. Sci. École Norm. Sup. (4) **8** (1975), no. 4, 487–507. MR0397619

Department of Mathematics, The Ohio State University, Columbus, Ohio 43210
Email address: paquette.30@osu.edu

Contemporary Mathematics
Volume **719**, 2018
https://doi.org/10.1090/conm/719/14471

Eternal Family Trees and dynamics on unimodular random graphs

Francois Baccelli, Mir-Omid Haji-Mirsadeghi, and Ali Khezeli

ABSTRACT. This paper is centered on covariant dynamics on unimodular random graphs and random networks (marked graphs), namely maps from the set of vertices to itself which are preserved by graph or network isomorphisms. Such dynamics are referred to as vertex-shifts here.

The first result of the paper is a classification of vertex-shifts on unimodular random networks. Each such vertex-shift partitions the vertices into a collection of connected components and foils. The latter are discrete analogues the stable manifold of the dynamics. The classification is based on the cardinality of the connected components and foils. Up to an event of zero probability, there are three classes of foliations in a connected component: F/F (with finitely many finite foils), I/F (infinitely many finite foils), and I/I (infinitely many infinite foils).

An infinite connected component of the graph of a vertex-shift on a random network forms an infinite tree with one selected end which is referred to as an Eternal Family Tree. Such trees can be seen as stochastic extensions of branching processes. Unimodular Eternal Family Trees can be seen as extensions of critical branching processes. The class of offspring-invariant Eternal Family Trees, which is introduced in the paper, allows one to analyze dynamics on networks which are not necessarily unimodular. These can be seen as extensions of not necessarily critical branching processes. Several construction techniques of Eternal Family Trees are proposed, like the joining of trees or moving the root to a far descendant.

Eternal Galton-Watson Trees and Eternal Multitype Galton-Watson Trees are also introduced as special cases of Eternal Family Trees satisfying additional independence properties. These examples allow one to show that the results on Eternal Family Trees unify and extend to the dependent case several well known theorems of the literature on branching processes.

1. Introduction

A network is a graph with marks on its vertices and edges. The marks can be used to assign weights to the vertices and lengths, capacities or directions to the edges. A rooted network is a network with a distinguished vertex called its root. The heuristic interpretation of unimodularity is that the root is *equally likely* to be any vertex, even though the network may have an infinite number of vertices.

2010 *Mathematics Subject Classification.* Primary 60C05; Secondary 60K99, 05C80.
Key words and phrases. Local weak convergence, networks, infinite graphs, branching process, criticality, offspring-invariance, point-shift, dynamical system, stable manifold, foliation.

This interpretation is made precise by assuming that the random rooted network satisfies the *mass transport principle.*

This paper studies general (that is, not necessarily continuous, not necessarily measure preserving) dynamical systems on unimodular networks. A dynamical system on a network G is a map $f : V(G) \to V(G)$, where $V(G)$ denotes the set of vertices of G. If f depends on the network in a measurable and isomorphism-covariant way, it will be called a *vertex shift.* The *stable manifolds* of f form a partition of $V(G)$ which will be referred to as the foliation of the vertices here: two vertices v and w are in the same *foil*[1] when $f^n(w) = f^n(v)$ for some $n \geq 0$. *Connected components* give another partition of $V(G)$ in the sense that v and w are in the same component if $f^n(v) = f^m(w)$ for some $m, n \geq 0$.

The first result of this paper is a classification of such dynamical systems on unimodular networks in terms of the cardinality of the connected components, the cardinality of the foils, and the limit of the images of f. It is established that, almost surely, there are three classes of components: (F/F), those where there are finitely many foils each of finite cardinality, and where the images of the component under f^n tend to a limit cycle (this is the only case where connected components can have a cycle); (I/F), those with infinitely many foils, all with finite cardinality, where the images under f^n tend to a single bi-infinite path (this is the only case where connected components form a two-ended tree); and (I/I), those with infinitely many foils, all with infinite cardinality, where the images under f^n converge to the empty set (this is the only case where connected components form a one-ended tree). In the last case, the set of f-pre-images of all orders of any vertex in the component is a.s. finite.

Palm probabilities of stationary point processes provide a subclass of unimodular networks, so that this classification generalizes that proposed in [**7**] for point-shifts on stationary point processes. Several other basic ideas from the theory of point processes are extended to general unimodular networks. In particular analogues of Mecke's point stationarity theorem for Palm probabilities and of Neveu's exchange formula between two Palm probabilities are derived. Also, as discussed in [**6**], there is an intimate connection between the theory of unimodular networks and that of *invariant measured equivalence relations* [**15**]. This will be explained in the bibliographical comments at the end of the relevant sections (in particular Subsection 2.3).

Any infinite connected component (or connected set of orbits) described above can be regarded as a directed tree, which is called an *Eternal Family Tree.* Such trees are studied in complement to the classification, with a focus on basic properties and construction methods. Such trees share similarities with branching processes, by regarding $f(v)$ as the *parent* of v, with the specificity that there is no vertex which is an ancestor of all other vertices, which explains the chosen name. For dynamics on unimodular networks, these branching like processes are always critical, that is, the expected number of children of the root of the associated Eternal Family Tree is one. The class of offspring-invariant Eternal Family Trees, defined in the paper, is used to analyze a class of dynamics on non-unimodular networks such that the associated Eternal Family Tree generalizes non-critical branching processes.

[1] Usually, in the context of foliations, the word "leaf" is used. But since the paper discusses several types of trees where the word leaf has another meaning, the authors preferred to use the word "foil" rather than "leaf".

A general way for obtaining unimodular and offspring-invariant Eternal Family Trees as limits is presented. Roughly speaking, it consists in moving the root of a given family tree to a *typical far vertex* (Theorem 4.11). In particular, any unimodular Eternal Family Tree of class $\mathcal{I}/\mathcal{I}$ is the limit of some random finite tree when moving the root to a typical far vertex (Proposition 4.15). Another general way for obtaining unimodular Eternal Family Trees is by joining the roots of a stationary sequence of rooted trees with finite mean population. Any unimodular Eternal Family Tree of class $\mathcal{I}/\mathcal{F}$ is the joining of a sequence of such trees (Theorem 4.17).

The *Eternal Galton-Watson Tree* is introduced in Section 6. It is obtained from the classical Galton-Watson Tree by moving the root to a typical far vertex. It is also characterized as the Eternal Family Trees which are offspring-invariant and enjoy a certain independence property. This framework is extended to the multi-type setting as well. These trees allow one to show in what sense the general results obtained on offspring-invariant Eternal Family Trees unify classical analytical results of the literature on branching processes and extend them to a non-independent framework.

The paper is structured as follows. Basic definitions on unimodular networks are recalled in Section 2. Vertex-shifts are discussed in Section 3 together with the classification theorem. Eternal Family Trees are discussed in Section 4 together with construction methods like tree joining and moving to a far vertex. Section 5 gathers results on offspring-invariance. The Eternal Galton-Watson Tree and related examples are discussed in Section 6.

The paper is based on the following literature: it leverages the framework for unimodular networks elaborated in [**6**], and it extends to all unimodular networks the classification of dynamics on the points of a stationary point process established in [**7**]. To the best of the authors' knowledge, the unimodular network dynamics classification theorem, the notions of Eternal Family Tree and Eternal Galton-Watson Tree, which are used to state and embody this classification, as well as the constructions proposed for such trees are new. Each section contains a bibliographical subsection which gathers the connections with earlier results.

2. Unimodular Networks

2.1. Definition. The following definitions are borrowed from [**6**]. A **network** is a (multi-) graph $G = (V, E)$ equipped with a complete separable metric space Ξ, called the **mark space** and with two maps from V and $\{(v, e) : v \in V, e \in E, v \sim e\}$ to Ξ (the symbol $\sim$ is used for adjacent pairs of vertices or edges). The image of v (resp. (v, e)) in Ξ is called its **mark**. Note that graphs and directed graphs are special cases of networks. In this paper, all networks are assumed to be locally finite; that is, the degree of each vertex is assumed to be finite. Moreover, a network is assumed to be connected except when explicitly mentioned. For $r \geq 0$, the closed ball of (graph-distance) radius $\lfloor r \rfloor$ with center $v \in V(G)$ is denoted by $N_r(G, v)$. An **isomorphism** between two networks is a graph isomorphism which also preserves the marks, namely the mark of the image of v (resp. (v, e)) by the bijection is the same as that of v (resp. (v, e)). A network isomorphism from G to itself is referred to as an **automorphism** of G.

A **rooted network** is a pair (G, o) in which G is a network and o is a distinguished vertex of G called the **root**. An **isomorphism** of rooted networks is a network isomorphism that takes the root of one to that of the other. Let $\mathcal{G}$ denote the set of isomorphism classes of connected and locally finite networks and $\mathcal{G}_*$ the set of isomorphism classes of rooted, connected and locally finite networks. The set $\mathcal{G}_{**}$ is defined similarly for networks with a pair of distinguished vertices. The isomorphism class of a network G (resp. (G,o) or (G,o,v)) is denoted by $[G]$ (resp. $[G,o]$ or $[G,o,v]$). The sets $\mathcal{G}_*$ and $\mathcal{G}_{**}$ can be equipped with a metric and its Borel sigma-field. The distance between (G,o) and (G',o') is $2^{-\alpha}$, where α is the supremum of those $r>0$ such that there is a rooted isomorphism between $N_r(G,o)$ and $N_r(G',o')$ such that the distance of the marks of the corresponding elements is at most $\frac{1}{r}$. This makes $\mathcal{G}_*$ a complete separable metric space which will be equipped with its Borel sigma-field in order to get a measure space. In this paper, several definitions will be provided which use non-rooted networks (e.g., Definitions 2.5, 2.7 and 3.1), but measurability issues are discussed using $\mathcal{G}_*$ and $\mathcal{G}_{**}$. In fact, it should be noted that $\mathcal{G}$ is not a standard space (when requiring measurability of the projection from $\mathcal{G}_*$ onto $\mathcal{G}$).

A **random network** is a measurable function from some probability space $(\Omega,\mathcal{F},\mathbb{P})$ to $\mathcal{G}_*$. This function will be denoted by $[\boldsymbol{G},\boldsymbol{o}]$ with bold symbols. A probability measure $\mathcal{P}$ on $\mathcal{G}_*$ can also be regarded as a random network when considering the identity map on the canonical probability space $(\mathcal{G}_*,\mathcal{P})$. When a richer probability space is needed, the blackboard bold notation $\mathbb{P}$ will be used instead.

For all measurable functions $g:\mathcal{G}_{**}\to\mathbb{R}^{\geq 0}$ and all $[G,o]\in\mathcal{G}_*$, let (throughout the paper, $[G,o]$ will be used in place of $([G,o])$ for the sake of light notation).

$$\begin{aligned} g_G^+(o) &:= g^+[G,o] := \sum_{v\in V(G)} g[G,o,v], \\ g_G^-(o) &:= g^-[G,o] := \sum_{v\in V(G)} g[G,v,o]. \end{aligned}$$

It is straightforward to see that g^+ and g^- are well-defined measurable functions on $\mathcal{G}_*$.

DEFINITION 2.1. A random network $[\boldsymbol{G},\boldsymbol{o}]$ is **unimodular** if for all measurable functions $g:\mathcal{G}_{**}\to\mathbb{R}^{\geq 0}$,

$$\mathbb{E}\left[g_{\boldsymbol{G}}^+(\boldsymbol{o})\right]=\mathbb{E}\left[g_{\boldsymbol{G}}^-(\boldsymbol{o})\right], \tag{2.1}$$

where the expectations may be infinite. A probability measure on $\mathcal{G}_*$ is called **unimodular** when, by considering it as a random network, one gets a unimodular network.

Equation (2.1) is referred to as the *mass transport principle*. Many examples of unimodular networks can be found in the literature, see for instance [**6**]. Here, only some elementary examples are recalled for the purpose of illustrating the new definitions. More elaborate examples will be discussed throughout the paper. In the following example, an **end** in a tree ([**13**]) is an equivalence class of semi-infinite simple paths (also known as *rays*), where two simple paths are equivalent when they have infinitely many common vertices.

EXAMPLE 2.2 (Regular Tree). The deterministic d-regular rooted tree T_d rooted at an arbitrary vertex is a unimodular network. In contrast, it can be shown that the d-regular rooted tree with one distinguished end is only unimodular in the case $d = 2$.

EXAMPLE 2.3 (Canopy Tree). The Canopy Tree with offspring cardinality $d \in \mathbb{N}$, introduced in [**1**], is the tree C_d whose vertices can be partitioned in infinitely many layers $L_0, L_1, L_2, \ldots$ and such that for each $i \geq 0$, each vertex $v \in L_i$ is connected to exactly one vertex $F(v)$ in L_{i+1}, and (for $i \geq 1$) to exactly d vertices in L_{i-1}. It is a one-ended tree. Assuming $d \geq 2$, let $\boldsymbol{o}$ be a random vertex such that $\mathbb{P}[\boldsymbol{o} \in L_i] = cd^{-i}$, where $c = \frac{d-1}{d}$ (note that $[C_d, v] = [C_d, w]$ for $v, w \in L_i$). Then, $[C_d, \boldsymbol{o}]$ is a unimodular network. This can be proved directly from the definition; another proof is given in Example 5.12 below.

EXAMPLE 2.4 (Marks: the i.i.d. Case). Let $[\boldsymbol{G}, \boldsymbol{o}]$ be a unimodular network. It is possible to enrich the marks of the vertices and edges of $[\boldsymbol{G}, \boldsymbol{o}]$ with i.i.d. marks which are also independent of the network itself. By doing so, the new random rooted network is also unimodular (Lemma 4.1 in [**9**]). If $[\boldsymbol{G}, \boldsymbol{o}]$ is the usual grid structure of $\mathbb{Z}^d$ with the origin as the root, the same holds if the marks are stationary (but not necessarily i.i.d.). i.e. if their joint distribution is invariant under the translations of $\mathbb{Z}^d$. More generally, graphs which are covariant with a stationary (marked) point process in $\mathbb{R}^d$ allow one to construct unimodular random networks in a systematic way by taking the root to be the origin under the Palm probability of the point process.

2.2. On Subset Selection. This subsection mainly formulates a preliminary result, the *no infinite/finite inclusion lemma* (Lemma 2.9 below), which will be extensively used in the proofs, and which is also of independent interest.

DEFINITION 2.5. A **covariant subset (of the set of vertices)** is a map S which associates to each network G a set $S_G \subseteq V(G)$ which is (1) covariant under network isomorphisms (that is, for all isomorphisms $\rho : G \to G'$, one has $\rho(S_G) = S_{G'}$), and (2) such that the function $[G, o] \mapsto 1_{\{o \in S_G\}}$ is measurable.

For instance, the set of vertices with a given degree defines a covariant subset.

LEMMA 2.6. *Let $[\boldsymbol{G}, \boldsymbol{o}]$ be a unimodular network and S be a covariant subset of the vertices. Then $\mathbb{P}[S_{\boldsymbol{G}} \neq \emptyset] > 0$ if and only if $\mathbb{P}[\boldsymbol{o} \in S_{\boldsymbol{G}}] > 0$.*

PROOF. Define $g[G, o, s] := 1_{\{s \in S_G\}}$ (which is well-defined and measurable by the definition of S_G). Assume $\mathbb{P}[S_{\boldsymbol{G}} \neq \emptyset] > 0$. By (2.1),

$$\begin{aligned} 0 < \mathbb{E}[\# S_{\boldsymbol{G}}] &= \mathbb{E}\left[\sum_{s \in V(\boldsymbol{G})} g[\boldsymbol{G}, \boldsymbol{o}, s]\right] \\ &= \mathbb{E}\left[\sum_{s \in V(\boldsymbol{G})} g[\boldsymbol{G}, s, \boldsymbol{o}]\right] = \mathbb{E}\left[1_{\{\boldsymbol{o} \in S_{\boldsymbol{G}}\}} \# V(\boldsymbol{G})\right]. \end{aligned}$$

Therefore, $\mathbb{P}[\boldsymbol{o} \in S_{\boldsymbol{G}}] > 0$ (note that the situation when $\# V(\boldsymbol{G}) = \infty$ poses no problem). The converse is clear. □

DEFINITION 2.7. A **covariant (vertex) partition** is a map Π which associates to all networks G a partition Π_G of $V(G)$ which is (1) covariant under networks isomorphisms (that is $\Pi_{\rho(G)} = \rho \circ \Pi_G$ for all isomorphisms ρ), and (2) such that the (well-defined) subset $\{[G, o, s] : s \in \Pi_G(o)\} \subseteq \mathcal{G}_{**}$ is measurable, where $\Pi_G(o)$ is the element of Π_G that contains o.

For examples, one can partition the vertices based on their degrees, which defines a covariant partition. As another example, it can be seen that by deleting all cut edges (i.e., edges the deletion of which disconnects the graph), the connected components of the remaining graph define a covariant partition.

REMARK 2.8. A covariant partition Π is not necessarily a collection of disjoint covariant subsets. Indeed, in some cases there is no covariant subset which is always an element of Π (in other words, one can not *select* an element of Π).

The following lemma will be used several times in Subsection 3.3 below.

LEMMA 2.9 (No Infinite/Finite Inclusion). *Let $[\boldsymbol{G}, \boldsymbol{o}]$ be a unimodular network, Π a covariant partition, and S a covariant subset. Almost surely, there is no infinite element E of $\Pi_{\boldsymbol{G}}$ such that $E \cap S_{\boldsymbol{G}}$ is finite and non-empty.*

PROOF. Let

$$g[G, o, s] := \frac{1}{\#(\Pi_G(o) \cap S_G)} 1_{\{s \in \Pi_G(o) \cap S_G, 0 < \#(\Pi_G(o) \cap S_G) < \infty\}}.$$

Suppose the claim is not true. Thus, by Lemma 2.6, with a positive probability $0 < \#(\Pi_{\boldsymbol{G}}(\boldsymbol{o}) \cap S_{\boldsymbol{G}}) < \infty$, $\#\Pi_{\boldsymbol{G}}(\boldsymbol{o}) = \infty$ and $\boldsymbol{o} \in S_{\boldsymbol{G}}$. Therefore, $\mathbb{E}\left[g_{\boldsymbol{G}}^-(\boldsymbol{o})\right] = \infty$ by the definition of g. But $0 \leq g^+ \leq 1$. This contradicts the mass transport relation (2.1). $\square$

COROLLARY 2.10. *Let $[\boldsymbol{G}, \boldsymbol{o}]$ be a unimodular network in which $V(\boldsymbol{G})$ is almost surely infinite. Then any covariant subset S of the vertices is almost surely either empty or infinite; that is,*

$$\mathbb{P}\left[\#S_{\boldsymbol{G}} \in \{0, \infty\}\right] = 1.$$

PROOF. This is a direct corollary of Lemma 2.9 by letting Π_G be the trivial partition $\{V(G)\}$. $\square$

The notion of subnetwork is a natural extension of that of subset.

DEFINITION 2.11. A **covariant subnetwork** H is a map $G \mapsto H_G$ defined for all networks G, where H_G is the restriction of G to a covariant subset. For a unimodular network $[\boldsymbol{G}, \boldsymbol{o}]$, the **intensity** of H relative to the $\boldsymbol{G}$ is defined by $\lambda_H^{\boldsymbol{G}} := \mathbb{P}[\boldsymbol{o} \in V(H_{\boldsymbol{G}})]$ (see Lemma 2.6 below). The conditioning of $\mathbb{P}$ on the event $\boldsymbol{o} \in V(H_{\boldsymbol{G}})$ will be denoted by $\mathbb{P}_H$.

REMARK 2.12. It is easy to see that $[H_{\boldsymbol{G}}, \boldsymbol{o}]$, conditioned on $\boldsymbol{o} \in V(H_{\boldsymbol{G}})$, is a unimodular network, provided $H_{\boldsymbol{G}}$ is connected a.s.

The following proposition gives a general connection between two covariant subnetworks.

PROPOSITION 2.13 (Exchange Formula). *Let $[\boldsymbol{G}, \boldsymbol{o}]$ be a unimodular network and H and H' be two covariant subnetworks. For all measurable functions $g : \mathcal{G}_{**} \to \mathbb{R}^{\geq 0}$,*

$$\lambda_H^{\boldsymbol{G}} \mathbb{E}_H \left[\sum_{v' \in V(H'_{\boldsymbol{G}})} g[\boldsymbol{G}, \boldsymbol{o}, v'] \right] = \lambda_{H'}^{\boldsymbol{G}} \mathbb{E}_{H'} \left[\sum_{v \in V(H_{\boldsymbol{G}})} g[\boldsymbol{G}, v, \boldsymbol{o}] \right].$$

PROOF. Let $\hat{g}(G,v,w) := 1_{\{v\in V(H_G)\}}1_{\{w\in V(H'_G)\}}g[G,v,w]$. The claim is a direct implication of (2.1) for $\hat{g}$. □

2.3. Bibliographical Comments. As already mentioned, the underlying framework for this section is that of [**6**]. In particular, a result analogous to Lemma 2.6 can be found in Lemma 2.3 in [**6**].

Example 9.9 of [**6**] shows that the theory of unimodular random graphs and networks has substantial overlap with that of *invariant measured equivalence relations* (developed in [**15**]), which is briefly recalled below. See also the literature cited in Example 9.9 of [**6**]. As will be seen, the existence of automorphisms makes a difference between the two theories (some issues are discussed below and in the bibliographical comments of Section 3). However, this is no major difference in the applications where generality is not lost by adding i.i.d. marks. Therefore, many definitions and results on unimodular networks can be stated in (or have analogous results in) the theory of measured equivalence relations and vice versa. The next two paragraphs do so for covariant subsets, covariant partitions and the no infinite/finite inclusion lemma (Lemma 2.9). However, covariant subnetworks (Definition 2.11) seem not to have a neat analogue. This is also discussed further in the bibliographical notes of Section 3. It should be noted that, to the best of the authors' knowledge, the analogues of main results of the present paper are new in the theory of measured equivalence relations; e.g., Theorem 3.9.

Adjacency of vertices provides a natural graph structure on $\mathcal{G}_*$ (for every network G and all adjacent pairs $x,y\in V(G)$, put an edge between $[G,x]$ and $[G,y]$). Connected components of this graph provide an equivalence relation on $\mathcal{G}_*$. Then, a probability measure on $\mathcal{G}_*$ creates an instance of a *measured equivalence relation* equipped with a *graphing* (the general definition is discussed below). If $[\boldsymbol{G},\boldsymbol{o}]$ is a random rooted network which has no non-rooted automorphisms other than the identity a.s., then the graph properties of $[\boldsymbol{G},\boldsymbol{o}]$ can be recovered from the graphing. See [**6**] for more details. In the general case, a **measured equivalence relation** is a tuple (X,R,μ), where X is a Polish space, μ is a Borel probability measure on X and $R\subseteq X^2$ is a Borel subset which is an equivalence relation such that each equivalence class is finite or countable. For $x\in X$, let $R[x]$ denote the equivalence class of x. Assume μ is *R-invariant*; i.e.,

$$\int \sum_{y\in R[x]} f(x,y)d\mu(x) = \int \sum_{y\in R[x]} f(y,x)d\mu(x),$$

for all Borel functions $f: X^2\to\mathbb{R}^{\geq 0}$ (this is analogous to unimodularity). The notion of covariant subsets (Definition 2.5) is analogous to Borel subsets $A\subseteq X$ (note that in the case of $\mathcal{G}_*$ described above, $\{v\in G : [G,v]\in A\}$ provides a covariant subset and vice versa). An equivariant partition (Definition 2.7) is analogous to a Borel equivalence relation R' on X such that $R'\subseteq R$. Then, Lemma 2.9 can be extended as follows with a similar proof.

Lemma 2.9': Let (X,R,μ) be an invariant measured equivalence relation, A a Borel subset of X, and R' a Borel sub-equivalence relation of R. Then for μ-a.e. $x\in X$, there is no element E of R' such that $E\subseteq R[x]$, E is infinite and $E\cap A$ is finite and non-empty.

Unimodular networks have many analogies with stationary point processes, and more precisely the Palm versions of stationary (marked) point processes [**12**]. Since random graphs and networks which are covariant with a stationary point process are unimodular [**23**], the main notions and results of this section can be seen as extensions of notions and results already known in the point process case. The notion of subnetwork in Definition 2.11 is analogous to that of thinning of a point process based on a selection by the marks [**12**]. The associated notion of intensity extends that of a sub-point process. The exchange formula (Proposition 2.13) is the analogue of Neveu's exchange formula [**26**]. The no infinite/finite inclusion lemma (Lemma 2.9) extends a result in [**7**] (in fact, the idea behind this lemma is present in some proofs in the literature in special cases). All these extensions and formalizations are new to the best of the authors' knowledge.

3. Vertex-Shifts and Foil Classification

Roughly speaking, vertex-shifts are dynamics on the vertices of networks and prescribe to what vertex to move from a given vertex.

3.1. Vertex-Shifts as Dynamical Systems.

DEFINITION 3.1. A **covariant vertex-shift** is a map f which associates to each network G a function $f_G : V(G) \to V(G)$ such that (i) f is covariant under isomorphisms (that is, for all isomorphisms $\rho : G \to G'$, one has $f_{G'} \circ \rho = \rho \circ f_G$) and (ii) the well-defined function $[G, o, s] \mapsto 1_{\{f_G(o)=s\}}$ on $\mathcal{G}_{**}$ is measurable. Below, a vertex-shift will always mean a covariant vertex-shift.

REMARK 3.2. Note that the definition of vertex-shifts (as well as covariant subsets and partitions) does not require a probability measure on $\mathcal{G}_*$. Also note that a vertex-shift should be defined for all networks. When defining a vertex-shift on a subclass of networks, the last requirement can be met by making the vertex-shift to be the identity on networks outside of this subclass.

EXAMPLE 3.3. For an infinite tree T with a distinguished end (for instance the Canopy Tree of Example 2.3), let $f_T(v)$ be the neighbor of v in the unique semi-infinite simple path that starts at v and passes through the distinguished end. Then f is a vertex-shift.

EXAMPLE 3.4. Consider the example of i.i.d. marks (Example 2.4). Assume the mark space is a subset of $\mathbb{R}$. Let $f(v)$ be neighbor of v with the smallest mark (let $f(v) = v$ if there is a tie). Then f is a vertex-shift.

Similarly, let $f'(v)$ be the neighbor w of v such that the mark of edge vw is the smallest (and $f'(v) = v$ if there is a tie). Then f' is also a vertex-shift.

EXAMPLE 3.5. Let G be a network with mark space $\mathbb{R}$ and assume the marks of the vertices are all distinct. Let $h : \mathcal{G}_* \to \mathbb{N}$ be a given measurable function. Define $f_G(v)$ to be the closest vertex w to v such that $h[G, w] > h[G, v]$, choosing the one with the smallest mark when there is a tie and let $f(v) = v$ when there is no such vertex. Then f is a vertex-shift.

In the following proposition, given a vertex-shift f, the dynamical system $\theta_f : \mathcal{G}_* \to \mathcal{G}_*$ is defined by

$$\theta_f[G, o] := [G, f(o)],$$

which is a well-defined and measurable function.

PROPOSITION 3.6 (Mecke's Point Stationarity). *Let f be a vertex-shift and $[\boldsymbol{G}, \boldsymbol{o}]$ be a unimodular network. Then θ_f preserves the distribution of $[\boldsymbol{G}, \boldsymbol{o}]$ if and only if $f_{\boldsymbol{G}}$ is almost surely bijective.*

For an instance of a bijective vertex-shift, see Example 3.7.

PROOF. Let $h : \mathcal{G}_* \to \mathbb{R}^{\geq 0}$ be a measurable function. By defining $g[G, o, s] := 1_{\{f_G(o)=s\}} h[G, s]$, (2.1) gives

$$\begin{aligned}\mathbb{E}\left[h(\theta_f[\boldsymbol{G}, \boldsymbol{o}])\right] &= \mathbb{E}\left[\sum_{v \in V(\boldsymbol{G})} g[\boldsymbol{G}, \boldsymbol{o}, v]\right] \\ &= \mathbb{E}\left[\sum_{v \in V(\boldsymbol{G})} g[\boldsymbol{G}, v, \boldsymbol{o}]\right] = \mathbb{E}\left[h[\boldsymbol{G}, \boldsymbol{o}] \# f_{\boldsymbol{G}}^{-1}(\boldsymbol{o})\right].\end{aligned}$$

If $f_{\boldsymbol{G}}$ is a.s. bijective, then it follows that $\theta_f[\boldsymbol{G}, \boldsymbol{o}]$ has the same distribution as $[\boldsymbol{G}, \boldsymbol{o}]$. Conversely, assume θ_f preserves the distribution of $[\boldsymbol{G}, \boldsymbol{o}]$. So the LHS is just $\mathbb{E}\left[h[\boldsymbol{G}, \boldsymbol{o}]\right]$. By letting $h[\boldsymbol{G}, \boldsymbol{o}]$ be the positive and negative parts of $(1 - \# f_{\boldsymbol{G}}^{-1}(\boldsymbol{o}))$ in the above equation, one gets $\mathbb{E}\left[(1 - \# f_{\boldsymbol{G}}^{-1}(\boldsymbol{o}))^2\right] = 0$, thus, $\# f_{\boldsymbol{G}}^{-1}(\boldsymbol{o}) = 1$ a.s. This in turn implies that $f_{\boldsymbol{G}}$ is bijective a.s. by Lemma 2.6. □

EXAMPLE 3.7. In the Canopy Tree of Example 3.3, assume the vertices have marks and that a total order on $F^{-1}(v)$ is obtained from the marks for every vertex v. The *Royal Line of Succession order* (see e.g. [**7**]) provides a total order on the vertices (isomorphic to the order on $\mathbb{Z}$) using the DFS (depth-first search) algorithm. Thus, one obtains a vertex-shift by mapping each vertex to its next vertex in this order. This vertex-shift is bijective and the orbit of each vertex coincides with the set of all vertices.

3.2. Foliation Associated with a Vertex-Shift. Let G be a network and f be an arbitrary vertex-shift. When there is no ambiguity, the symbol f will also be used for the function $f_G : V(G) \to V(G)$. The following definitions are adapted from [**7**]. Let $\sim_f$ be the equivalence relation on $V(G)$ defined by

$$x \sim_f y \Leftrightarrow \exists n \in \mathbb{N}; f^n(x) = f^n(y).$$

The symbol $\sim_f$ should not be confused with $\sim$ used for adjacency of vertices and edges.

DEFINITION 3.8. Each equivalence class of $\sim_f$ is called a **foil**. The partition of $V(G)$ generated by the foils is called the **f-foliation of** G and is denoted by $\mathcal{L}_G^f$. For $x \in V(G)$, denote by $L^f(x)$ the foil which contains x.

The (not-necessarily simple) graph $G^f = (V(G), E^f(G))$ with set of vertices $V(G)$ and set of directed edges $E^f(G) = \{(x, f(x)), x \in V(G)\}$ will be called the **f-graph** of G. Directed paths and directed cycles in G^f are called **f-paths** and **f-cycles**, respectively. For $x \in V(G)$, denote by $C^f(x)$ the (undirected) connected component of G^f which contains x; that is,

$$C^f(x) := \{y \in V(G) : \exists m, n : f^m(x) = f^n(y)\}.$$

The set of connected components of G^f will be denoted by $\mathcal{C}_G^f$. If $x \sim_f y$, then x and y are in the same connected component of $C^f(x)$. In other words, the foliation $\mathcal{L}_G^f$ is a refinement of $\mathcal{C}_G^f$.

For an acyclic connected component C of G^f, one can put a total order on the foils in C: each foil L is **older** than $f^{-n}(L)$ (which is a foil if it is non-empty) for any $n > 0$. The justification of this order will be discussed in Section 4. With this terminology, C does not have an *oldest* foil, but it may have a *youngest* foil; that is, a foil L such that $f^{-1}(L) = \emptyset$. The order on the foils of C is hence similar to that of $\mathbb{Z}$ or $\mathbb{N}$.

For $x \in V(G)$ and $n \geq 0$, let

$$D_n(x) := f^{-n}(x) = \{y \in V(G) : f^n(y) = x\}, \qquad d_n(x) := \#D_n(x).$$

Similarly, define

$$D(x) := \bigcup_{n=1}^{\infty} D_n(x) = \{y \in V(G) : \exists\, n \geq 0 : f^n(y) = x\}, \qquad d(x) := \#D(x).$$

The sequence of sets $f^n(V(G))$ is decreasing in n. Its limit (which may be the empty set) is denoted by $f^{\infty}(V(G))$. One can define $f^{\infty}(C)$ for a connected component C of G^f similarly; that is,

$$f^{\infty}(C) = \{x \in C : \forall n \geq 0 : D_n(x) \neq \emptyset\}.$$

3.3. Unimodular Classification Theorem.

THEOREM 3.9 (Foil Classification in Unimodular Networks). *Let $[\boldsymbol{G}, \boldsymbol{o}]$ be a unimodular network and f be a vertex-shift. Almost surely, every vertex has finite degree in the graph $\boldsymbol{G}^f$. In addition, each component C of $\boldsymbol{G}^f$ has at most two ends and it belongs to one of the following three classes:*

(i) *Class $\mathcal{F}/\mathcal{F}$: C and all its foils are finite. If $n = n(C)$ is the number of foils in C $(1 \leq n < \infty)$, then*
  - *C has a unique f-cycle and its length is n;*
  - *$f^{\infty}_{\boldsymbol{G}}(C)$ is the set of vertices of the cycle;*
  - *Each foil of C contains exactly one vertex of the cycle.*

(ii) *Class $\mathcal{I}/\mathcal{F}$: C is infinite but all its foils are finite. In this case,*
  - *The (undirected) f-graph on C is a tree;*
  - *There is a unique bi-infinite f-path in C, each foil in C contain exactly one vertex of the path, and $f^{\infty}_{\boldsymbol{G}}(C)$ coincides with the set of vertices of the path;*
  - *The order of the foils of C is of type $\mathbb{Z}$; that is, there is no youngest foil in C.*

(iii) *Class $\mathcal{I}/\mathcal{I}$: C and all foils of C are infinite. In this case,*
  - *The (undirected) f-graph on C is a tree;*
  - *C has one end, there is no bi-infinite f-path in C, $D(v)$ is finite for every vertex $v \in C$, and $f^{\infty}_{\boldsymbol{G}}(C) = \emptyset$;*
  - *The order of the foils of C is of type $\mathbb{N}$ or $\mathbb{Z}$, that is, there may or may not be a youngest foil in C.*

Before going though the proof of this theorem, here are examples of the different classes in Theorem 3.9. In a unimodular network with finitely many vertices, all components of any vertex-shift are of class $\mathcal{F}/\mathcal{F}$. The same holds in Example 3.4 provided the marks are i.i.d. uniform random numbers in $[0, 1]$. Subsection 4.3 provides a general example of class $\mathcal{I}/\mathcal{F}$. The unimodular Eternal Galton-Watson

Tree (introduced in Subsection 6.1), the Canopy Tree (Example 3.3) and Proposition 5.11 provide instances of class $\mathcal{I}/\mathcal{I}$. There is no youngest foil in the first one, and there is one in the two other instances.

The proof of Theorem 3.9 requires a few preliminary results.

PROPOSITION 3.10. *Under the assumptions of Theorem 3.9, almost surely, $d_n(x)$ is finite for all $n \geq 0$ and all $x \in V(\boldsymbol{G})$. Moreover,*

$$\forall n \geq 0,\ \mathbb{E}\left[d_n(\boldsymbol{o})\right] = 1.$$

If in addition $\boldsymbol{G}^f$ is acyclic a.s., then $\mathbb{E}\left[d(\boldsymbol{o})\right] = \infty$.

PROOF. Define $g[G,x,y] := 1_{\{y=f^n(x)\}}$, which is well-defined and measurable. Therefore, the mass transport principle (2.1) holds for g and the first claim follows. For the second claim, note that when $\boldsymbol{G}^f$ is acyclic, $D(\boldsymbol{o})$ is the disjoint union of $D_n(\boldsymbol{o})$ for $n \geq 0$, and hence $\mathbb{E}\left[d(\boldsymbol{o})\right] = \sum_{n=0}^{\infty} \mathbb{E}\left[d_n(\boldsymbol{o})\right] = \infty$. □

PROPOSITION 3.11. *Let $[\boldsymbol{G},\boldsymbol{o}]$ be a unimodular network and g be a vertex-shift.*

(i) *If g is injective a.s., then it is bijective a.s.*

(ii) *If g is surjective a.s., then it is bijective a.s.*

PROOF. Define $h[G,o,s] := 1_{\{g_G(o)=s\}}$. By (2.1), $\mathbb{E}\left[\#g_{\boldsymbol{G}}^{-1}(\boldsymbol{o})\right] = \mathbb{E}\left[1\right] = 1$.

(i). One has $\#g_{\boldsymbol{G}}^{-1}(\boldsymbol{o}) \leq 1$. So $\#g_{\boldsymbol{G}}^{-1}(\boldsymbol{o}) = 1$ a.s. By Lemma 2.6, it follows that $\{x : \#g_{\boldsymbol{G}}^{-1}(x) \neq 1\} = \emptyset$ a.s.

(ii). One has $\#g_{\boldsymbol{G}}^{-1}(\boldsymbol{o}) \geq 1$. So $\#g_{\boldsymbol{G}}^{-1}(\boldsymbol{o}) = 1$ a.s. The rest of the proof is as above. □

COROLLARY 3.12. *Under the assumptions of Theorem 3.9, almost surely, if $L(x)$ is infinite, then each $L(f^i(x))$, $i \geq 0$, is infinite too.*

PROOF. Note that $f^i(L(x)) \subseteq L(f^i(x))$ for all $i \geq 0$. Now, the claim follows from the fact that the degrees of all vertices in $\boldsymbol{G}^f$ are finite a.s. (Proposition 3.10). □

PROPOSITION 3.13. *Under the assumptions of Theorem 3.9, almost surely, all infinite (undirected) components of $\boldsymbol{G}^f$ are acyclic (and hence are trees).*

PROOF. It is a classical (deterministic) fact that each component of $\boldsymbol{G}^f$ is either acyclic (and hence a tree) or it has a unique f-cycle.

Let S be the union of the f-cycles, which is a covariant subset. The intersection of each component with S is either empty or a unique f-cycle. Therefore, the claim follows by using Lemma 2.9 for the partition $\mathcal{C}_{\boldsymbol{G}}^f$ consisting of the components of $\boldsymbol{G}^f$ and the covariant subset S. □

PROPOSITION 3.14. *Under the assumptions of Theorem 3.9, almost surely, each component of $\boldsymbol{G}^f$ contains at most one bi-infinite f-path.*

PROOF. Let S be the union of bi-infinite f-paths, which is a covariant subset. Define

$$g_G(x) := \begin{cases} f_G(x), & x \in S, \\ x, & x \notin S, \end{cases}$$

which is a vertex-shift. By the definition of S, g is surjective. Therefore, by Proposition 3.11 g is injective too a.s. It means that no two bi-infinite f-paths collide. Therefore, the bi-infinite paths are in different components a.s. and the claim is proved. □

PROOF OF THEOREM 3.9. It is again a classical deterministic fact that finite connected components satisfy the properties of class $\mathcal{F}/\mathcal{F}$, as shown by Lemma 5 in [7]. So it is enough to prove that each infinite component is of class $\mathcal{I}/\mathcal{F}$ or $\mathcal{I}/\mathcal{I}$ a.s.

By Proposition 3.13, almost surely, each infinite component is acyclic and its f-graph is a tree. Since the degree of each vertex is finite a.s., König's infinity lemma implies that $v \in f_{\boldsymbol{G}}^{\infty}(V(\boldsymbol{G}))$ if and only if v is in a bi-infinite f-path. So, $f_{\boldsymbol{G}}^{\infty}(V(\boldsymbol{G}))$ is the union of the bi-infinite f-paths a.s.

Assume that, with positive probability, there exists a component of $\boldsymbol{G}^f$ that has both finite and infinite foils. Let P be the partition consisting of such components (and one more element which is the set of remaining vertices). By Corollary 3.12, each component in P has an oldest finite foil. Define

$$S := \{v \in V(\boldsymbol{G}) : \#L(v) < \infty, \#L(f(v)) = \infty\},$$

which is a covariant subset. The intersection of S with each component in P is the oldest finite foil in the component. Now, one gets a contradiction by using Lemma 2.9 for the covariant partition P and the covariant subset S.

In other words, almost surely, each connected component has either only finite foils or only infinite foils.

Let C be an infinite component containing some finite foils. As already said, almost surely, all foils in C are finite. By Lemma 2.9, almost surely, C does not have a youngest foil (use Lemma 2.9 for the partition $\mathcal{C}_{\boldsymbol{G}}^f$ and the union of the youngest foils $\{L \in \mathcal{L}_{\boldsymbol{G}}^f : f^{-1}(L) = \emptyset\}$). So the order of foils in C is of type $\mathbb{Z}$ a.s. Let $L \subseteq C$ be an arbitrary foil and consider $\cup_{n=0}^{\infty} f^{-n}(L)$, which is the union of the foils younger than L. Since the degree of each vertex is finite a.s., König's infinity lemma implies that there is a bi-infinite path in C a.s. By Proposition 3.14, C contains exactly one bi-infinite f-path a.s. Now, C satisfies all properties of class $\mathcal{I}/\mathcal{F}$.

Finally, Let C be a component containing some infinite foils. As mentioned above, all foils in C are infinite a.s. It remains to prove that C has no bi-infinite f-path a.s.

Let S be the union of the bi-infinite f-paths. It can be seen that S is a covariant subset. By Proposition 3.14, each foil includes at most one vertex in S almost surely. Therefore, Lemma 2.9 implies that almost surely no infinite foil intersects S, which is the desired property. Thus, C is of class $\mathcal{I}/\mathcal{I}$ a.s. □

REMARK 3.15. For a connected component C of G^f, the vertex-shift f is said to *evaporate* C if $f_G^{\infty}(C) = \emptyset$, that is, for all $x \in C$, there exists some $n > 0$ such that $D_n(x) = \emptyset$ [7]. Thus, under the assumptions of Theorem 3.9, almost surely, the vertex-shift f evaporates C if and only if C is of class $\mathcal{I}/\mathcal{I}$.

EXAMPLE 3.16. Let $[\boldsymbol{G}, \boldsymbol{o}]$ be a unimodular graph equipped with i.i.d. uniform random marks in $[0, 1]$ (Example 2.4). Consider the vertex-shift of Example 3.5. Assuming $h[\boldsymbol{G}, \cdot]$ takes infinitely many values a.s., it can be seen that $\boldsymbol{G}^f$ is acyclic and $D(v)$ is finite for each vertex v. Therefore, Theorem 3.9 implies that each connected component is of class $\mathcal{I}/\mathcal{I}$ almost surely.

EXAMPLE 3.17 (Stationary Drainage Networks). Let $V = \{(x, y) \in \mathbb{Z}^2 : x + y \text{ is even}\}$. Define the following random directed graph $\boldsymbol{G}$ with vertex set V: for each point $(x, y) \in V$, add a directed edge from (x, y) to one of the two vertices

$(x-1, y-1)$ and $(x+1, y-1)$ randomly. Assume the joint law of the choices is invariant under translations (i.e. is stationary, but not necessarily i.i.d.). Say (x, y) flows into (x', y') if there is a directed path from the former to the latter. It is shown below that Theorem 3.9 implies the following: $\boldsymbol{G}$ is connected a.s. if and only if the number of points flowing into the origin is finite a.s. By the stationarity assumption, the connected component of $\boldsymbol{G}$ containing the origin is unimodular (Example 2.4). In addition, the point $f(x, y)$ chosen for (x, y) as indicated above is a vertex-shift. Each foil is a set of consecutive points in a horizontal line. It follows from Lemma 2.9, that no foil is a half-line a.s. Therefore, almost surely, either there is a foil which is a full horizontal line, or all foils are finite horizontal intervals. It is easily seen that the former is equivalent to the connectedness of $\boldsymbol{G}$ and implies that all foils are full horizontal lines. Therefore, Theorem 3.9 implies that the former is equivalent to the condition that there is no bi-infinite directed path in $\boldsymbol{G}$. By Lemma 2.6, one can show that this hold a.s. if and only if the number of points flowing into the origin is finite a.s. So the claim is proved.

In this example, one can replace V with any stationary point process in $\mathbb{Z}^2$ and $f(x, y)$ by $(\tau(x, y), y-1)$. One should assume that the joint law of V and $\tau(\cdot, \cdot)$ is stationary and that τ is monotonic on each horizontal line.

EXAMPLE 3.18. Let G be a transitive graph with a unimodular automorphism group. Definition 5.2 of [**19**] defines the *oriented wired spanning forest* `OWSF` of G, which is a random directed subgraph of G. It has the property that there is a unique outgoing edge from each vertex. So, this naturally defines a vertex-shift on (every connected component of) `OWSF`. Theorem 10.1 of [**19**] and Theorem 3.9 above imply that every connected component of `OWSF` is of class $\mathcal{I}/\mathcal{F}$ or $\mathcal{I}/\mathcal{I}$ almost surely, the former only when the component is *roughly isometric to* $\mathbb{Z}$.

The next two results are applications of the classification theorem. The proofs are only sketched. In particular, the technicalities about ends of trees are not discussed. These two results will not be used below and can hence been skipped at the first reading.

Proposition 3.19 provides another proof, based on Theorem 3.9, of a result in [**6**] in the special case of trees.

PROPOSITION 3.19. *The number of ends of a unimodular random tree is almost surely either 0, 1, 2 or uncountable.*

PROOF. For any given locally finite tree, there is a well-known metric on the set of ends of the tree that makes it a compact and complete metric space (see [**13**]). Therefore, it is enough to show that almost surely, if the tree has at least three ends, then it has no isolated end. Recall that there exists an isolated end if and only if by deleting a finite subset of the vertices, one can get a component of the remaining graph which has only one end.

Let T be a tree with at least three ends. Fix a vertex $v \in V(T)$. Call a neighbor w of v good if by deleting the edge vw, the connected component containing v has only one end. It follows from the assumption of at least three ends that there is at most one good neighbor. Let $f_T(v) := w$ if w is a good neighbor of v and $f_T(v) := v$ if v has no good neighbor. It can be seen that f defines a vertex shift. By the assumption of at least three ends, it can be seen that there is $n \in \mathbb{N}$ such that $f^n(v) = f^{n+1}(v)$; i.e. each connected component of the f-graph T^f has a cycle of length one. Moreover, if T has an isolated end, then there is a vertex v such that $f(v) \neq v$. In this case, $f^{-n}(v)$ is a single vertex for each $n \geq 0$.

Let $[\boldsymbol{T}, \boldsymbol{o}]$ be a unimodular random tree and assume that with positive probability, $\boldsymbol{T}$ has at least three ends and an isolated end. As stated above, on this event there is a vertex v such that $f(v) \neq v$. Now, the connected component of v in $\boldsymbol{T}^f$ is infinite and has a cycle, which contradicts the classification theorem (Theorem 3.9). This completes the proof. □

THEOREM 3.20. *Let $[\boldsymbol{T}, \boldsymbol{o}]$ be a unimodular network whose underlying graph is a tree a.s. Then, on the event that the number of ends of $\boldsymbol{T}$ is not 2, there is no measurable way of selecting a non-empty and non-dense subset of the ends of $\boldsymbol{T}$.*

PROOF. The proof is similar to the proof of Proposition 3.19. Given a tree T, if T has at most one end, the claim is trivial. So, by the assumption, assume T has at least three ends. Let D_T be the set of ends of T and S_T be the selected subset of D_T. By considering the closure of S_T, one may assume S_T is closed from the beginning. Assume $\emptyset \neq S_T \neq D_T$. For $v \in V(T)$, let $f_T(v) := w$ if by deleting the edge vw, the component containing v has none of the ends in S_T and $f(v) := v$ if no such neighbor exists. It can be seen that f is a well-defined vertex-shift and $f(v) \neq v$ for at least one vertex v.

It can be seen that if S_T has only one element, then the f-graph T^f is connected and has at least three ends. If not, each connected component of T^f has a cycle with length one and some of these components are infinite. By Theorem 3.9, both cases happen with zero probability for $[\boldsymbol{T}, \boldsymbol{o}]$ and the claim is proved. □

3.4. Bibliographical Comments. The notion of vertex-shift is an analogue of that of point-shifts for point processes [**17**]. Theorem 3.6 extends Mecke's point stationarity theorem [**18**]. The notion of foliation and Theorem 3.9 extend results for point-shifts in [**7**] to unimodular networks. Proposition 3.19 is stated in [**6**] where it is proved for general unimodular random graphs.

The observations of Example 3.17 can be applied to various examples of drainage networks in the literature. In the case where the choices are i.i.d. with probability $\frac{1}{2}$, one obtains the river model of [**27**], which is studied in various manners in the literature [**3**]. If V is replaced by a Bernoulli point process and τ selects the closest point in the line, one obtains the Howard model [**29**].

The results of this section have analogues in the context of measured equivalence relations, as introduced in Subsection 2.3. Consider the equivalence relation on $\mathcal{G}_*$ discussed in Subsection 2.3. Let $\mathcal{G}_\# \subseteq \mathcal{G}_*$ be the set of rooted networks which have no non-rooted automorphisms other than the identity. If $[G, o] \in \mathcal{G}_\#$, then the map $v \mapsto [G, v]$, defined for $v \in V(G)$, is a bijection between G and the equivalence class of $[G, o]$. Therefore, it is easy to see that any vertex-shift f induces naturally a measurable function from $\mathcal{G}_\#$ to itself. Conversely, every such function gives rise to a vertex-shift (which is only defined on networks with no nontrivial automorphisms). For networks having nontrivial automorphisms, some care is needed. Usually (including the cases in this section), generality is not lost by adding i.i.d. marks. Moreover, one can extend the results of this section to all invariant measured equivalence relations as follows.

Let (X, R, μ) be an invariant measured equivalence relation. The analogue of a vertex-shift is a measurable function $f : X \to X$ such that $\forall x : (x, f(x)) \in R$ (the differences are discussed above). For instance, if R is generated by a measurable action of a group H on X (i.e., it is an *orbit equivalence relation* [**15**]), then for every $h \in H$, the map $x \mapsto h.x$ is analogous to a vertex-shift. Returning to the general case, Proposition 3.6 is analogous to Corollary 1 of [**15**]. Also, a function f as above gives a graph structure on X by considering the directed edges $(x, f(x))$. Note that every connected component C of this graph is equipped with a function from C to itself. So, foils on C can be defined similar to Definition 3.8. Then, the classification theorem (Theorem 3.9) can be extended as follows with a similar proof (note also that connected components and foils in this context provide two Borel sub-equivalence relations of R).

Theorem 3.9': Let (X, R, μ) be an invariant measured equivalence relation and $f : X \to X$ be a measurable function such that $\forall x : (x, f(x)) \in R$. Define the graph structure on X as above. Then, for μ-a.e. $x \in X$ and for every connected component C of this graph contained in $R[x]$, the graph structure on C belongs to one of the classes $\mathcal{F}/\mathcal{F}$, $\mathcal{I}/\mathcal{F}$ or $\mathcal{I}/\mathcal{I}$ mentioned in Theorem 3.9.

By adding i.i.d. marks if necessary, this result implies Theorem 3.9.

4. Eternal Family Trees

Consider a random network $[\boldsymbol{G}, \boldsymbol{o}]$ and the directed graph $\boldsymbol{G}^f$ of a vertex-shift f. When regarding $f(x)$ as the *parent* of x and $f^{-1}(x)$ as its *children* for $x \in V(\boldsymbol{G})$, one may regard an acyclic connected component of $\boldsymbol{G}^f$ as an extension of a branching process. A first difference with such a process is that, in a connected component of $\boldsymbol{G}^f$, there is no vertex which is an ancestor of all other vertices. This is formalized in the definition of *Eternal* Family Trees below, abbreviated as EFTs, which are objects of independent interest. In this context, the foils of $\boldsymbol{G}^f$ represent vertices of the same *generation*. A second major difference is that there are no independence assumptions as in classical branching processes. The concept is further discussed in Subsection 4.4).

Subsection 4.1 gives the definition and basic properties of EFTs. It is shown that if $[\boldsymbol{G}, \boldsymbol{o}]$ is a unimodular network, then, conditioned on being infinite, the connected component of $\boldsymbol{G}^f$ containing $\boldsymbol{o}$ is a unimodular EFT. Proposition 4.4 classifies components in two classes, $\mathcal{I}/\mathcal{I}$ and $\mathcal{I}/\mathcal{F}$. Subsection 4.2 introduces a method for constructing general unimodular EFTs which is more interesting for class $\mathcal{I}/\mathcal{I}$. Subsection 4.3 describes the structure of EFTs of class $\mathcal{I}/\mathcal{F}$ and provides another general construction. Several examples stemming from branching processes will be discussed in forthcoming sections.

4.1. Family Trees.

A **Family Tree** (abbreviated as FT) is a *directed* tree T in which the out-degree of each vertex is at most 1. For a vertex $v \in V(T)$ which has one outgoing edge vw, let $F(v) := w$, and call $F(v)$ the **parent** of v. Note that there may be vertices without parent. When the out-degrees of all vertices are exactly 1, T is called an **Eternal Family Tree**.

It is straightforward that in a Family Tree, (1) there is at most one vertex without parent and (2) for all pairs of vertices (v, w), there exist $m, n \geq 0$ such that $F^n(v) = F^m(w)$. This implies that Family Trees are in one-to-one correspondence with undirected trees with one selected vertex (not to be confused with the root) or end (the latter in the eternal case).

A **rooted Family Tree** is a pair (T, o) in which T is a FT and o is a distinguished vertex called the root. The root may or may not have a parent, even if the graph has a vertex without parent.

A Family Tree T can be considered as a network by choosing a mark for each pair (v, e) according to the direction of e. Also note that when F is defined for all vertices (i.e. in the eternal case), F is a covariant vertex-shift which is called **the parent vertex-shift**, and T coincides with the F-graph T^F (see Definition 3.8).

In line with the definitions used for networks, the neighbors of x, other than $F(x)$, are called its **children**. An **ordered** Family Tree is a Family Tree with an underlying total order on the children of each vertex (note that this order can be obtained by putting marks as in the definition of networks). The vertices $(F^n(x))_{n=0}^{\infty}$ are called the **ancestors** of x. The set of **descendants of order** n of x is $D_n(x) := D_n(T, x) := \{y : F^{(n)}(y) = x\}$, with $d_n(x) := d_n(T, x) := \#D_n(x)$. The (tree of) **descendants** $D(x)$ of x is the subtree with vertices $\cup_{n=0}^{\infty} D_n(x)$.

Let $l(\cdot, \cdot) = l_T(\cdot, \cdot)$ be the function which assigns to each pair (v, w) of vertices of a Family Tree T, the number of generations between v and w w.r.t. F, which is defined by

$$l(v, v) = 0, \qquad l(v, F(w)) = l(v, w) - 1 \tag{4.1}$$

for all $v, w \in V(T)$. Note that this function is invariant under isomorphisms. One also has $l(w, v) = -l(v, w)$ and $l(v, w) + l(w, z) = l(v, z)$. For a rooted Eternal Family Tree $[T, o]$, the level sets of $l(o, \cdot)$ are just the foils of T for the parent vertex-shift F. The level set $\{w \in V(T) : l(o, w) = n\}$ is called the ***n*-th generation** of $[T, o]$.

The results listed below and in Subsection 4.2 are valid in both the ordered and the non-ordered cases.

DEFINITION 4.1. Let $\mathcal{T}$ denote the set of isomorphism classes of Family Trees, and define $\mathcal{T}_*$ and $\mathcal{T}_{**}$ similarly (they form closed subspaces of $\mathcal{G}_*$ and $\mathcal{G}_{**}$, respectively for a suitable mark space). A **random Family Tree** is a random network with values in $\mathcal{T}_*$ almost surely. A **unimodular** FT is defined as in Definition 2.1. A **proper** random FT is a random FT in which

$$0 < \mathbb{E}\left[d_n(\boldsymbol{o})\right] < \infty,$$

for all $n \geq 0$. The same notation will be used for ordered Family Trees, as there will be no possible confusion from the context.

REMARK 4.2. All the results in this section and Section 5 are also valid for Family Trees in which the vertices and edges are equipped with marks; i.e. networks where the underlying graph is a Family Tree.

PROPOSITION 4.3. *Let $[\boldsymbol{T}, \boldsymbol{o}]$ be a unimodular FT.*

(i) *If $\boldsymbol{T}$ has infinitely many vertices a.s., then it is eternal a.s. Moreover, $[\boldsymbol{T}, \boldsymbol{o}]$ is a proper random EFT, with $\mathbb{E}\left[d_n(\boldsymbol{o})\right] = 1$ for all $n \geq 0$, and $\mathbb{E}\left[d(\boldsymbol{o})\right] = \infty$.*

(ii) *If $\boldsymbol{T}$ is finite with positive probability, then $\mathbb{E}\left[d_n(\boldsymbol{o})\right] < 1$ for all $n > 0$.*

PROOF. (i). For all Family Trees T, let S_T be the set of vertices of T without parent. The set S_T has at most one element. Now, $S_{\boldsymbol{T}}$ is a finite covariant subset of the infinite unimodular network $[\boldsymbol{T}, \boldsymbol{o}]$. Lemma 2.6 gives that $S_{\boldsymbol{T}} = \emptyset$ a.s.; that is, $\boldsymbol{T}$ is an Eternal Family Tree a.s. Now, the second part is a direct consequence of Proposition 3.10.

(ii). Using the definition of (i) above, the finiteness of T implies $S_{\boldsymbol{T}} \neq \emptyset$ with positive probability. Therefore, $\mathbb{P}[\boldsymbol{o} \in S_{\boldsymbol{T}}] > 0$ by Lemma 2.6. Send unit mass from each vertex v to $F^n(v)$ provided $F^n(v)$ exists; that is, $g[G,v,w] := 1_{\{w=F^n(v)\}}$. By (2.1),

$$\mathbb{E}[d_n(\boldsymbol{o})] = \mathbb{E}\left[g^-(\boldsymbol{o})\right] = \mathbb{E}\left[g^+(\boldsymbol{o})\right] \leq \mathbb{E}\left[1_{\{\boldsymbol{o} \notin S_{\boldsymbol{T}}\}}\right] < 1.$$

□

The subtree of descendants of the root of an EFT can be seen as some generalized branching processes, where the generalization lies in the fact that there are no independence assumptions. A unimodular EFT is always critical in the sense that the mean number of children of the root is 1 (Proposition 4.3).

PROPOSITION 4.4 (Classification of Unimodular EFTs). *For the parent vertex-shift, a unimodular FT with infinitely many vertices almost surely belongs to one of the following classes:*

(i) *Class $\mathcal{I}/\mathcal{I}$: every generation is infinite, each vertex has finitely many descendants and there is no bi-infinite F-path; i.e., the tree has only one end.*
(ii) *Class $\mathcal{I}/\mathcal{F}$: every generation is finite and the set of vertices with infinitely many descendants form a unique bi-infinite F-path; i.e., the tree has two ends.*

PROOF. The claim is a direct corollary of Proposition 4.3 and Theorem 3.9. □

Proposition 4.15 and part (ii) of Theorem 4.17 below provide specific constructions for the classes $\mathcal{I}/\mathcal{I}$ and $\mathcal{I}/\mathcal{F}$. Some examples of these classes are also mentioned after Theorem 3.9.

EXAMPLE 4.5. Let $[\boldsymbol{G}, \boldsymbol{o}]$ be a unimodular network and f a vertex-shift. Let $C_{(\boldsymbol{G},\boldsymbol{o})}$ be the connected component of the f-graph $\boldsymbol{G}^f$ that contains $\boldsymbol{o}$. Then $[C_{(\boldsymbol{G},\boldsymbol{o})}, \boldsymbol{o}]$, conditioned on being infinite, is a unimodular EFT. This can be proved by verifying (2.1) directly and using the fact that $[G,o] \mapsto [C_{(G,o)}, o]$ is a measurable map.

4.2. Moving the Root to a Typical Far Descendant. This subsection introduces a method for constructing general unimodular EFTs. The idea is to start with a (not necessarily unimodular) random FT and to move the root to a *typical* n-th descendant for large n as defined below. In fact, this method constructs a larger class of EFTs, namely offspring-invariant EFTs, which will be studied in detail in Section 5. The main focus of the present section is on the unimodular case.

DEFINITION 4.6. Let $[\boldsymbol{T}, \boldsymbol{o}]$ be a random (ordered or non-ordered) Family Tree, not necessarily unimodular, and let $n \in \mathbb{Z}^{\geq 0}$. Assume $0 < \mathbb{E}[d_n(\boldsymbol{o})] < \infty$. By **moving the root to a typical n-descendant**, one means considering the following measure on $\mathcal{T}_*$:

$$\mathcal{P}_n(A) := \frac{1}{\mathbb{E}[d_n(\boldsymbol{o})]} \mathbb{E}\left[\sum_{v \in D_n(\boldsymbol{o})} 1_A[\boldsymbol{T}, v]\right]. \tag{4.2}$$

Whenever $(\mathcal{P}_n)$ converges weakly to some probability distribution, define

$$\mathcal{P}_\infty := \lim_{n\to\infty} \mathcal{P}_n.$$

The probability measure $\mathcal{P}_n$ defines a new random network. Below, expectation w.r.t. $\mathcal{P}$ and $\mathcal{P}_n$ are denoted by $\mathcal{E}$ and $\mathcal{E}_n$ respectively. The probability $\mathcal{P}_n$ defined in (4.2) is just the image of $\mathcal{P}$ by the operator σ_n defined below.

DEFINITION 4.7. Let σ_n be the operator which associates to any probability measure $\mathcal{Q}$ on $\mathcal{T}_*$ the probability measure

$$\sigma_n\mathcal{Q}[A] = \frac{1}{\int_{\mathcal{T}_*} d_n(T,o)\mathrm{d}\mathcal{Q}([T,o])}\int_{\mathcal{T}_*}\sum_{v\in D_n(T,o)} 1_A[T,v]\mathrm{d}\mathcal{Q}([T,o])$$

on $\mathcal{T}_*$, given that the denominator is positive and finite. Let $\sigma := \sigma_1$ and $\sigma_\infty\mathcal{Q}$ be the weak limit of $\sigma_n\mathcal{Q}$ as $n\to\infty$, assuming the limit exists.

For a random Family Tree $[\boldsymbol{T},\boldsymbol{o}]$ with distribution $\mathcal{P}$, one has $\mathcal{P}_n = \sigma_n\mathcal{P}$ for each $n\le\infty$. Denote by $\sigma_n[\boldsymbol{T},\boldsymbol{o}]$ a random FT with this distribution. The random Family Tree $[\boldsymbol{T},\boldsymbol{o}]$ is called **offspring-invariant** if its distribution is invariant under σ.

EXAMPLE 4.8. Consider a semi-infinite path $v_0, v_1, \ldots$ and attach to each of its vertices disjoint semi-infinite paths to form a deterministic tree T. By letting v_0 be the vertex without father, one may regard $[T, v_0]$ as a FT. Then, since $d_n(v_0)$ is deterministic, the probability measure $\mathcal{P}_n$ defined in (4.2) is obtained by choosing a new root uniformly at random in $D_n(v_0)$. Therefore, under $\mathcal{P}_n$, the distance of the root to the path $\{v : d_1(v) = 2\}$ is uniformly at random in $\{0, 1, \ldots, n\}$. It can be seen that $\mathcal{P}_\infty$ exists here and is just the distribution of a bi-infinite path rooted at an arbitrary vertex. In other words, the vertices having two children vanish in the limit.

Two other examples of $\mathcal{P}_\infty$ are described in Subsections 4.3 and 6.1 below. See Propositions 4.20 and 6.5.

The following three results establish the basic results of the operator σ alluded to above. The proofs are postponed to the end of the subsection.

LEMMA 4.9. *The operators $(\sigma_n)_n$ form a semigroup on the space of proper probability measures on $\mathcal{T}_*$. In other words, $\sigma_n\mathcal{P} = \sigma^{(n)}\mathcal{P}$, for every proper probability measure $\mathcal{P}$. More generally,*

$$\sigma_m \circ \sigma_n(\cdot) = \sigma_{m+n}(\cdot) \tag{4.3}$$

whenever $\sigma_n(\cdot)$ and $\sigma_{m+n}(\cdot)$ are defined.

PROPOSITION 4.10. *A random Family Tree $[\boldsymbol{T},\boldsymbol{o}]$ is a unimodular EFT a.s. if and only if it is offspring-invariant and $\mathbb{E}[d_1(\boldsymbol{o})] = 1$. In this case, the classification of Proposition 4.4 holds.*

The following is a continuity-like result that studies when the limit $\mathcal{P}_\infty$ of iterates of σ on $\mathcal{P}$ is offspring-invariant or unimodular (see also Lemma 4.14 below).

THEOREM 4.11. *Let $[\boldsymbol{T},\boldsymbol{o}]$ be a random FT. Assume that $\mathcal{P}_\infty$ exists and that*

$$\limsup_{n\to\infty} \frac{\mathbb{E}[d_{n+1}(\boldsymbol{o})]}{\mathbb{E}[d_n(\boldsymbol{o})]} > 0. \tag{4.4}$$

Then

(i) *The sequence* $\mathbb{E}[d_{n+1}(\boldsymbol{o})]/\mathbb{E}[d_n(\boldsymbol{o})]$ *is convergent and its limit* $\mathcal{E}_\infty[d_1(\boldsymbol{o})]$ *is positive and finite.*
(ii) $\mathcal{P}_\infty$ *is offspring-invariant, and hence proper;*
(iii) $\mathcal{P}_\infty$ *is unimodular if and only if* $\lim_{n\to\infty}\mathbb{E}[d_{n+1}(\boldsymbol{o})]/\mathbb{E}[d_n(\boldsymbol{o})] = 1$.

REMARK 4.12. If $\mathcal{P}_\infty$ exists but (4.4) fails, one can still obtain a result similar to (though a bit weaker than) Theorem 4.11. In this case, under $\mathcal{P}_\infty$, the root is a.s. in the last generation, thus $\mathcal{P}_\infty$ is not proper, but the mass transport principle holds *along the generation of the root.* See Proposition 5.3 for a precise definition of this property.

REMARK 4.13. Condition (4.4) in Theorem 4.11 is equivalent to the condition that $\sigma\mathcal{P}_\infty$ is defined. This can be proved with a small change in the proof of Theorem 4.11.

The following lemma is a more general continuity-like property of the operator σ. Its proof is similar to that of Theorem 4.11 and is skipped here.

LEMMA 4.14. *Let* $P_1, P_2, \ldots$ *be an arbitrary sequence of probability measures on* $\mathcal{T}_*$ *that converges weakly to a probability measure* P. *Assume* σP_n *is defined for each* n. *If the sequence* σP_n *converges weakly to a probability measure* Q *and* σP *is defined, then* $\sigma P = Q$.

Here is a result on the construction of unimodular EFTs of class $\mathcal{I}/\mathcal{I}$.

PROPOSITION 4.15. *Any offspring-invariant random Family Tree* $[\boldsymbol{T}, \boldsymbol{o}]$ *can be constructed by applying* σ_∞ *to the (non-eternal) Family Tree* $[D(\boldsymbol{o}), \boldsymbol{o}]$ *of the descendants of* $\boldsymbol{o}$ *in* $\boldsymbol{T}$*; i.e.* $[\boldsymbol{T}, \boldsymbol{o}] \sim \sigma_\infty[D(\boldsymbol{o}), \boldsymbol{o}]$. *In particular, any unimodular EFT of class* $\mathcal{I}/\mathcal{I}$ *can be constructed by applying* σ_∞ *to a random finite FT.*

The following lemma is needed to prove the above results.

LEMMA 4.16. *Under the assumptions of Definition 4.6, for all* $n \geq 0$,

$$\mathcal{E}_n[d_m(\boldsymbol{o})] = \frac{\mathbb{E}[d_{m+n}(\boldsymbol{o})]}{\mathbb{E}[d_n(\boldsymbol{o})]}, \tag{4.5}$$

where in the left hand side $\boldsymbol{o}$ *denotes the root of a random FT with distribution* $\mathcal{P}_n$.

PROOF. Using the fact that the $(m+n)$-descendants of the root are just the disjoint union of the m-descendants of the n-descendants of the root, one gets that

$$\mathcal{E}_n[d_m(\boldsymbol{o})] = \frac{1}{\mathbb{E}[d_n(\boldsymbol{o})]}\mathbb{E}\left[\sum_{v\in D_n(\boldsymbol{o})} d_m(v)\right] = \frac{\mathbb{E}[d_{m+n}(\boldsymbol{o})]}{\mathbb{E}[d_n(\boldsymbol{o})]},$$

which proves the claim. □

PROOF OF LEMMA 4.9. Let $\mathcal{P}$ be a probability distribution on $\mathcal{T}_*$ such that $\sigma_n\mathcal{P}$ and $\sigma_{m+n}\mathcal{P}$ are defined. By (4.5), one obtains that $\sigma_m \circ \sigma_n\mathcal{P}$ is also defined. Denote the expectation operators w.r.t. $\mathcal{P}$ and w.r.t. $\sigma_m \circ \sigma_n\mathcal{P}$ by $\mathcal{E}$ and

$\mathcal{E}'$ respectively. For all measurable functions $h : \mathcal{T}_* \to R^{\geq 0}$,

$$\begin{aligned}
\mathcal{E}'[h] &= \frac{1}{\mathcal{E}_n[d_m(\boldsymbol{o})]}\mathcal{E}_n\left[\sum_{y\in D_m(\boldsymbol{o})} h(\boldsymbol{T}, y)\right] \\
&= \frac{1}{\mathcal{E}_n[d_m(\boldsymbol{o})]\mathcal{E}\left[d_n(\boldsymbol{o})\right]}\mathcal{E}\left[\sum_{x\in D_n(\boldsymbol{o})}\sum_{y\in D_m(x)} h(\boldsymbol{T}, y)\right] \\
&= \frac{1}{\mathcal{E}[d_{m+n}(\boldsymbol{o})]}\mathcal{E}\left[\sum_{y\in D_{m+n}(\boldsymbol{o})} h(\boldsymbol{T}, y)\right] = \mathcal{E}_{m+n}[h],
\end{aligned}$$

where the first two equations use (4.2) and the third one uses (4.5). This proves the claim. □

PROOF OF PROPOSITION 4.10. Let $\mathcal{P}$ be the distribution of $[\boldsymbol{T}, \boldsymbol{o}]$.

First, assume $[\boldsymbol{T}, \boldsymbol{o}]$ is a unimodular EFT. Therefore, Proposition 4.3 gives $\mathbb{E}\left[d_1(\boldsymbol{o})\right] = 1$. By applying the mass transport principle (2.1) to the function $[G, o, x] \mapsto 1_{\{x\in D_1(o)\}}1_A([G, x])$, one gets $\mathcal{P}_1[A] = \mathcal{P}[A]$, which proves that $\sigma\mathcal{P} = \mathcal{P}$.

Conversely, assume $\sigma\mathcal{P} = \mathcal{P}$ and $\mathbb{E}\left[d_1(\boldsymbol{o})\right] = 1$. It is easy to see that $\boldsymbol{T}$ is eternal a.s. (see also Lemma 5.2). To prove unimodularity, by Proposition 2.2 in [**6**], it is enough to prove that $\mathcal{P}$ is *involution invariant*; that is, (2.1) holds for all functions supported on doubly-rooted networks in which the two roots are neighbors.

Note that invariance under σ implies that $\boldsymbol{o}$ has a parent a.s. Now, by the definition of $\mathcal{E}_1$ and $\mathbb{E}\left[d_1(\boldsymbol{o})\right] = 1$ one gets that for all measurable functions $g : \mathcal{G}_{**} \to \mathbb{R}^{\geq 0}$,

$$\begin{aligned}
\mathbb{E}\left[\sum_{v\sim \boldsymbol{o}} g(\boldsymbol{T}, \boldsymbol{o}, v)\right] &= \mathbb{E}\left[\sum_{v\in D_1(\boldsymbol{o})} g(\boldsymbol{T}, \boldsymbol{o}, v)\right] + \mathbb{E}\left[g(\boldsymbol{T}, \boldsymbol{o}, F(\boldsymbol{o}))\right] \\
&= \mathcal{E}_1\left[g(\boldsymbol{T}, F(\boldsymbol{o}), \boldsymbol{o})\right] + \mathbb{E}\left[g(\boldsymbol{T}, \boldsymbol{o}, F(\boldsymbol{o}))\right],
\end{aligned}$$

where the symbol $\sim$ means adjacency of vertices. Similarly,

$$\mathbb{E}\left[\sum_{v\sim \boldsymbol{o}} g(\boldsymbol{T}, v, \boldsymbol{o})\right] = \mathcal{E}_1\left[g(\boldsymbol{T}, \boldsymbol{o}, F(\boldsymbol{o}))\right] + \mathbb{E}\left[g(\boldsymbol{T}, F(\boldsymbol{o}), \boldsymbol{o})\right].$$

Now, $\mathcal{P}_1 = \mathcal{P}$ implies that the right-hand-sides of the above two equations are equal. Therefore, so are the left hand sides and thus $\mathcal{P}$ is involution invariant. This proves that $[\boldsymbol{T}, \boldsymbol{o}]$ is unimodular. □

PROOF OF THEOREM 4.11. Let $b(v) := b_T(v) := d_1(F(v))$ be the number of siblings of v including v itself, which is defined whenever v has a parent. The key point for proving properness is that the distribution of $b(\boldsymbol{o})$ under $\mathcal{P}_{n+1}$ is the size-biased version of the distribution of $d_1(\boldsymbol{o})$ under $\mathcal{P}_n$ (see (4.6)). Convergence of the former implies properness of the limit of the latter (see also Lemma 4.3 in [**25**]). This is discussed in the following in detail.

(i). Fix $k \in \mathbb{Z}^{\geq 0}$. By (4.3) (for $m := 1$),

$$\mathcal{P}_{n+1}[b(\boldsymbol{o}) = k] = \frac{\mathcal{E}_n\left[\sum_{v\in D_1(\boldsymbol{o})} 1_{\{b(v)=k\}}\right]}{\mathcal{E}_n[d_1(\boldsymbol{o})]} = \frac{\mathbb{E}\left[d_n(\boldsymbol{o})\right]}{\mathbb{E}\left[d_{n+1}(\boldsymbol{o})\right]}k\mathcal{P}_n\left[d_1(\boldsymbol{o}) = k\right], \tag{4.6}$$

where (4.5) was used (again for $m := 1$) to get the last expression. The indicator functions $[T, o] \mapsto 1_{\{b(o)=k\}}$ and $[T, o] \mapsto 1_{\{d_1(o)=k\}}$ are bounded continuous functions on $\mathcal{T}_*$. Therefore, one has

$$\text{(4.7)} \qquad \begin{cases} \lim_{n\to\infty} \mathcal{P}_{n+1}[b(\boldsymbol{o}) = k] & = & \mathcal{P}_\infty[b(\boldsymbol{o}) = k], \\ \lim_{n\to\infty} \mathcal{P}_n[d_1(\boldsymbol{o}) = k] & = & \mathcal{P}_\infty[d_1(\boldsymbol{o}) = k]. \end{cases}$$

Choose $k > 0$ such that $\mathcal{P}_\infty[b_1(\boldsymbol{o}) = k] > 0$. By taking lim inf in (4.6), and using (4.7), one gets $\mathcal{P}_\infty[d_1(\boldsymbol{o}) = k] > 0$ (note that $\liminf_{n\to\infty} \frac{\mathbb{E}[d_n(\boldsymbol{o})]}{\mathbb{E}[d_{n+1}(\boldsymbol{o})]} < \infty$ by assumption). Now, by (4.6) and (4.7) again, one gets that the limit

$$c := \lim_{n\to\infty} \frac{\mathbb{E}\left[d_n(\boldsymbol{o})\right]}{\mathbb{E}\left[d_{n+1}(\boldsymbol{o})\right]}$$

exists and for each $k \geq 0$, one has $\mathcal{P}_\infty[b(\boldsymbol{o}) = k] = ck\mathcal{P}_\infty\left[d_1(\boldsymbol{o}) = k\right]$. In other words, under $\mathcal{P}_\infty$, the distribution of $b(\boldsymbol{o})$ is the size-biased version of that of $d_1(\boldsymbol{o})$. By summing over k, one gets $0 < c < \infty$ and

$$\mathcal{E}_\infty[d_1(\boldsymbol{o})] = \frac{1}{c} = \lim_{n\to\infty} \frac{\mathbb{E}\left[d_{n+1}(\boldsymbol{o})\right]}{\mathbb{E}\left[d_n(\boldsymbol{o})\right]}.$$

(ii). By the previous part, one has $\mathcal{E}_\infty[d_1(\boldsymbol{o})] \in (0, \infty)$, thus, $\sigma\mathcal{P}_\infty$ is defined. Let $g : \mathcal{T}_* \to \mathbb{R}^{\geq 0}$ be a bounded continuous function and $k \geq 0$ be arbitrary. By (4.3), one has $\mathcal{P}_{n+1} = \sigma\mathcal{P}_n$. Therefore, by (4.2) and (4.5) (for $m := 1$), one gets

$$\begin{aligned} \mathcal{E}_{n+1}\left[1_{\{b(\boldsymbol{o})\leq k\}} g[\boldsymbol{T}, \boldsymbol{o}]\right] &= \frac{1}{\mathcal{E}_n[d_1(\boldsymbol{o})]} \mathcal{E}_n\left[\sum_{v\in D_1(\boldsymbol{o})} 1_{\{b(v)\leq k\}} g[\boldsymbol{T}, v]\right] \\ &= \frac{\mathbb{E}\left[d_n(\boldsymbol{o})\right]}{\mathbb{E}\left[d_{n+1}(\boldsymbol{o})\right]} \mathcal{E}_n\left[1_{\{d_1(\boldsymbol{o})\leq k\}} \sum_{v\in D_1(\boldsymbol{o})} g[\boldsymbol{T}, v]\right]. \end{aligned}$$

In both sides, the functions under the expectation operator are bounded and continuous. Therefore,

$$\mathcal{E}_\infty\left[1_{\{b(\boldsymbol{o})\leq k\}} g[\boldsymbol{T}, \boldsymbol{o}]\right] = \frac{1}{\mathcal{E}_\infty[d_1(\boldsymbol{o})]} \mathcal{E}_\infty\left[1_{\{d_1(\boldsymbol{o})\leq k\}} \sum_{v\in D_1(\boldsymbol{o})} g[\boldsymbol{T}, v]\right].$$

By letting $k \to \infty$, monotone convergence and $g \geq 0$ imply

$$\mathcal{E}_\infty\left[g[\boldsymbol{T}, \boldsymbol{o}]\right] = \frac{1}{\mathcal{E}_\infty[d_1(\boldsymbol{o})]} \mathcal{E}_\infty\left[\sum_{v\in D_1(\boldsymbol{o})} g[\boldsymbol{T}, v]\right].$$

This means that $\sigma\mathcal{P}_\infty = \mathcal{P}_\infty$; i.e. $\mathcal{P}_\infty$ is offspring-invariant. In particular, $\sigma^n\mathcal{P}_\infty$ is defined for each $n > 0$, which implies that $\mathcal{P}_\infty$ is proper.

(iii). The claim is a direct consequence of parts (i) and (ii) and Proposition 4.10.

□

PROOF OF PROPOSITION 4.15. For any proper random FT $[\boldsymbol{T}, \boldsymbol{o}]$, let $[\boldsymbol{T}_n, \boldsymbol{o}_n]$ be a random FT with the same distribution as $\sigma_n[\boldsymbol{T}, \boldsymbol{o}]$. Using (4.2) directly, one obtains that $\sigma_n[D(\boldsymbol{o}), \boldsymbol{o}]$ has the same distribution as $[D(F^n(\boldsymbol{o}_n)), \boldsymbol{o}_n]$. By assuming offspring-invariance of $[\boldsymbol{T}, \boldsymbol{o}]$, one gets that $\sigma_n[D(\boldsymbol{o}), \boldsymbol{o}]$ has the same distribution as $[D(F^n(\boldsymbol{o})), \boldsymbol{o}]$. Since the trees $D(F^n(\boldsymbol{o}))$ cover any neighborhood of $\boldsymbol{o}$, the distribution of $[D(F^n(\boldsymbol{o})), \boldsymbol{o}]$ tends weakly to that of $[\boldsymbol{T}, \boldsymbol{o}]$. Therefore, the distribution of $\sigma_n[D(\boldsymbol{o}), \boldsymbol{o}]$ also tends to that of $[\boldsymbol{T}, \boldsymbol{o}]$ and the claim is proved. The last claim is a corollary of the the first one and Proposition 4.10. □

4.3. Joining a Stationary Sequence of Trees. Consider a stationary sequence of random rooted trees $([\boldsymbol{T}_i, \boldsymbol{o}_i])_{i=-\infty}^{\infty}$ defined on a common probability space. One may regard each $[\boldsymbol{T}_i, \boldsymbol{o}_i]$ as a (non-ordered) Family Tree by directing the edges towards $\boldsymbol{o}_i$. Add a directed edge $\boldsymbol{o}_i\boldsymbol{o}_{i+1}$ for each $i \in \mathbb{Z}$. By letting $\boldsymbol{o} := \boldsymbol{o}_0$, the resulting random rooted EFT, denoted by $[\boldsymbol{T}, \boldsymbol{o}]$, will be referred to as the *joining* of the sequence $([\boldsymbol{T}_i, \boldsymbol{o}_i])_{i=-\infty}^{\infty}$.

If $\mathbb{E}[\#V(\boldsymbol{T}_0)] < \infty$, then one can move the root of $\boldsymbol{T}$ to a *typical vertex of* $\boldsymbol{T}_0$ as in Definition 4.6. More precisely, consider the following measure on $\mathcal{T}_*$ associated with $[\boldsymbol{T}, \boldsymbol{o}]$:

$$\mathcal{P}'[A] := \frac{1}{\mathbb{E}[\#V(\boldsymbol{T}_0)]}\mathbb{E}\left[\sum_{v\in V(\boldsymbol{T}_0)} 1_A([\boldsymbol{T}, v])\right]. \tag{4.8}$$

It is easy to see that $\mathcal{P}'$ is a probability measure.

THEOREM 4.17. *Let $[\boldsymbol{T}, \boldsymbol{o}]$ be the joining of a stationary sequence of trees $([\boldsymbol{T}_i, \boldsymbol{o}_i])_{i=-\infty}^{\infty}$ such that $\mathbb{E}[\#V(\boldsymbol{T}_0)] < \infty$, as defined above. Let $[\boldsymbol{T}', \boldsymbol{o}']$ be a random rooted* EFT *with distribution $\mathcal{P}'$ defined by* (4.8).

(i) *$[\boldsymbol{T}', \boldsymbol{o}']$ is a unimodular* EFT *and it is of class $\mathcal{I}/\mathcal{F}$ a.s. As a result, all generations of $\boldsymbol{T}$ and $\boldsymbol{T}'$ are finite a.s.*
(ii) *Any unimodular non-ordered* EFT *of class $\mathcal{I}/\mathcal{F}$ can be constructed by joining a stationary sequence of trees as in the previous part.*

REMARK 4.18. Every ordered EFT of class $\mathcal{I}/\mathcal{F}$ can also be constructed in a way similar to that of Part (ii) of Theorem 4.17. Here, only the non-ordered case was discussed for simplicity.

PROOF OF THEOREM 4.17. (i) Let $m := \mathbb{E}[\#V(\boldsymbol{T}_0)]$ and $g : \mathcal{G}_{**} \to \mathbb{R}^{\geq 0}$ be a measurable function. One has

$$\mathbb{E}\left[\sum_{w\in V(\boldsymbol{T}')} g[\boldsymbol{T}', \boldsymbol{o}', w]\right] = \frac{1}{m}\mathbb{E}\left[\sum_{v\in V(\boldsymbol{T}_0)}\sum_{w\in V(\boldsymbol{T})} g[\boldsymbol{T}, v, w]\right]$$
$$= \frac{1}{m}\mathbb{E}\left[\sum_{j=-\infty}^{\infty}\sum_{v\in V(\boldsymbol{T}_0)}\sum_{w\in V(\boldsymbol{T}_j)} g[\boldsymbol{T}, v, w]\right] = \frac{1}{m}\sum_{j=-\infty}^{\infty}\hat{g}(0, j),$$

where $\hat{g}(i, j) := \mathbb{E}\left[\sum_{v\in V(\boldsymbol{T}_i)}\sum_{w\in V(\boldsymbol{T}_j)} g[\boldsymbol{T}, v, w]\right]$. Similarly, one obtains

$$\mathbb{E}\left[\sum_{w\in V(\boldsymbol{T}')} g[\boldsymbol{T}', w, \boldsymbol{o}']\right] = \frac{1}{m}\sum_{j=-\infty}^{\infty}\hat{g}(j, 0).$$

The stationarity of the sequence $\boldsymbol{T}_i$ implies that $\hat{g}(0, j) = \hat{g}(-j, 0)$. Therefore, the right-hand-sides of the above equations are equal. This implies that $[\boldsymbol{T}', \boldsymbol{o}']$ is unimodular.

Now, since there is a bi-infinite path in $\boldsymbol{T}$ a.s., the same holds for $\boldsymbol{T}'$ almost surely. Therefore, Proposition 4.4 implies that $[\boldsymbol{T}', \boldsymbol{o}']$ is of class $\mathcal{I}/\mathcal{F}$ and thus all generations of $\boldsymbol{T}'$ are finite a.s. This implies that all generations of $\boldsymbol{T}$ are finite a.s. too.

(ii) Let $[\tilde{\boldsymbol{T}}, \tilde{\boldsymbol{o}}]$ be a non-ordered unimodular EFT of class $\mathcal{I}/\mathcal{F}$. For all Eternal Family Trees T with a unique bi-infinite F-path, let S_T be the bi-infinite F-path. So $S_{\tilde{\boldsymbol{T}}}$ is almost surely defined. It is easy to see that S is a covariant subset of vertices. For simplicity, denote by $[\boldsymbol{T}, \boldsymbol{o}]$ the random EFT obtained by conditioning on $\boldsymbol{o} \in S_{\tilde{\boldsymbol{T}}}$. Denote the bi-infinite F-path in $\boldsymbol{T}$ by $(\boldsymbol{o}_i)_{i=-\infty}^{\infty}$, where $\boldsymbol{o}_0 = \boldsymbol{o}$ and $F(\boldsymbol{o}_i) = \boldsymbol{o}_{i+1}$. Let $\boldsymbol{T}_i$ be the Family Tree spanned by $D(\boldsymbol{o}_i) \setminus D(\boldsymbol{o}_{i-1})$. It will be shown below that the sequence $\boldsymbol{T}_i$ is the desired random FT sequence.

The first step consists in proving that $(\boldsymbol{T}_i)_i$ is a stationary sequence, that is, its distribution is invariant under the shifts $i \mapsto i+1$ and $i \mapsto i-1$. Only invariance w.r.t. the first shift is proved. The other one can be proved similarly. For this, it is enough to show that the distribution of $[\boldsymbol{T}, \boldsymbol{o}]$ (which is not unimodular) is invariant under the map $\theta_F[T, o] := [T, F(o)]$. Note that $[\boldsymbol{T}, \boldsymbol{o}] \mapsto ([\boldsymbol{T}_i, \boldsymbol{o}_i])_i$ is a measurable bijective map. Define

$$f_T(v) := \begin{cases} F(v), & v \in S_T \\ v, & v \notin S_T \end{cases}.$$

Note that f_T is bijective. So, by Proposition 3.6, θ_f preserves the distribution of $[\tilde{\boldsymbol{T}}, \tilde{\boldsymbol{o}}]$. Since, on $[\boldsymbol{T}, \boldsymbol{o}]$ f and F almost surely agree on $\boldsymbol{o}$, it follows that θ_F preserves the distribution of $[\boldsymbol{T}, \boldsymbol{o}]$, and the first claim is proved.

Next, one proves that for the stationary sequence $\boldsymbol{T}_i$, the construction (4.8) gives back the distribution of $[\tilde{\boldsymbol{T}}, \tilde{\boldsymbol{o}}]$. First, note that by joining the roots of $[\boldsymbol{T}_i, \boldsymbol{o}_i]$, one obtains the same $[\boldsymbol{T}, \boldsymbol{o}]$ here. For all $v \in V(T)$, let $\tau_T(v)$ be the first ancestor of v in S_T. For a given event A, define

$$g[T, v, w] := 1_A([T, v]) 1_{\{w = \tau_T(v)\}}.$$

By the exchange formula (Proposition 2.13) for the subnetworks S_T and T itself, by letting $\mathbb{P}_S$ be the distribution of $[\boldsymbol{T}, \boldsymbol{o}]$ conditionally on $\boldsymbol{o} \in S_{\boldsymbol{T}}$, one gets

$$\mathbb{P}\left[\tilde{\boldsymbol{o}} \in S_{\tilde{\boldsymbol{T}}}\right] \cdot \mathbb{E}_S \left[\sum_{v \in V(\boldsymbol{T}_0)} 1_A[\boldsymbol{T}, v] \right] = \mathbb{P}\left[[\tilde{\boldsymbol{T}}, \tilde{\boldsymbol{o}}] \in A \right].$$

By letting $A := \mathcal{T}_*$, one obtains $\mathbb{P}\left[\tilde{\boldsymbol{o}} \in S_{\tilde{\boldsymbol{T}}}\right] \cdot \mathbb{E}_S\left[\#V(\boldsymbol{T}_0)\right] = 1$. By substituting this in the above equation, it follows that the distribution constructed in (4.8) coincides with the distribution of $[\tilde{\boldsymbol{T}}, \tilde{\boldsymbol{o}}]$ and the claim is proved. □

LEMMA 4.19. *Let $\boldsymbol{T}$ be the joining of a stationary sequence of trees $([\boldsymbol{T}_i, \boldsymbol{o}_i])_{i=-\infty}^{\infty}$ rooted at $\boldsymbol{o} := \boldsymbol{o}_0$. Then $\boldsymbol{T}$ satisfies the following properties:*

(i) *For all measurable functions $h : \mathcal{T}_* \to \mathbb{R}^{\geq 0}$,*

$$\mathbb{E}\left[\sum_{v \in L(\boldsymbol{o})} h[\boldsymbol{T}, v] \right] = \mathbb{E}\left[\sum_{v \in V(\boldsymbol{T}_0)} h[\boldsymbol{T}, v] \right].$$

(ii) *By moving the root of $\boldsymbol{T}$ to a typical vertex in $\boldsymbol{L}(\boldsymbol{o})$ (an operation defined as in (4.8)) one gets the same same distribution as by moving the root of $\boldsymbol{T}$ to a typical vertex in $\boldsymbol{T}_0$.*

PROOF. (i). The key point is that $L(\boldsymbol{o}_0)$ is the disjoint union of $D_i(\boldsymbol{T}_i, \boldsymbol{o}_i)$ for $i \geq 0$. Therefore,

$$\text{LHS} = \sum_{i=0}^{\infty} \mathbb{E}\left[\sum_{v \in D_i(\boldsymbol{T}_i, \boldsymbol{o}_i)} h[\boldsymbol{T}, v]\right] = \sum_{i=0}^{\infty} \mathbb{E}\left[\sum_{v \in D_i(\boldsymbol{T}_0, \boldsymbol{o}_0)} h[\boldsymbol{T}, v]\right] = \text{RHS},$$

where the second equality is implied by stationarity of the sequence $\boldsymbol{T}_i$.

(ii). By (i), one gets $\mathbb{E}[\#L(\boldsymbol{o}_0)] = \mathbb{E}[\#V(\boldsymbol{T}_0)]$. This, together with (i) readily imply the claim. □

The next proposition shows that the distribution obtained by joining a stationary sequence of trees is a special case of moving the root to a typical far descendant.

PROPOSITION 4.20. *Let $[\boldsymbol{T}, \boldsymbol{o}_0]$ be the joining defined in Theorem 4.17 and $\mathcal{P}$ be its distribution. Then the distribution $\mathcal{P}_\infty$ (defined in Subsection 4.2) exists and $\mathcal{P}_\infty = \mathcal{P}'$ (defined in (4.8)).*

PROOF. Note that $D_n(\boldsymbol{T}, \boldsymbol{o}_0)$ is the disjoint union of $D_{n-i}(\boldsymbol{T}_{-i}, \boldsymbol{o}_{-i})$ for $0 \leq i \leq n$. Therefore,

$$\mathbb{E}[d_n(\boldsymbol{o}_0)] = \sum_{i=0}^{n} \mathbb{E}[d_{n-i}(\boldsymbol{T}_{-i}, \boldsymbol{o}_{-i})] = \sum_{i=0}^{n} \mathbb{E}[d_{n-i}(\boldsymbol{T}_0, \boldsymbol{o}_0)] = \mathbb{E}[\#N_n(\boldsymbol{T}_0, \boldsymbol{o}_0)],$$

where $N_n(\boldsymbol{T}_0, \boldsymbol{o}_0)$ is the n-neighborhood of $\boldsymbol{o}_0$ in T_0. Similarly, for any bounded continuous function $h : \mathcal{T}_* \to \mathbb{R}^{\geq 0}$,

$$\begin{aligned}
\mathcal{E}_n[h] &= \frac{1}{\mathbb{E}[d_n(\boldsymbol{o}_0)]}\mathbb{E}\left[\sum_{v \in D_n(\boldsymbol{T}, \boldsymbol{o}_0)} h[\boldsymbol{T}, v]\right] \\
&= \frac{1}{\mathbb{E}[d_n(\boldsymbol{o}_0)]}\sum_{i=0}^{n}\mathbb{E}\left[\sum_{v \in D_{n-i}(\boldsymbol{T}_{-i}, \boldsymbol{o}_{-i})} h[\boldsymbol{T}, v]\right] \\
&= \frac{1}{\mathbb{E}[d_n(\boldsymbol{o}_0)]}\sum_{i=0}^{n}\mathbb{E}\left[\sum_{v \in D_{n-i}(\boldsymbol{T}_0, \boldsymbol{o}_0)} h[\boldsymbol{T}, v]\right] \\
&= \frac{1}{\mathbb{E}[\#N_n(\boldsymbol{T}_0, \boldsymbol{o}_0)]}\mathbb{E}\left[\sum_{v \in N_n(\boldsymbol{T}_0, \boldsymbol{o}_0)} h[\boldsymbol{T}, v]\right].
\end{aligned}$$

By monotone convergence, both the numerator and the denominator converge and

$$\lim_{n\to\infty} \mathcal{E}_n[h] = \frac{1}{\mathbb{E}[\#V(\boldsymbol{T}_0)]}\mathbb{E}\left[\sum_{v \in V(\boldsymbol{T}_0)} h[\boldsymbol{T}, v]\right] = \mathcal{E}'[h].$$

This proves the claim. □

4.4. Bibliographical Comments. The function l defined in (4.1) is a special case of a *Busemann function* (see e.g. [**30**]) which is used in a variety of contexts (see e.g. [**21**]). Roughly speaking, a Busemann function on a graph is a *level structure* on the set of vertices such that the edges are between consecutive levels only (it is usually assumed that there are infinitely many levels on both sides). Such a function gives rise to the notion of *generations* as in this section. Several instances of trees with a level structure are considered in the literature; e.g., in [**20**], [**2**], and [**21**]. The connections are discussed in Subsection 6.3. However, to the best of the authors' knowledge, the general notion of an EFT and the fact that unimodular EFTs are in a sense equivalent to vertex shifts on unimodular networks are new. It should be noted that the main emphasis of this paper is on the unimodular and offspring-invariant cases. The general definition of $\mathcal{P}_\infty$ and the other constructions of EFTs given in the present section are also believed to be new (the precise connections with known definitions and results are discussed in the next paragraph).

An operator similar to σ is defined in [**2**] and its invariant probability distributions is studied therein. In the language of the present work, this operator can be rephrased as follows. A *Sin-Tree* is a rooted EFT with only one end (equivalently, the number of descendants of each vertex is finite). The kernel Q^∞ is defined by $Q^\infty([T,o],A) = \sum_{v\in D_1(o)} 1_A[T,v]$, where $[T,o]$ is a rooted Sin-Tree and A is an event. This kernel acts on the space of measures on the set of (isomorphism classes of) rooted Sin-Trees. An *invariant Sin-Tree* is a random rooted Sin-Tree whose distribution is invariant under the action of Q^∞.

In the setting of the present paper, for probability measures with an average number of children equal to one, the above action is identical to that of the operator σ. It can be seen that invariant Sin-Trees are precisely unimodular EFTs of class $\mathcal{I}/\mathcal{I}$. More generally, (the distributions of) offspring-invariant EFTs with the property that $D(v)$ is finite for all vertices v, are precisely the probability measures that are eigenvectors of Q^∞ corresponding to non-zero eigenvalues.

The statement of Lemma 4.14 is related to Lemma 4.3 in [**25**], in which P_n and σP_n should be replaced by a probability measure on $\mathbb{R}^{\geq 0}$ and its *size-biased* version (i.e., the probability measure $A \mapsto \frac{1}{c}\int_A x dP_n(x)$) respectively.

5. Trees and Networks Beyond Unimodularity

The offspring-invariant setting introduced in Section 4 relaxes the unimodularity assumption and is hence, in this sense, a generalization of the unimodular setting. A simple instance of non-unimodular offspring-invariant EFTs is the d-regular tree with one distinguished end of Example 2.2, when $d > 2$. Subsection 5.1 focuses on offspring-invariant EFTs. In particular, it is shown that they satisfy a modified version of the mass transport principle and a cardinality classification similar to Theorem 3.9. The offspring-invariant setting is also extended to general networks in Subsection 5.2.

5.1. On Offspring-Invariant Random EFTs.

5.1.1. *Offspring-Invariant Mass Transport and Classification.* Here are general properties of offspring-invariant random EFTs which will be used later.

PROPOSITION 5.1. *The distribution of an offspring-invariant random Family Tree $[\boldsymbol{T},\boldsymbol{o}]$ is uniquely determined by that of $[D(\boldsymbol{o}),\boldsymbol{o}]$.*

PROOF. This proposition is a direct corollary of Proposition 4.15. □

LEMMA 5.2. *Let $[\boldsymbol{T},\boldsymbol{o}]$ be an offspring-invariant random FT.*

(i) *$\boldsymbol{T}$ is eternal almost surely.*
(ii) *By letting $m := \mathbb{E}[d_1(\boldsymbol{o})]$, one has for all $n \geq 0$,*

$$\mathbb{E}[d_n(\boldsymbol{o})] = m^n. \tag{5.1}$$

In particular,

$$\mathbb{E}[d(\boldsymbol{o})] = \begin{cases} \frac{1}{1-m}, & m < 1 \\ \infty, & m \geq 1 \end{cases}.$$

PROOF. By (4.3), σ_n preserves the distribution of $[\boldsymbol{T}, \boldsymbol{o}]$ for each n. Now, (4.2) easily implies that $F^n(\boldsymbol{o})$ is defined a.s. This proves that $\boldsymbol{T}$ is eternal a.s.

By (4.5), one gets $\mathbb{E}[d_{n+1}(\boldsymbol{o})] = m\mathbb{E}[d_n(\boldsymbol{o})]$. This implies the second claim by induction on n. □

PROPOSITION 5.3 (Offspring-Invariant Mass Transport Principle). *Let $[\boldsymbol{T}, \boldsymbol{o}]$ be an offspring-invariant random EFT. Then, for all measurable functions $g : \mathcal{T}_{**} \to \mathbb{R}^{\geq 0}$,*

$$\mathbb{E}\left[\sum_{v \in V(\boldsymbol{T})} g[\boldsymbol{T}, \boldsymbol{o}, v]\right] = \mathbb{E}\left[\sum_{v \in V(\boldsymbol{T})} m^{l(v,\boldsymbol{o})} g[\boldsymbol{T}, v, \boldsymbol{o}]\right], \tag{5.2}$$

where $l(v, \boldsymbol{o}) = l_{\boldsymbol{T}}(v, \boldsymbol{o})$ is defined in (4.1). *In particular, the mass transport principle holds along the generation of the root in the sense that*

$$\mathbb{E}\left[\sum_{v \in L(\boldsymbol{o})} g[\boldsymbol{T}, \boldsymbol{o}, v]\right] = \mathbb{E}\left[\sum_{v \in L(\boldsymbol{o})} g[\boldsymbol{T}, v, \boldsymbol{o}]\right]. \tag{5.3}$$

PROOF. By additivity of both sides w.r.t. g, one can assume that for all $v, w \in V(G)$, $g(G, v, w)$ is zero except when $l(v, w) = i$ for a given $i \in \mathbb{Z}$. In this case, (5.2) can be written as $\mathbb{E}[g^+[\boldsymbol{G}, \boldsymbol{o}]] = m^i\mathbb{E}[g^-[\boldsymbol{T}, \boldsymbol{o}]]$. Let $g_n(T, o, v) := g(T, o, v)1_{\{F^n(o)=F^{n+i}(v)\}}$. By (4.2) and (5.1), one gets

$$\begin{aligned} \mathbb{E}\left[g_n^+[\boldsymbol{T}, \boldsymbol{o}]\right] &= \mathcal{E}_n\left[g_n^+\right] &= \frac{1}{m^n}\mathbb{E}\left[\sum_{w \in D_n(\boldsymbol{o})} \sum_{v \in D_{n+i}(\boldsymbol{o})} g(\boldsymbol{T}, w, v)\right], \\ \mathbb{E}\left[g_n^-[\boldsymbol{T}, \boldsymbol{o}]\right] &= \mathcal{E}_{n+i}\left[g_n^-\right] &= \frac{1}{m^{n+i}}\mathbb{E}\left[\sum_{w \in D_{n+i}(\boldsymbol{o})} \sum_{v \in D_n(\boldsymbol{o})} g(\boldsymbol{T}, v, w)\right]. \end{aligned}$$

Therefore, $\mathbb{E}[g_n^+[\boldsymbol{T}, \boldsymbol{o}]] = \mathbb{E}\left[m^i g_n^-[\boldsymbol{T}, \boldsymbol{o}]\right]$. Since $g_n \uparrow g$, Fatou's lemma implies that (5.2) holds for g and the claim is proved. □

Proposition 5.3 can be stated in the language of measured equivalence relations as below (see Example 9.9 of [**6**] for the connections of this subject with random networks). The following definitions are translated from [**15**] to the current setting (see Subsection 2.3).

DEFINITION 5.4. Call a random rooted network $[\boldsymbol{G}, \boldsymbol{o}]$ **quasi-invariant** when the measures $\mathcal{P}^{(r)}(A) := \mathbb{E}[\sum_v 1_A[\boldsymbol{G}, \boldsymbol{o}, v]]$ and $\mathcal{P}^{(l)}(A) := \mathbb{E}[\sum_v 1_A[\boldsymbol{G}, v, \boldsymbol{o}]]$ on $\mathcal{G}_{**}$ are mutually absolutely continuous. In this case, let Δ be a version of the Radon-Nikodym derivative

$$\Delta(o, v) := \Delta[G, o, v] := \frac{d\mathcal{P}^{(r)}}{d\mathcal{P}^{(l)}}[G, o, v]$$

and call it the **Radon-Nikodym cocycle** of $[\boldsymbol{G}, \boldsymbol{o}]$.

The Radon-Nikodym cocycle almost surely satisfies $\Delta[\boldsymbol{G}, v, v] = 1$ for all $v \in V(\boldsymbol{G})$, and $\Delta[\boldsymbol{G}, v, w]\Delta[\boldsymbol{G}, w, z] = \Delta[\boldsymbol{G}, v, z]$ for all triples of vertices $v, w, z \in V(\boldsymbol{G})$. Moreover, the network is unimodular if and only if $\Delta[\boldsymbol{G}, \cdot, \cdot] \equiv 1$ almost surely. For quasi-invariant networks, a version of the mass transport principle holds similar to (5.2) when replacing $m^{l(v,\boldsymbol{o})}$ by $\Delta(\boldsymbol{T}, v, \boldsymbol{o})$. Therefore, Proposition 5.3 can be restated as follows:

COROLLARY 5.5. *Any offspring-invariant* EFT *with mean offspring cardinality* m *is quasi-invariant with Radon-Nikodym cocycle* $\Delta[T, o, v] = m^{l(o,v)}$.

The following is analogous to Lemma 2.6 for offspring-invariant EFTs. Indeed, it holds for all quasi-invariant random networks with the same proof.

LEMMA 5.6. *Let* $[\boldsymbol{T}, \boldsymbol{o}]$ *be an offspring-invariant* EFT *and* S *a covariant subset as in Definition* 2.5. *Then* $\mathbb{P}[S_{\boldsymbol{T}} \neq \emptyset] > 0$ *if and only if* $\mathbb{P}[\boldsymbol{o} \in S_{\boldsymbol{T}}] > 0$.

PROOF. Let $g[T, o, v] := 1_{\{v \in S_T\}}$. By (5.2), one gets

$$\begin{aligned}\mathbb{E}\left[\#S_{\boldsymbol{T}}\right] &= \mathbb{E}\left[\sum_{v \in V(\boldsymbol{T})} g[\boldsymbol{T}, \boldsymbol{o}, v]\right] \\ &= \mathbb{E}\left[\sum_{v \in V(\boldsymbol{T})} m^{l(v,\boldsymbol{o})} g[\boldsymbol{T}, v, \boldsymbol{o}]\right] = \mathbb{E}\left[1_{\{\boldsymbol{o} \in S_{\boldsymbol{T}}\}} \sum_{v \in V(\boldsymbol{T})} m^{l(v,\boldsymbol{o})}\right].\end{aligned}$$

The LHS is nonzero if and only if $S_{\boldsymbol{T}} \neq \emptyset$ with positive probability. On the other hand, the RHS is nonzero if and only $\mathbb{P}\left[\boldsymbol{o} \in S_{\boldsymbol{T}}\right] > 0$. □

Lemma 5.7 and Proposition 5.8 below provide a classification of offspring-invariant EFTs which extends Theorem 3.9.

LEMMA 5.7. *Let* $[\boldsymbol{T}, \boldsymbol{o}]$ *be an offspring-invariant random* EFT *and* $m := \mathbb{E}\left[d_1(\boldsymbol{o})\right]$. *Then, the generation* $L(\boldsymbol{o})$ *of the root is infinite a.s. if and only if*

$$\lim_{n \to \infty} \frac{\mathbb{P}\left[d_n(\boldsymbol{o}) > 0\right]}{m^n} = 0. \tag{5.4}$$

PROOF. One has $L(\boldsymbol{o}) = \bigcup_{n=1}^{\infty} L_n(\boldsymbol{o})$, where $L_n(\boldsymbol{o}) := F^{-n}(F^n(\boldsymbol{o}))$. By offspring-invariance, (4.2) and (5.1), one gets

$$\mathbb{E}\left[\frac{1}{\#L_n(\boldsymbol{o})}\right] = \frac{1}{m^n}\mathbb{E}\left[\sum_{v \in D_n(\boldsymbol{o})} \frac{1}{\#L_n(v)}\right] = \frac{1}{m^n}\mathbb{E}\left[1_{\{d_n(\boldsymbol{o})>0\}}\right] = \frac{1}{m^n}\mathbb{P}\left[d_n(\boldsymbol{o}) > 0\right].$$

By monotone convergence, one obtains

$$\mathbb{E}\left[\frac{1}{\#L(\boldsymbol{o})}\right] = \lim_{n \to \infty} \frac{\mathbb{P}\left[d_n(\boldsymbol{o}) > 0\right]}{m^n}.$$

On the other hand, $L(\boldsymbol{o})$ is infinite a.s. if and only if the LHS is zero. This proves the claim. □

PROPOSITION 5.8 (Cardinality Classification of Offspring-Invariant EFTs). *Let* $[\boldsymbol{T}, \boldsymbol{o}]$ *be an offspring-invariant random* EFT *and* $m := \mathbb{E}\left[d_1(\boldsymbol{o})\right]$. *Then almost surely, either all generations of* $\boldsymbol{T}$ *are finite or all are infinite. Moreover, almost surely,*

(i) *If $m > 1$, then all generations of $\boldsymbol{T}$ are infinite. Moreover, $\boldsymbol{T}$ has either one or infinitely many ends.*

(ii) *If $m = 1$, then the following are equivalent.*
- *All generations of $\boldsymbol{T}$ are finite (resp. infinite).*
- *There is a (resp. no) vertex v such that $D(v)$ is infinite.*
- *$\boldsymbol{T}$ has two ends (resp. one end).*

(iii) *If $m < 1$, then $D(v)$ is finite for each $v \in V(\boldsymbol{T})$ and $\boldsymbol{T}$ has one end. Moreover, each generation is infinite a.s. if and only if (5.4) holds.*

The case $m = 1$ above is just Proposition 4.4, which is restated here.

Before presenting the proof, here are examples for the different cases in Proposition 5.8. In part (i), both cases of one end or infinitely many ends are possible. The biased Canopy Tree for $d_1 > d$ (Example 5.13 below) is an example with one end. Super-critical Eternal Galton-Watson Trees introduced in Subsection 6.1 are examples of having infinitely many ends (see Proposition 6.8). Moreover, offspring-invariant EFTs can have countably many ends, in contrast with unimodular trees (Proposition 3.19). This is illustrated by Example 5.15 below. In part (iii), the generations (i.e. foils) of $\boldsymbol{T}$ can be all finite or all infinite. The biased Canopy Tree for $d_1 < d$ (Example 5.13 below) is an example of the infinite case. Subcritical Eternal Galton-Watson Trees introduced in Subsection 6.1 provide examples of the finite case (see Proposition 6.8).

PROOF OF PROPOSITION 5.8. Note that all networks are assumed to have all their vertices with finite degrees. This implies that if the k-th generation is infinite, then so are the k'-th generations for all $k' < k$. Therefore, if there are both finite and infinite generations in an EFT, then there is a *first* finite generation.

Assume that with positive probability, there are both finite and infinite generations. In this case, let $S_{\boldsymbol{T}}$ be the first finite generation (let it be empty otherwise). S is a covariant subset (Definition 2.5). Therefore, Lemma 5.6 implies that $\mathbb{P}[\boldsymbol{o} \in S_{\boldsymbol{T}}] > 0$. Since $S_{\boldsymbol{T}}$ is finite, there is $N < \infty$ such that

$$\mathbb{P}[\boldsymbol{o} \in S_{\boldsymbol{T}}, \#S_{\boldsymbol{T}} < N] > 0.$$

Let S' be the last generation before S_T, which is infinite whenever $S_T \neq \emptyset$. Let $h[T, o, v] := 1$ if $o \in S_T, v \in S'_T$ and $\#S_T < N$ and let it be 0 otherwise. By the above inequality and infiniteness of $S'_{\boldsymbol{T}}$ (when nonempty), one gets $\mathbb{E}\left[h^+_{\boldsymbol{T}}(\boldsymbol{o})\right] = \infty$. On the other hand, $h^-_{\boldsymbol{T}}(\boldsymbol{o}) \leq N$ and thus, $\mathbb{E}\left[h^-_{\boldsymbol{T}}(\boldsymbol{o})\right] \leq N$. This contradicts (5.2) (note that here, only the vertices with $l(\boldsymbol{o}, v) = 1$ matter in the RHS of (5.2)). Therefore, almost surely, either all generations are finite or all are infinite.

(i) Equation (5.4) holds trivially. Therefore, by Lemma 5.7, $L(\boldsymbol{o})$ is infinite a.s. So, the argument at the beginning of the proof shows that all generations are infinite a.s. For the second claim, let

$$S_T := \{v \in V(T) : d(v) = \infty\}.$$

Similarly to the proof of Corollary 2.10, by using the mass transport principle along the generation of the root (5.3), one can show that $\#S_{\boldsymbol{T}} \cap L(\boldsymbol{o})$ is in $\{0, \infty\}$ a.s. When it is zero, $\boldsymbol{T}$ has one end and when it is ∞, $\boldsymbol{T}$ has infinitely many ends. This proves the claim.

(ii) Proposition 4.10 implies that $[\boldsymbol{T}, \boldsymbol{o}]$ is unimodular. Now, the claim follows by Proposition 4.4.

(iii) The second claims is proved in Lemma 5.7. For the first claim, Lemma 5.2 shows that $D(\boldsymbol{o})$ is finite a.s. By Lemma 5.6 for the covariant subset S defined above, one gets that almost surely, $D(v)$ is finite for all vertices v. This implies that $\boldsymbol{T}$ has one end a.s. and the claim is proved. □

5.1.2. *Sub-EFTs.* This subsection shows two construction methods regarding offspring-invariant EFTs. In particular, the construction by covariant sub-EFTs in Proposition 5.9, which is analogous to that in Remark 2.12 in the unimodular case, is used in the examples of Subsection 5.1.3.

PROPOSITION 5.9. *Let $[\boldsymbol{T}, \boldsymbol{o}]$ be an offspring-invariant EFT and S be a covariant subset of the vertices. If the subgraph induced by $S_{\boldsymbol{T}}$ is a sub-EFT of $\boldsymbol{T}$ a.s., then $[S_{\boldsymbol{T}}, \boldsymbol{o}]$, conditioned on $\boldsymbol{o} \in S_{\boldsymbol{T}}$, is an offspring-invariant EFT with the same mean number of children as $[\boldsymbol{T}, \boldsymbol{o}]$.*

Here, the notation $S_{\boldsymbol{T}}$ is used both for a subset of the vertices and for the sub-EFT induced by the subset.

PROOF OF PROPOSITION 5.9. Let $g : \mathcal{T}_* \to \mathbb{R}^{\geq 0}$ be a measurable function. Define

$$\hat{g}[T, v] := g[S_T, v]1_{S_T}(v) = g[S_T, v]1_{S_T}(v)1_{S_T}(F(v)),$$

where the last equation used that $v \in S_{\boldsymbol{T}}$ implies $F(v) \in S_{\boldsymbol{T}}$ almost surely. It is easy to see that $\hat{g}$ is measurable. One has

$$\begin{aligned} \mathbb{E}\left[1_{S_{\boldsymbol{T}}}(\boldsymbol{o}) \sum_{v \in D_1(\boldsymbol{o}) \cap S_{\boldsymbol{T}}} g[S_{\boldsymbol{T}}, v]\right] &= \mathbb{E}\left[\sum_{v \in D_1(\boldsymbol{o})} \hat{g}[\boldsymbol{T}, v]\right] \\ &= m\mathbb{E}\left[\hat{g}[\boldsymbol{T}, \boldsymbol{o}]\right] = m\mathbb{E}\left[g[S_{\boldsymbol{T}}, \boldsymbol{o}]1_{S_{\boldsymbol{T}}}(\boldsymbol{o})\right]. \end{aligned}$$

Therefore,

$$\mathbb{E}\left[\sum_{v \in D_1(\boldsymbol{o}) \cap S_{\boldsymbol{T}}} g[S_{\boldsymbol{T}}, v] \,|\, \boldsymbol{o} \in S_{\boldsymbol{T}}\right] = m\mathbb{E}\left[g[S_{\boldsymbol{T}}, \boldsymbol{o}] \,|\, \boldsymbol{o} \in S_{\boldsymbol{T}}\right].$$

By letting $g \equiv 1$, one gets $m = \mathbb{E}\left[\#(D_1(\boldsymbol{o}) \cap S_{\boldsymbol{T}}) \,|\, \boldsymbol{o} \in S_{\boldsymbol{T}}\right]$. Now the claim is obtained by substituting this value of m in the above equation and using (4.2). □

The next construction method is pruning.

DEFINITION 5.10. Let (T, o) be a rooted FT. The **pruning** of (T, o) from generation $z \geq 0$ is the rooted Family Tree $K(T, o, z)$ which is the restriction of T to the set $\{v \in V(T) : l(o, v) \leq z\}$ rooted at o.

It is easily seen that pruning induces a measurable map $K : \mathcal{T}_* \times \mathbb{Z}^{\geq 0} \to \mathcal{T}_*$.

PROPOSITION 5.11 (Pruning). *Let $[\boldsymbol{T}, \boldsymbol{o}]$ be an offspring-invariant random EFT. Assume $m := \mathbb{E}\left[d_1(\boldsymbol{o})\right] > 1$. Let Z be a random variable such that $Z + 1$ is geometric with parameter $1 - \frac{1}{m}$, independent of $[\boldsymbol{T}, \boldsymbol{o}]$. Then, by pruning $[\boldsymbol{T}, \boldsymbol{o}]$ from generation Z, one gets a unimodular EFT of class $\mathcal{I}/\mathcal{I}$.*

PROOF. Denote $K(\boldsymbol{T}, \boldsymbol{o}, Z)$ by $[\boldsymbol{T}', \boldsymbol{o}]$, which is a random EFT. First, note that

$$\mathbb{E}\left[d_1(\boldsymbol{T}', \boldsymbol{o})\right] = \mathbb{E}\left[d_1(\boldsymbol{T}, \boldsymbol{o})1_{\{Z>0\}}\right] = \mathbb{E}\left[d_1(\boldsymbol{T}, \boldsymbol{o})\right]\mathbb{P}\left[Z > 0\right] = 1.$$

The next step is to show that the distribution $\mathcal{P}'$ of $[\boldsymbol{T}', \boldsymbol{o}]$ is offspring-invariant. For a vertex v in $K(T, o, z)$, let $(K(T, o, z), v)$ be the tree obtained from $K(T, o, z)$ by considering v as the root. One has

$$\begin{aligned}
\sigma\mathcal{P}'[A] &= \mathbb{E}\left[\sum_{v \in D_1(\boldsymbol{T}', \boldsymbol{o})} 1_A[\boldsymbol{T}', v]\right] = \mathbb{E}\left[1_{\{Z>0\}} \sum_{v \in D_1(\boldsymbol{T}, \boldsymbol{o})} 1_A[K(\boldsymbol{T}, o, Z), v]\right] \\
&= \mathbb{E}\left[1_{\{Z>0\}} \sum_{v \in D_1(\boldsymbol{T}, \boldsymbol{o})} 1_A[K(\boldsymbol{T}, v, Z-1)]\right] \\
&= \frac{1}{m}\mathbb{E}\left[\sum_{v \in D_1(\boldsymbol{T}, \boldsymbol{o})} 1_A(K(\boldsymbol{T}, v, Z))\right] = \mathbb{P}\left[K(\boldsymbol{T}, \boldsymbol{o}, Z) \in A\right],
\end{aligned}$$

where the fourth equation used the fact that Z is independent from $[\boldsymbol{T}, \boldsymbol{o}]$ and $Z-1$ conditioned on $Z > 0$ has the same distribution as Z and the last equation holds because of offspring-invariance of $[\boldsymbol{T}, \boldsymbol{o}]$. This shows that $\sigma\mathcal{P}' = \mathcal{P}'$. Now, unimodularity of $[\boldsymbol{T}', \boldsymbol{o}]$ follows by Proposition 4.10. Since there is a youngest foil in $\boldsymbol{T}'$, Proposition 4.4 implies that $[\boldsymbol{T}', \boldsymbol{o}]$ is of class $\mathcal{I}/\mathcal{I}$. □

5.1.3. *Examples.* Here are examples of offspring-invariant EFTs, some of which are obtained from the results of the last subsection. More elaborate examples are provided by the Eternal Galton-Watson Tree and its multi-type version which will be introduced in Subsections 6.1 and 6.2.

EXAMPLE 5.12 (Canopy Tree). As already mentioned, the d-regular tree with one distinguished end is offspring-invariant. For $d > 2$, one can prune it according to Proposition 5.11. The resulting random EFT is just the Canopy Tree with offspring cardinality $d-1$.

EXAMPLE 5.13 (Biased Canopy Tree). Consider the Canopy Tree in Example 2.3 with offspring cardinality $d > 1$, and choose the root such that $\mathbb{P}[\boldsymbol{o} \in L_i]$ is proportional to $\tilde{d}^{-i}$ for an arbitrary $\tilde{d} > 1$. It is not difficult to check that this gives an offspring-invariant random EFT with $\mathbb{E}\left[d_1(\boldsymbol{o})\right] = \frac{d}{\tilde{d}}$. Therefore, this EFT is unimodular if and only if $\tilde{d} = d > 1$ (by Proposition 4.3). Moreover, for $\tilde{d} > d$, by pruning the biased Canopy Tree as in Proposition 5.11, one obtains the usual (unimodular) Canopy Tree with offspring cardinality $d-1$.

See also Example 6.18 below for another example of pruning.

EXAMPLE 5.14. Let $d > 2$ and $[\boldsymbol{T}, \boldsymbol{o}]$ be the d-regular tree with one distinguished end (Example 2.2). Attach to each vertex $v \in V(\boldsymbol{T})$ a path of additional vertices $g_0(v), g_1(v), g_2(v), \ldots$, where $g_0(v) = v$ to obtain an Eternal Family Tree $\boldsymbol{T}'$. The additional vertices associated with different $v \in V(\boldsymbol{T})$ are assumed disjoint and the edges of each such path are directed towards the d-regular tree. Given $[\boldsymbol{T}, \boldsymbol{o}]$, Let $\boldsymbol{o}' := g_{Z-1}(\boldsymbol{o})$, where Z is a geometric random variable with parameter $1 - \frac{1}{m}$, independent of $[\boldsymbol{T}, \boldsymbol{o}]$ and where $m = d-1$. Then, it can be seen that $[\boldsymbol{T}', \boldsymbol{o}']$ is an offspring-invariant EFT with mean number of children m (see also Example 6.17 below).

It can be seen that in this example, one can replace $[\boldsymbol{T}, \boldsymbol{o}]$ by any offspring-invariant EFT with mean number of children $m > 1$. Moreover, one can deduce from Proposition 5.9 that the condition $m > 1$ is necessary.

The following example shows that offspring-invariant EFTs can have countably many ends in contrast with unimodular trees (Proposition 3.19).

EXAMPLE 5.15 (Offspring-Invariant Comb). A **comb** is an Eternal Family Tree T with the following property: For every vertex $v \in V(T)$ with $d_1(v) = k$, one has

$$\{d_1(w) : w \in D_1(v)\} = \{1, 2, \ldots, k\}$$

Note that for every $i \leq d_1(v)$, there is exactly one child w of v such that $d_1(w) = i$. The name 'comb' is chosen because, when $d_1(v) = 2$, the descendant subtree $D(v)$ of v looks like a comb (see the bibliographical comments at the end of this section). A rooted comb $[T, o]$ is uniquely characterized by the sequence $c_i := d_1(F^i(o))$ for $i \geq 0$. Choose a random sequence $(C_i)_{i \geq 0}$ such that the sequence $(C_0, C_1 - C_0 + 1, C_2 - C_1 + 1, \ldots)$ is i.i.d. with a geometric distribution. Its corresponding random rooted comb $[\boldsymbol{T}, \boldsymbol{o}]$ is called an **offspring-invariant comb**. It can be shown that it is indeed offspring-invariant (see also Example 6.17 below).

Here, except the end of the path $\boldsymbol{o}, F(\boldsymbol{o}), F^2(\boldsymbol{o}), \ldots$, each end is realized by a path of vertices with constant degree. It follows that $\boldsymbol{T}$ has countably many ends.

5.2. Generalization of Offspring-Invariance to Random Networks. In this subsection, the framework of offspring-invariant EFTs is generalized to networks, where the parent vertex-shift of EFTs is replaced by an arbitrary vertex-shift on networks.

DEFINITION 5.16. Let f be a given vertex-shift and $[\boldsymbol{G}, \boldsymbol{o}]$ be a random rooted network with distribution $\mathcal{P}$. Then, $[\boldsymbol{G}, \boldsymbol{o}]$ is called **offspring-invariant w.r.t.** f or just f**-offspring-invariant** if $\sigma^{(f)}\mathcal{P} = \mathcal{P}$, where

$$(\sigma^{(f)}\mathcal{P})(A) := \frac{1}{m}\mathbb{E}\left[\sum_{v \in D_1(\boldsymbol{o})} 1_A[\boldsymbol{G}, v]\right], \ \forall A \tag{5.5}$$

and $m := \mathbb{E}[d_1(\boldsymbol{o})]$ (see the notation in Definition 3.8).

Note that the RHS of (5.5) is identical to that of (4.2) for $n = 1$. This suggests that some of the results in Subsection 4.2 hold in the new setting. The n-fold iteration of $\sigma^{(f)}$ on $\mathcal{P}$, whenever define, has a similar equation to (4.2) with the same proof as Lemma 4.9. Also, the f-offspring-invariance of $\mathcal{P}_\infty$ (defined similarly) holds with the same conditions as in Theorem 4.11. Below, some properties of offspring-invariant random networks are discussed beyond those for EFTs.

LEMMA 5.17. *A unimodular network is offspring-invariant w.r.t. any vertex-shift and always* $m = 1$. *Conversely, if* $[\boldsymbol{G}, \boldsymbol{o}]$ *is* f*-offspring-invariant with* $m = 1$*, then the connected component of* $\boldsymbol{G}^f$ *containing* $\boldsymbol{o}$ *is unimodular.*

PROOF. For the first claim, apply (2.1) to $g[G, o, v] := 1_A[G, o]1_{\{v = f_G(o)\}}$. The second claim can also be proved similar to Proposition 4.10. □

According to this lemma, the focus is on the case $m \neq 1$ from now on. It should be noted that in the converse of the lemma, one cannot deduce that $[\boldsymbol{G}, \boldsymbol{o}]$ is unimodular.

LEMMA 5.18. *If* $[\boldsymbol{G}, \boldsymbol{o}]$ *is* f*-offspring-invariant and* $m \neq 1$*, then the connected component of* $\boldsymbol{G}^f$ *containing* $\boldsymbol{o}$ *is acyclic a.s. and is an offspring-invariant* EFT.

Before the proof, it is good to mention that if $[\boldsymbol{G},\boldsymbol{o}]$ is in addition quasi-invariant, then all connected components of $\boldsymbol{G}^f$ are acyclic (see Lemma 5.6).

PROOF. Let $C = C_G$ be the union of the f-cycles of G^f. By (5.5), one gets

$$\sigma^{(f)}P[\boldsymbol{o} \in C] = \frac{1}{m}\mathbb{E}\left[\#D_1(\boldsymbol{o}) \cap C\right] = \frac{1}{m}\mathbb{P}\left[\boldsymbol{o} \in C\right],$$

where the last equality holds because $\#D_1(\boldsymbol{o}) \cap C$ is $\{0,1\}$-valued depending on whether $\boldsymbol{o} \in C$ or not. Therefore, by f-offspring-invariance and $m \neq 1$, one gets $\mathbb{P}\left[\boldsymbol{o} \in C\right] = 0$. Assume it is proved that $\mathbb{P}\left[f^n(\boldsymbol{o}) \in C\right] = 0$ for some $n \geq 0$. By (5.5) again, one has

$$\begin{aligned}\mathbb{P}\left[f^{n+1}(\boldsymbol{o}) \in C\right] &= \sigma^{(f)}\mathcal{P}[\boldsymbol{o} \in f^{-n-1}(C)] = \frac{1}{m}\mathbb{E}\left[\#D_1(\boldsymbol{o}) \cap f^{-n-1}(C)\right] \\ &= \frac{1}{m}\mathbb{E}\left[1_{\{\boldsymbol{o} \notin f^{-n}(C)\}}\#D_1(\boldsymbol{o}) \cap f^{-n-1}(C)\right] = 0,\end{aligned}$$

where the last two equations use the fact that $\#D_1(\boldsymbol{o}) \cap f^{-n-1}(C)$ is zero whenever $\boldsymbol{o} \notin f^{-n}(C)$ and the latter happens with probability one. Inductively, this proves that $\mathbb{P}\left[f^n(\boldsymbol{o}) \in C\right] = 0$ for all $n \geq 0$ and the claim is obtained. □

The following proposition gives a criterion for verifying f-offspring-invariance. See Definition 5.4.

PROPOSITION 5.19. *Let f be a vertex-shift and $[\boldsymbol{G},\boldsymbol{o}]$ be a quasi-invariant random network with Radon-Nikodym cocycle Δ. Then $[\boldsymbol{G},\boldsymbol{o}]$ is f-offspring-invariant if and only if there is a constant c such that $\Delta(\boldsymbol{o}, f(\boldsymbol{o})) = c$ a.s. Moreover, if this holds, then $c = \frac{1}{m}$, where $m := \mathbb{E}\left[d_1(\boldsymbol{o})\right]$.*

PROOF. Note that the latter condition is equivalent to $\Delta(f(\boldsymbol{o}),\boldsymbol{o}) = m$ a.s. Also, by the definition of Δ, one gets

$$\mathbb{E}\left[\sum_{v \in D_1(\boldsymbol{o})} 1_A[\boldsymbol{G}, v]\right] = \mathbb{E}\left[1_A[\boldsymbol{G},\boldsymbol{o}]\Delta(f(\boldsymbol{o}),\boldsymbol{o})\right], \tag{5.6}$$

for all events $A \subseteq \mathcal{G}_*$. First, suppose $\Delta(f(\boldsymbol{o}),\boldsymbol{o}) = c$ a.s. By substituting this in (5.6), (5.5) gives that $[\boldsymbol{G},\boldsymbol{o}]$ is f-offspring-invariant and $c = m$. Conversely, assume the latter holds. Since Equations (5.5) and (5.6) hold for any event A, one gets $\Delta(f(\boldsymbol{o}),\boldsymbol{o}) = m$ a.s. and the claim is proved. □

In Example 5.20 below, the *horocyclic product* (also called *horospheric product*) of two independent offspring-invariant EFTs is considered. First, the definition of horocyclic products (see **[21]** and the references therein) is recalled in the case of EFTs. Here, the notation (F, l and D_n of Subsection 4.1) is used without reference to the underlying EFT as the context always indicates what is meant. Let (T_1, o_1) and (T_2, o_2) be rooted family trees. According to Subsection 4.4, the function l can be regarded as a Buesmann function on T_1 and T_2. So the **horocyclic product** (G, o) of the two EFTs can be defined as follows **[21]**: the vertex set is

$$V(G) := \{(v_1, v_2) \in V(T_1) \times V(T_2) : l(o_1, v_1) + l(o_2, v_2) = 0\}$$

and $o := (o_1, o_2)$. Also, add a directed edge from each vertex $(v_1, v_2) \in V(G)$ to each vertex in $\{F(v_1)\} \times F^{-1}(v_2)$. That is, (v_1, v_2) has $d_1(v_2)$ outgoing edges and $d_1(v_1)$ incoming edges. It can be seen that the graph is connected. Also, the horocyclic product of random rooted EFTs makes sense (see **[21]**).

EXAMPLE 5.20 (Horocyclic Products). Let $[\boldsymbol{T}_1, \boldsymbol{o}_1]$ and $[\boldsymbol{T}_2, \boldsymbol{o}_2]$ be two independent offspring-invariant EFTs with average offspring cardinalities m_1 and m_2 respectively. Let $[\boldsymbol{G}, \boldsymbol{o}]$ be their horocyclic product (see above). Note that $[\boldsymbol{T}_1, \boldsymbol{o}_1]$ and $[\boldsymbol{T}_2, \boldsymbol{o}_2]$ are quasi-invariant and their Radon-Nikodym cocycles are given in Corollary 5.5. The arguments before Theorem 2.18 of [**21**] imply that $[\boldsymbol{G}, \boldsymbol{o}]$ is also quasi-invariant, with Radon-Nikodym cocycle $\Delta((o_1, o_2), (v_1, v_2)) = (m_1/m_2)^{l(o_1, v_1)}$. Therefore, $[\boldsymbol{G}, \boldsymbol{o}]$ is unimodular if and only if $m_1 = m_2$. It should be noted that when each $\boldsymbol{T}_i$ is a $(m_i + 1)$-regular tree with one distinguished end, $\boldsymbol{G}$ is just the Diestel-Leader graph [**14**].

A natural vertex-shift is now defined on this horocyclic product $[\boldsymbol{G}, \boldsymbol{o}]$ using i.i.d. extra marks (assuming that each vertex in $\boldsymbol{T}_2$ has at least one child a.s.) Roughly speaking, it consists in picking a member $f(v_1, v_2)$ of $\{F(v_1)\} \times F^{-1}(v_2)$ uniformly at random, and independently for all vertices (v_1, v_2). This can be made rigorous by adding i.i.d. marks $t(v_1, v_2) \in [0,1] \times [0,1]$ to the vertices to obtain a new random network (the Radon-Nikodym cocycle is not changed by adding i.i.d. marks). Using the first coordinate of $t(v_1, v_2)$, one samples a number in $\{1, 2, \ldots, \#F^{-1}(v_2)\}$ uniformly at random. Then, one uses this number and the natural order of the second coordinates of the marks on $\{F(v_1)\} \times F^{-1}(v_2)$ to pick one element of this set. This defines a vertex-shift f with the desired property.

Finally, let $[\boldsymbol{T}, \boldsymbol{o}]$ be the connected component containing $\boldsymbol{o}$ of the graph $\boldsymbol{G}^f$ of the vertex-shift (Definition 3.8) rooted at $\boldsymbol{o}$. One gets $\Delta(\boldsymbol{o}, f(\boldsymbol{o})) = (m_1/m_2)^{-1}$. Therefore, Proposition 5.19 implies that $[\boldsymbol{T}, \boldsymbol{o}]$ is an offspring-invariant EFT with average offspring cardinality m_1/m_2.

The above construction is used in Example 6.10 below, which constructs a nontrivial explicit example of offspring-invariant EFTs.

5.3. Bibliographical Comments.

The name 'comb' is used in [**28**] for a different tree which is a subtree of the two dimensional lattice on $\mathbb{Z}^2$. This tree is related to the descendant subtree $D(v)$ in Example 5.15 in the case $d_1(v) = 2$. Similar to the construction in [**28**], the EFT of Example 5.15 can be naturally embedded in $\mathbb{Z}^k$ for some $1 \leq k \leq \infty$. So, it could be called a **multi-dimensional comb** as well.

Quasi-invariance and the Radon-Nikodym cocycle are originally defined for measured equivalence relations [**15**]. According to the discussions in Subsection 2.3, these definitions are translated to random rooted networks in Definition 5.4. This translation is likely to be known before. For example, [**8**] does something similar, but its definition is slightly different from that of [**15**]. The notion of offspring-invariance is believed to be new.

Offspring-invariant EFTs are related to *conformal trees* [**21**] as discussed below (the following is a translation of the notions in [**21**] according to the discussion in Subsection 2.3). Assume that for every undirected tree T and every vertex $x \in V(T)$, a finite measure ν_x is given on ∂T (where ∂T is the set of ends of T) which is invariant under rooted-isomorphisms of (T, x). A measurability condition should also be satisfied (see Definition 2.6 of [**21**]). This system of finite measures is called **conformal** if, for all trees T, and all pairs of vertices $x, y \in V(T)$, one has

$$\frac{d\nu_x}{d\nu_y}(\gamma) = m^{l_\gamma(x,y)}$$

for some constant m, where $l_\gamma(\cdot,\cdot)$ is the Buesmann function associated to the end γ (see Subsection 4.4). In fact, given a random rooted undirected tree $[\boldsymbol{T},\boldsymbol{o}]$, one can assume that the above condition is satisfied for almost all realizations of $\boldsymbol{T}$. Below, it is shown that if $[\boldsymbol{T},\boldsymbol{o}]$ is unimodular and every ν_x is a probability measure, an offspring-invariant EFT is obtained.

Assume that almost surely, for every $x \in V(\boldsymbol{T})$, ν_x is a probability measure. Select a random end of $\boldsymbol{T}$ with distribution $\nu_{\boldsymbol{o}}$ (conditional on $[\boldsymbol{T},\boldsymbol{o}]$). By the arguments at the beginning of this section, this gives a random rooted EFT, namely $[\boldsymbol{T}',\boldsymbol{o}']$. If in addition, $[\boldsymbol{T},\boldsymbol{o}]$ is unimodular, then Remark 2.15 of **[21]** implies that $[\boldsymbol{T}',\boldsymbol{o}']$ is quasi-invariant with Radon-Nikodym cocycle $\Delta[T,o,x] = m^{l(o,x)}$. Therefore, Proposition 5.19 implies that $[\boldsymbol{T}',\boldsymbol{o}']$ is offspring-invariant with mean offspring cardinality m. Note that the converse does not hold in general, since offspring-invariant EFTs have generally no presumed system of finite measures. Also, their underlying undirected tree is not necessarily unimodular

In the case when $\nu_{(\cdot)}$ is not a probability measure, or $[\boldsymbol{T},\boldsymbol{o}]$ is non-unimodular, the above construction does not work and there seems no direct connection between the two notions. As an example, **[21]** considers a conformal system of finite measures for the *Augmented Galton-Watson tree* (AGW), but it does not satisfy the above conditions in general (otherwise, it would be connected to the *Eternal Galton-Watson tree* defined in the next section). The same construction can be done for the *Unimodular Galton-Watson tree* (UGW), but does not satisfy the first condition in general. More properties of the AGW and UGW are discussed in Subsection 6.3.

6. Eternal Branching Processes

This section introduces special cases of Eternal Family Trees satisfying certain mutual independence assumptions and pertaining hence to branching processes. The main object is the Eternal Galton-Watson Tree introduced in Subsection 6.1. Such trees have connections with many objects in the literature which are reviewed in detail in Subsection 6.3. These connections show how general properties established for EFTs unify several concepts and results previously known. For instance, the classification of offspring-invariant EFTs under the independence assumptions boils down (but with a new non-analytic proof) to classical results on branching processes with immigration (Proposition 6.8). Although the critical case has been defined previously, some of the results are new. In particular, Theorem 6.6 provides a characterization of Eternal Galton-Watson Trees as offspring-invariant EFTs satisfying a specific independence property. Also, the multi-type version in Subsection 6.2 appears to be new.

6.1. Eternal Galton-Watson Trees. Let π be a probability distribution on $\mathbb{Z}^{\geq 0}$. Recall that the ordinary **Galton-Watson Tree** (abbreviated GWT) with offspring distribution π is a rooted tree defined by a branching process: starting from a single vertex, each vertex v gives birth to a random number $d_1(v)$ of new vertices, where $d_1(v)$ has distribution π, and the same process is repeated recursively for the new vertices independently. One obtains a Family Tree by connecting each vertex to its children. Denote by $\mathcal{P}_{\mathrm{GW}}$ the law of the GWT.

Let $\widehat{\pi}$ be the size-biased version of π; that is, $\widehat{\pi}(k) := \frac{k\pi(k)}{m}$, where m is the expected value of π, assuming $0 < m < \infty$.

DEFINITION 6.1. The **(ordered) Eternal Galton-Watson Tree** (EGWT) with offspring distribution π (abbreviated as EGW(π)) is a random EFT constructed as follows. Start from a path $(\boldsymbol{o}_i)_{i=0}^{\infty}$. For each $i > 0$, regard $\boldsymbol{o}_i$ as the parent of $\boldsymbol{o}_{i-1}$. Then, choose an independent random number z_i with the size-biased distribution $\hat{\pi}$ and add $z_i - 1$ new vertices as children of $\boldsymbol{o}_i$ (so that $\boldsymbol{o}_i$ has a total of z_i children). Choose a uniform random order between the children of $\boldsymbol{o}_i$. For $\boldsymbol{o}_0$ and each new vertex, generate their descendants as in an ordinary Galton-Watson Tree with offspring distribution π. Finally, add a directed edge from each vertex to its parent and let $\boldsymbol{o} := \boldsymbol{o}_0$ be its root. Denote by $\mathcal{P}_{\mathrm{EGW}}$ the law of the EGWT. The **non-ordered** EGWT is obtained by forgetting the order of the vertices in the ordered EGWT. The same symbols will be used for the ordered and non-ordered cases.

REMARK 6.2. The arguments in this section are valid for both ordered and non-ordered cases except when explicitly mentioned.

EXAMPLE 6.3. The d-regular tree with one distinguished end (Example 2.2) is an example of EGWT. For this, π is concentrated on $d-1$.

The proofs of the following results are postponed to the end of the subsection.

PROPOSITION 6.4. *The Eternal Family Tree EGW(π),*

(i) *is offspring-invariant.*
(ii) *is a unimodular EFT if and only if it is critical, that is, $m = 1$.*

PROPOSITION 6.5. *If $[\boldsymbol{T}, \boldsymbol{o}]$ is an ordinary GW(π) tree, then $\sigma_\infty[\boldsymbol{T}, \boldsymbol{o}]$ is well defined and is an EGW(π) tree. In other words, EGWTs are obtained from ordinary Galton-Watson Trees by moving the root to a typical far descendant.*

The next theorem provides a characterization of EGWTs.

THEOREM 6.6 (Characterization of EGWTs). *A random Family Tree $[\boldsymbol{T}, \boldsymbol{o}]$ is an EGWT (resp. a critical EGWT) if and only if*

(i) *it is offspring-invariant (resp. unimodular), and*
(ii) *the number of children $d_1(\boldsymbol{o})$ of the root is independent of the non-descendants $D^c(\boldsymbol{o})$ of the root, namely, the subtree induced by $\{\boldsymbol{o}\} \cup (V(\boldsymbol{T}) \setminus D(\boldsymbol{o}))$ rooted at $\boldsymbol{o}$.*

REMARK 6.7. The strong assumption of offspring-invariance in Theorem 6.6 can be somewhat relaxed. The more general result is that a random EFT is an EGWT if and only if it is quasi-invariant and $d_1(\boldsymbol{o})$ is independent (jointly) of $D^c(\boldsymbol{o})$ and $\Delta[\boldsymbol{T}, \boldsymbol{o}, F(\boldsymbol{o})]$ (see Definition 5.4 and Corollary 5.5). The same proof works except for the terms of Equation (6.3) which should be adapted with the new assumptions.

The following proposition classifies EGWTs beyond the classification of Theorem 3.9.

PROPOSITION 6.8 (Foil Classification of EGWTs). *The EGWT almost surely satisfies the following properties:*

(i) *When $m = 1$, in the non-degenerate case (that is, when $\pi(1) \neq 1$), the EGWT is a unimodular EFT of class $\mathcal{I}/\mathcal{I}$ and there is no youngest generation. In the degenerate case (when $\pi(1) = 1$), it is of class $\mathcal{I}/\mathcal{F}$ and each generation (i.e. foil) has only one vertex.*

(ii) *When $m > 1$, all generations are infinite and there is no youngest generation. Moreover, $\mathbb{P}[\#D(\boldsymbol{o}) = \infty] > 0$ and there are uncountably many bi-infinite F-paths; i.e., the tree has uncountably many ends.*

(iii) *When $m < 1$, the descendant tree of each vertex is finite; i.e., the tree has only one end. Moreover, all generations are finite (resp. infinite) if and only if $\sum_{k\geq 1}(k \log k)\pi_k < \infty$ (resp. the sum is infinite).*

It follows from part (i) of Proposition 6.8 that in a non-degenerate critical EGWT, almost surely the number of vertices in each generation is infinite, which is a known result (see Subsection 6.3).

REMARK 6.9. The existence of a youngest foil can be interpreted as the extinction of the EFT and is an analogue of extinction in branching processes. Therefore, by Proposition 6.8, if $m \geq 1$, then the eternal branching process almost surely does not suffer of extinction, although the descendants of the root may suffer of extinction. For $m < 1$, extinction is equivalent to $\mathbb{E}[d_1(\boldsymbol{o}) \log d_1(\boldsymbol{o})] < \infty$.

EXAMPLE 6.10 (Horocyclic product of EGWTs). Let $[\boldsymbol{T}_i, \boldsymbol{o}_i]$ be an EGWT with offspring distribution $\pi^{(i)}$ (for $i = 1, 2$). Example 5.20 defines a vertex-shift f on the horocyclic product of $[\boldsymbol{T}_1, \boldsymbol{o}_1]$ and $[\boldsymbol{T}_2, \boldsymbol{o}_2]$. Let $[\boldsymbol{T}, \boldsymbol{o}]$ be the connected component of the graph of f containing the root. It has been shown that $[\boldsymbol{T}, \boldsymbol{o}]$ is quasi-invariant. Below, a random rooted EFT is explicitly constructed that has the same distribution as $[\boldsymbol{T}, \boldsymbol{o}]$. This construction shows that $[\boldsymbol{T}, \boldsymbol{o}]$ is a *height-dependent EGWT* as described below.

First, construct an independent sequence of random variables $(b_i)_{i=-\infty}^{\infty}$ such that b_i has distribution $\widehat{\pi}^{(2)}$ for $i \geq 0$ (corresponding to $d_1(F^{i+1}(\boldsymbol{o}_2))$ above) and has distribution $\pi^{(2)}$ for $i < 0$ (corresponding to $d_1(F(v_i))$, where v_i is the second coordinate of $f^{-i}(\boldsymbol{o})$ above). Then, start with a path, namely $\boldsymbol{o}'_0, \boldsymbol{o}'_1, \ldots$, and set $F(\boldsymbol{o}'_i) := \boldsymbol{o}'_{i+1}$. For each $i \geq 1$, independently add some potential children to $\boldsymbol{o}'_i$ such that their cardinality (by counting the child $\boldsymbol{o}'_{i-1}$) has distribution $\widehat{\pi}^{(1)}$. Keep each of the new vertices independently with probability $\frac{1}{b_{-i}}$ and delete it otherwise. Then, for $\boldsymbol{o}'_0$ and each newly added vertex, given it is on level j, add some potential children with distribution $\pi^{(1)}$ and keep each of them independently with probability $\frac{1}{b_j}$. Continuing this process for the new vertices iteratively gives a random rooted EFT (rooted at $\boldsymbol{o}'_0$). It can be seen that it has the same distribution as $[\boldsymbol{T}, \boldsymbol{o}]$.

PROPOSITION 6.11. *In Example 6.10 above, $[\boldsymbol{T}, \boldsymbol{o}]$ is an EGWT if and only if either $[\boldsymbol{T}_2, \boldsymbol{o}_2]$ is a regular tree with a distinguished end (i.e., $\pi^{(2)}$ is concentrated on a single number), or $\boldsymbol{T}_1$ is a path a.s. (i.e., $\pi^{(1)}$ is concentrated on $\{0, 1\}$).*

For proving this proposition, one can use either the characterization of Theorem 6.6 (see the ArXiv version) or the explicit construction in Example 6.10. The proof is skipped for brevity.

PROOF OF PROPOSITION 6.4. Let $m := \mathbb{E}[d_1(\boldsymbol{o})]$. It is enough to prove the claim in the ordered case. The non-ordered case is obtained by forgetting the order. To prove offspring-invariance of the EGWT, one should prove that for all A,

$$\mathbb{P}[A] = \frac{1}{m}\mathbb{E}\left[\sum_{v \in D_1(\boldsymbol{o})} 1_A[\boldsymbol{T}, v]\right]. \tag{6.1}$$

For this, it is enough to prove the last result when A is a subset of $B := \{[T, o] : d_1(F(o)) = k, o = c_j(F(o))\}$, where $c_j(v)$ denotes the j-th child of v using the assumed order on the children and $1 \leq j \leq k$ are arbitrary. One has

$$\mathbb{P}[A] = \mathbb{P}[B]\,\mathbb{P}[A\,|B] = \frac{k\pi_k}{m}\cdot\frac{1}{k}\mathbb{P}[A\,|B].$$

On the other hand, by the definition of A, the summation in (6.1) is non-zero only when $d_1(\boldsymbol{o}) = k$ and only the summand for $v = c_j(\boldsymbol{o})$ can be non-zero. Hence

$$\frac{1}{m}\mathbf{E}\sum_{v\in D_1(\boldsymbol{o})} 1_A[\boldsymbol{T}, v] = \frac{1}{m}\mathbb{P}[d_1(\boldsymbol{o}) = k, [T, c_j(\boldsymbol{o})] \in A] = \frac{\pi_k}{m}\mathbb{P}[[T, c_j(\boldsymbol{o})] \in A\,|d_1(\boldsymbol{o}) = k].$$

The explicit construction of EGWT implies that $[\boldsymbol{T}, \boldsymbol{o}]$ conditioned on B has the same distribution as $[\boldsymbol{T}, c_j(\boldsymbol{o})]$ conditioned on $d_1(\boldsymbol{o}) = k$. Therefore, the above equations imply that (6.1) holds and the claim is proved. □

PROOF OF PROPOSITION 6.5. The descendants $[D(\boldsymbol{o}), \boldsymbol{o}]$ of the root form an ordinary GW(π) tree. Thus, the claim follows from Propositions 4.15 and 6.4. □

PROOF OF THEOREM 6.6. If $[\boldsymbol{T}, \boldsymbol{o}]$ is an EGWT, the claim follows by Proposition 6.4 and the construction of EGWT. Conversely, suppose $[\boldsymbol{T}, \boldsymbol{o}]$ is offspring-invariant and it has the mentioned independence property. If $[\boldsymbol{T}, \boldsymbol{o}]$ is non-ordered, add a uniformly random order on the children of each vertex independently. It is not difficult to verify that the resulting ordered EFT is also offspring-invariant. Therefore, one may assume $[\boldsymbol{T}, \boldsymbol{o}]$ is an ordered EFT without loss of generality.

Let $m := \mathbb{E}[d_1(\boldsymbol{o})]$. For a vertex v, let $c_i(v)$ be the i-th child of v for $1 \leq i \leq d_1(v)$ using the assumed order on the children. Let $A', A_1, A_2, \ldots A_k$ be a sequence of events in $\mathcal{T}_*$ and let $A = A(A'; A_1, \ldots, A_k)$ be the event that $d_1(o) = k$, $D^c(o) \in A'$ and $D(c_i(o)) \in A_i$ for all i. The next step consists in showing that

$$\mathbb{P}[A] = \mathbb{P}[d_1(\boldsymbol{o}) = k]\left(\prod_{j=1}^{k}\mathbb{P}[D(\boldsymbol{o}) \in A_j]\right)\mathbb{P}[D^c(\boldsymbol{o}) \in A'].\tag{6.2}$$

First, assume (6.2) holds. By letting $A' := \mathcal{T}_*$, it follows that conditional on $d_1(\boldsymbol{o}) = k$, the subtrees $D(c_1(\boldsymbol{o})), \ldots, D(c_k(\boldsymbol{o}))$ are i.i.d. with the same distribution as $D(\boldsymbol{o})$. This means that $D(\boldsymbol{o})$ is an ordinary (ordered) Galton-Watson Tree. Therefore, Propositions 5.1 and 6.4 imply that $[\boldsymbol{T}, \boldsymbol{o}]$ is an EGWT, which is the desired property.

To prove (6.2), it is enough to assume that the events $D(o) \in A_i$ for $1 \leq i \leq k$ depend only on the generations $0, 1, 2, \ldots, n$ of $D(o)$, which is the subtree induced by $D_0(o) \cup \cdots \cup D_n(o)$. The proof is by induction on n. For $n = 0$, (6.2) follows by the independence of $d_1(\boldsymbol{o})$ and $D^c(\boldsymbol{o})$. For $n \geq 1$, assume the claim holds for $n-1$. Let j be arbitrary such that $1 \leq j \leq k$ and let $h_j[T, v] := 1_A[T, F(v)]1_{\{v=c_j(F(v))\}}$. One can write $1_A[T, o] = \sum_{v\in D_1(o)} h_j[T, v]$ (note that only the summand for $v := c_j(o)$ can be non-zero). Therefore,

$$\mathbb{P}[A] = \mathbb{E}\left[\sum_{v\in D_1(\boldsymbol{o})} h_j[T, v]\right].$$

Now, one can use this equation for $j = 1$ together with invariance under σ and (4.2) to obtain

$$\begin{aligned}
\mathbb{P}[A] &= m\mathbb{E}[h_1(\boldsymbol{o})] \\
&= m\mathbb{P}[\boldsymbol{o} = c_1(F(\boldsymbol{o})), [\boldsymbol{T}, F(\boldsymbol{o})] \in A] \\
&= m\mathbb{P}[D(\boldsymbol{o}) \in A_1, \boldsymbol{o} = c_1(F(\boldsymbol{o})), [\boldsymbol{T}, F(\boldsymbol{o})] \in A(A'; \mathcal{T}_*, A_2, \dots, A_k)]. \qquad (6.3)
\end{aligned}$$

The induction hypothesis in (6.2) implies that the event $D(o) \in A_1$ is independent of $D^c(o)$ (note that $D(o) \in A_1$ depends on one less generation of $D(o)$ than the event $[T, o] \in A$). Therefore, (6.3) and the definition of the event A imply

$$\begin{aligned}
\mathbb{P}[A] &= m\mathbb{P}[D(\boldsymbol{o}) \in A_1] \cdot \mathbb{P}[\boldsymbol{o} = c_1(F(\boldsymbol{o})), [\boldsymbol{T}, F(\boldsymbol{o})] \in A(A'; \mathcal{T}_*, A_2, \dots, A_k)] \\
&= \mathbb{P}[D(\boldsymbol{o}) \in A_1] \cdot \mathbb{P}[A(A'; \mathcal{T}_*, A_2, \dots, A_k)], \qquad (6.4)
\end{aligned}$$

where in the last equation, (6.3) is used again for the event $A(A'; \mathcal{T}_*, A_2, \dots, A_k)$. By applying the same argument as above for $j = 2$ and the event $A(A'; \mathcal{T}_*, A_2, \dots, A_k)$, one gets

$$\mathbb{P}[A] = \mathbb{P}[D(\boldsymbol{o}) \in A_1] \cdot \mathbb{P}[D(\boldsymbol{o}) \in A_2] \cdot \mathbb{P}[A(A'; \mathcal{T}_*, \mathcal{T}_*, A_3, \dots, A_k)].$$

Continuing inductively, one obtains

$$\begin{aligned}
\mathbb{P}[A] &= \left(\prod_{j=1}^{k} \mathbb{P}[D(\boldsymbol{o}) \in A_j]\right) \mathbb{P}[A(A'; \mathcal{T}_*, \dots, \mathcal{T}_*)] \\
&= \left(\prod_{j=1}^{k} \mathbb{P}[D(\boldsymbol{o}) \in A_j]\right) \mathbb{P}[d_1(\boldsymbol{o}) = k, D^c(\boldsymbol{o}) \in A'] \\
&= \left(\prod_{j=1}^{k} \mathbb{P}[D(\boldsymbol{o}) \in A_j]\right) \mathbb{P}[d_1(\boldsymbol{o}) = k] \, \mathbb{P}[D^c(\boldsymbol{o}) \in A'],
\end{aligned}$$

where in the last equation, the independence assumption is used. Therefore, (6.2) is proved and the proof is complete. □

PROOF OF PROPOSITION 6.8. Let $[\boldsymbol{T}, \boldsymbol{o}]$ be the EGWT. The event $\#D(\boldsymbol{o}) = \infty$ is the event of non-extinction in a Galton-Watson process. Therefore, it happens with positive probability if and only if $m > 1$ or $\pi(1) = 1$.

(i) As mentioned above, $D(\boldsymbol{o})$ is finite a.s. Therefore, Proposition 5.8 implies that $[\boldsymbol{T}, \boldsymbol{o}]$ is of class $\mathcal{I}/\mathcal{I}$. For the second claim, consider $K(\boldsymbol{T}, \boldsymbol{o}, 0)$ which is obtained by pruning $[\boldsymbol{T}, \boldsymbol{o}]$ from generation 0. Since $\boldsymbol{T}$ has class $\mathcal{I}/\mathcal{I}$, the generation $L(\boldsymbol{o})$ is infinite a.s. Conditioned on $K(\boldsymbol{T}, \boldsymbol{o}, 0)$, the descendants $D(v)$ for $v \in L(\boldsymbol{o})$ are i.i.d. GWTs. Now the claim follows by the Borel-Cantelli lemma.

(ii) Propositions 5.8 and 6.4 show that the generations are infinite a.s. Like the previous case, conditioned on $K(\boldsymbol{T}, \boldsymbol{o}, 0)$, the descendants $D(v)$ for $v \in L(\boldsymbol{o})$ are i.i.d. supercritical GWTs, each of them is infinite with positive probability. Now, by the Borel-Cantelli lemma, one obtains that there is no youngest generation and there is more than one end. Therefore, Proposition 5.8 implies that there are infinitely many ends.

It is easy to see that the probability that $D(\boldsymbol{o})$, which is an ordinary GWT, has exactly one end, is zero. Therefore, by Lemma 5.6 and Proposition 6.4, almost surely there is no vertex $v \in V(\boldsymbol{T})$ such that $D(v)$ has exactly one end.

This implies that $\boldsymbol{T}$ has no isolated end a.s. Since the space of ends of a tree is complete ([**13**]), this implies that $\boldsymbol{T}$ has uncountably many ends a.s.

(iii) By Propositions 5.8 and 6.4, $\boldsymbol{T}$ has only one end, and each generation is finite (resp. infinite) if and only if (5.4) holds (resp. doesn't hold). It is proved in [**16**] that for subcritical GWTs, (5.4) is equivalent to $(\sum k \log k)\pi_k < \infty$. □

6.2. Eternal Multi-Type Galton-Watson Trees. In this subsection, the EGWTs constructed in Subsection 6.1 are extended to multi-type branching processes. These are EFTs where each vertex is equipped with a mark called its **type**.

First, the definition and notation of (ordinary) multi-type Galton-Watson Trees are recalled using the notation in [**22**]. In the construction, an initial vertex is considered and some new vertices are added as its children such that the distribution of their cardinality and types depends on the type of the initial vertex. Then, the same process is repeated for each newly added vertex independently (depending only on the type of the vertex).

Let J be a finite or countable set describing the possible types of vertices and $t(v)$ denote the type of vertex v. For $j \in J$, let $d_1^{(j)}(v)$ denote the number of children of v that have type j. For each $i \in J$, let $\pi^{(i)}$ be a probability distribution on $(\mathbb{Z}^{\geq 0})^J$, which represents the joint distribution of $(d_1^{(j)}(v))_{j\in J}$, for each vertex v of type i. So, the notation $\pi^{(i)}(k)$ for $k = (k_j)_{j\in J}$ expresses the probability that a given vertex of type i has k_j children of type j for each $j \in J$.

Assume the means $m_{i,j} := \mathbb{E}\left[d_1^{(j)}(v)\,|t(v) = i\right]$ and $m_i := \mathbb{E}\left[d_1(v)\,|t(v) = i\right] = \sum_{j\in J} m_{i,j}$ are finite for each $i, j \in J$.

DEFINITION 6.12 (EMGWT). The **Eternal Multi-Type Galton-Watson Tree** (EMGWT) is a random rooted EFT equipped with marks defined as follows. For $i, j \in J$ such that $m_{i,j} > 0$, let $\hat{\pi}^{(i,j)}$ be the version of $\pi^{(i)}$ biased by the number of children of type j, which is defined by $\hat{\pi}^{(i,j)}(k) := \frac{k_j}{m_{i,j}}\pi^{(i)}(k)$ for $k \in (\mathbb{Z}^{\geq 0})^J$. Assume the matrix $M := (m_{i,j})$ has a non-negative left-eigenvector b with eigenvalue $\rho > 0$ and assume $\sum_i b_i = 1$, i.e.

$$\sum_{j\in J} b_j m_{j,i} = \rho b_i, \quad ,\forall i \in J.$$

Start with a path $(\boldsymbol{o}_n)_{n=0}^{\infty}$. Let $\boldsymbol{o} := \boldsymbol{o}_0$ and $F(\boldsymbol{o}_n) := \boldsymbol{o}_{n+1}$ for $n \geq 0$. Choose the type of $\boldsymbol{o}$ such that $\mathbb{P}[t(\boldsymbol{o}) = i] = b_i$ for each $i \in J$. Inductively for $n \geq 0$, choose the type of $\boldsymbol{o}_{n+1}$ such that $\mathbb{P}[t(\boldsymbol{o}_{n+1}) = j \mid t(\boldsymbol{o}_n) = i] = \frac{1}{\rho b_i} b_j m_{j,i}$ for each $i, j \in J$. Then, given $t(\boldsymbol{o}_{n+1}) = j$ and $t(\boldsymbol{o}_n) = i$, add some children to $\boldsymbol{o}_{n+1}$ by choosing their cardinality and types (including the child $\boldsymbol{o}_n$) with the biased distribution $\hat{\pi}^{(j,i)}$. For $\boldsymbol{o}_0$ and all newly added vertices, sample their descendants as i.i.d. ordinary multi-type GWTs with offspring distribution $\pi^{(\cdot)}$. The choices in each step are independent of the previous steps.

REMARK 6.13. In Definition 6.12, it is neither assumed that J is finite, nor that M is positive regular (i.e. that M^n is positive for some $n \in \mathbb{N}$), nor that ρ is the largest eigenvalue.

Here are some properties of EMGWTs. The proofs are similar to, but lengthier than the ones for EGWTs in Subsection 6.1 and are skipped for brevity.

PROPOSITION 6.14. *The EMGWT constructed in Definition 6.12 is offspring-invariant with* $\mathbb{E}[d_1(\boldsymbol{o})] = \rho$. *Therefore, it is unimodular if and only if* $\rho = 1$.

PROPOSITION 6.15. *Let $\mathcal{P}$ be the distribution of the ordinary multi-type* GWT *with the parameters described above. If the distribution of the type of the initial vertex is a left-eigenvector of M as above, then $\mathcal{P}_\infty$ exists and is the distribution of the* EMGWT *constructed in Definition* 6.12.

PROPOSITION 6.16. *A random Family Tree $[\boldsymbol{T}, \boldsymbol{o}]$ equipped with types in a countable set is an* EMGWT *if and only if (i) it is offspring-invariant, and (ii) conditional to the type of the root, the cardinality and types of the children of the root is independent of the non-descendants of the root.*

The following example shows that some of the EFTs constructed in the previous examples are instances of EMGWTs (of course, by forgetting the types). Therefore, Proposition 6.14 implies that all of them are offspring-invariant.

EXAMPLE 6.17. Consider Definition 6.12 with the following sets of parameters:
(i) Let $J := \mathbb{Z}^{\geq 0}$ and $d \geq 2$. Assume each vertex of type j has exactly d children of type $j-1$ when $j > 0$ and no children when $j = 0$. Let b_i be proportional to $\tilde{d}^{-i}$ for $i \in J$ and $\rho := d/\tilde{d}$. It can be seen that these parameters satisfy the assumptions of Definition 6.12 and the resulting EMGWT is just the biased Canopy Tree of Example 5.13. In particular, when $\tilde{d} = d$, one gets the Canopy Tree of Example 2.3.
(ii) Let $J := \{1, 2\}$ and $d \geq 1$. Let $\pi^{(1)}$ and $\pi^{(2)}$ be concentrated on $(d, 1)$ and $(0, 1)$ respectively. Let $\rho := d$ and $b := (1 - \frac{1}{d}, \frac{1}{d})$. One gets an EMGWT. When $d \geq 2$, this is just the EFT of Example 5.14. For $d = 1$, it is a single bi-infinite path where all vertices have type 2.
(iii) Let $J := \mathbb{N}$ and $\pi^{(j)}$ be concentrated on $(1, 1, \ldots, 1, 0, 0, \ldots)$, where $j \in J$ and there are precisely j ones in the vector. For an arbitrary $0 < p < 1$, the vector b in which $b_i = p(1-p)^{i-1}$ is a left eigenvector of the corresponding matrix M with eigenvalue $\rho := \frac{1}{p}$. The resulting EMGWT is just the comb of Example 5.15.

EXAMPLE 6.18. It can be seen that by pruning an (single-type) EGWT as in Definition 5.10, one obtains (the underlying EFT of) an EMGWT. Here, the type of each vertex can be set as its distance from the last generation.

6.3. Bibliographical Comments. This subsection gathers a comprehensive list of connections between EGWTs and random trees of the literature. In spite of these connections, the main mathematical objects and several results of this section appear to be new, like for instance Theorem 6.6 and the results stressed as such below.

First note that the unimodular case of EGWTs (i.e. the critical case $m = 1$) should not be confused with the Unimodular Galton-Watson Tree (UGW) [**10**] which is an important example of unimodular graphs. The latter is an undirected tree with a different construction. The underlying undirected tree of a EGWT is related to the UGW tree (with the same set of parameters $(\pi_k)_k$) as follows: The distribution of the former is absolutely continuous w.r.t. the latter if and only if $m > 1$ and $\sum k \log k \pi_k < \infty$. The converse holds if and only if $\pi_0 = 0$ in addition to the previous conditions. The same holds for the Augmented Galton-Watson tree (AGW) [**24**] since its distribution is mutually absolutely continuous w.r.t. that of UGW.

The critical case of EGWT (where $m = 1$) was introduced in [**2**] as an example of *invariant Sin-Trees* (see Subsection 4.4). The **skeleton tree** considered in [**4, 11**] is the special case of the latter with offspring distribution Poisson of parameter 1.

The critical case of EGWTs is also described in a different context in [**20**] as discussed below and part (i) of Proposition 6.8 is proved therein. However, the characterization theorem (Theorem 6.6) is believed to be new even when $m = 1$. The cases $m > 1$ and $m < 1$ are also new, to the best of the authors' knowledge.

The EGWT is related to the **size-biased Galton-Watson tree** [**25**]. Both are obtained from the ordinary GWT by biasing the probability distribution by the population of the n-th generation and letting n tend to infinity, but the latter keeps the root at the initial vertex and the former moves the root to an n-descendant (as in Definition 4.6). Another similarity is their undirected trees, where the only difference is the degree distribution of the root: the probability that the root has k neighbors is π_{k-1} in the former and $\frac{k\pi_k}{m}$ in the latter. Therefore, the distributions of the two undirected trees are mutually absolutely continuous (resp. identical) if and only if $\pi_k > 0$ for all $k \geq 0$ (resp. π is a Poisson distribution). However, this does not hold for the multi-type versions of EGWTs and Size-Biased Galton-Watson Trees [**22**] and the laws of the undirected trees are generally non-equivalent.

EGWTs can also be connected to the stationary regime of **branching processes with immigration** as discussed in [**5**], when deleting the ancestors $\{F^i(\boldsymbol{o}) : i \geq 0\}$ of the root and regarding the (other) children of $F^i(\boldsymbol{o})$ as the immigrants at time $-(i-1)$ for each $i \geq 0$. Then, the part up to level 0 of the EFT is converted to a branching processes with immigration. With this adaptation, the properties of the stationary regime of this type of processes (Lemma 6.6 in [**5**]) are exactly the foil classification of EGWTs. Hence, Proposition 6.8 gives a new (non analytical) proof of these results, and the classification of offspring-invariant EFTs (Proposition 5.8) can be seen as an extension of these classical results beyond branching processes.

In [**20**], successive iterates of a *clustering mechanism* are considered starting from a given stationary point process (or a single point). Such a mechanism is defined by replacing each point in the point process with a random cluster of points independently and simultaneously for all points. The question studied is the *stability* of this, i.e. the convergence of the Palm distribution of the n-th step (as n tends to infinity), which is similar to the operator σ_n in (4.2) above. However, to disregard trivial cases, the expected number of points in the clusters is assumed to be 1. To study the limit, the genealogy of the points in all steps is considered. This is called the *method of backward trees* in [**20**]. Note that the descendants of every given point form a GWT. Some criteria for the convergence are studied in [**20**]. Moreover, by considering the limit, the *backward tree* is considered, which in our language is the EFT obtained by applying σ_∞ to the GWT. This tree is identical with the case $m = 1$ of the EGWT. See also Section 13.5 of [**12**] for a concise introduction to this and other related works.

Conclusion

The main connections between the objects discussed in the present paper are summarized in the figure.

Acknowledgements

This work was supported by a grant of the Simons Foundation (#197982 to the University of Texas at Austin). The second author thanks the *Research and Technology Vice-presidency* of Sharif University of Technology for its support. The third author thanks the University of Texas at Austin where he was a visiting

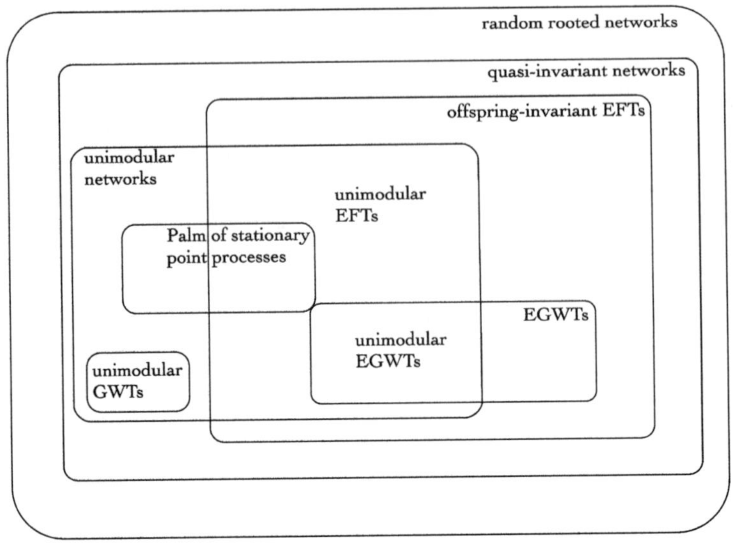

scholar at when this work was initiated. The authors thank Charles Bordenave and Lewis Bowen for stimulating discussions on the topics presented in this paper. They also thank the anonymous reviewer and the editor for their comments and suggestions.

References

[1] M. Aizenman and S. Warzel, *The canopy graph and level statistics for random operators on trees*, Math. Phys. Anal. Geom. **9** (2006), no. 4, 291–333 (2007), DOI 10.1007/s11040-007-9018-3. MR2329431

[2] D. Aldous, *Asymptotic fringe distributions for general families of random trees*, Ann. Appl. Probab. **1** (1991), no. 2, 228–266. MR1102319

[3] D. Aldous, *The continuum random tree. III*, Ann. Probab. **21** (1993), no. 1, 248–289. MR1207226

[4] D. Aldous and J. M. Steele, *The objective method: probabilistic combinatorial optimization and local weak convergence*, Probability on discrete structures, Encyclopaedia Math. Sci., vol. 110, Springer, Berlin, 2004, pp. 1–72, DOI 10.1007/978-3-662-09444-0_1. MR2023650

[5] S. Asmussen and H. Hering, *Branching processes*, Progress in Probability and Statistics, vol. 3, Birkhäuser Boston, Inc., Boston, MA, 1983. MR701538

[6] D. Aldous and R. Lyons, *Processes on unimodular random networks*, Electron. J. Probab. **12** (2007), no. 54, 1454–1508, DOI 10.1214/EJP.v12-463. MR2354165

[7] F. Baccelli and M.-O. Haji-Mirsadeghi (2016). *Point-Shift Foliation of a Point Process*, arxiv.org/abs/1601.03653

[8] I. Benjamini and N. Curien, *Ergodic theory on stationary random graphs*, Electron. J. Probab. **17** (2012), no. 93, 20, DOI 10.1214/EJP.v17-2401. MR2994841

[9] I. Benjamini, R. Lyons, and O. Schramm, *Unimodular random trees*, Ergodic Theory Dynam. Systems **35** (2015), no. 2, 359–373, DOI 10.1017/etds.2013.56. MR3316916

[10] I. Benjamini and S. Müller, *On the trace of branching random walks*, Groups Geom. Dyn. **6** (2012), no. 2, 231–247, DOI 10.4171/GGD/156. MR2914859

[11] C. Bordenave (2016) *Lecture notes on random graphs and probabilistic combinatorial optimization*, /www.math.univ-toulouse.fr/ bordenave/coursRG.pdf.

[12] D. J. Daley and D. Vere-Jones, *An introduction to the theory of point processes. Vol. II*, 2nd ed., Probability and its Applications (New York), Springer, New York, 2008. General theory and structure. MR2371524

[13] R. Diestel, *Graph theory*, 4th ed., Graduate Texts in Mathematics, vol. 173, Springer, Heidelberg, 2010. MR2744811

[14] R. Diestel and I. Leader, *A conjecture concerning a limit of non-Cayley graphs*, J. Algebraic Combin. **14** (2001), no. 1, 17–25, DOI 10.1023/A:1011257718029. MR1856226

[15] J. Feldman and C. C. Moore, *Ergodic equivalence relations, cohomology, and von Neumann algebras. I*, Trans. Amer. Math. Soc. **234** (1977), no. 2, 289–324, DOI 10.2307/1997924. MR0578656

[16] C. R. Heathcote, E. Seneta, and D. Vere-Jones, *A refinement of two theorems in the theory of branching processes* (English, with Russian summary), Teor. Verojatnost. i Primenen. **12** (1967), 341–346. MR0217889

[17] M. Heveling and G. Last, *Characterization of Palm measures via bijective point-shifts*, Ann. Probab. **33** (2005), no. 5, 1698–1715, DOI 10.1214/009117905000000224. MR2165576

[18] M. Heveling and G. Last, *Point shift characterization of Palm measures on abelian groups*, Electron. J. Probab. **12** (2007), no. 5, 122–137, DOI 10.1214/EJP.v12-394. MR2280261

[19] T. Hutchcroft (2018). *Universality of high-dimensional spanning forests and sandpiles.* https://arxiv.org/abs/1804.04120.

[20] O. Kallenberg, *Stability of critical cluster fields*, Math. Nachr. **77** (1977), 7–43, DOI 10.1002/mana.19770770102. MR0443078

[21] V. A. Kaimanovich and F. Sobieczky, *Stochastic homogenization of horospheric tree products*, Probabilistic approach to geometry, Adv. Stud. Pure Math., vol. 57, Math. Soc. Japan, Tokyo, 2010, pp. 199–229. MR2648261

[22] T. Kurtz, R. Lyons, R. Pemantle, and Y. Peres, *A conceptual proof of the Kesten-Stigum theorem for multi-type branching processes*, Classical and modern branching processes (Minneapolis, MN, 1994), IMA Vol. Math. Appl., vol. 84, Springer, New York, 1997, pp. 181–185, DOI 10.1007/978-1-4612-1862-3_14. MR1601737

[23] G. Last and H. Thorisson, *Invariant transports of stationary random measures and mass-stationarity*, Ann. Probab. **37** (2009), no. 2, 790–813, DOI 10.1214/08-AOP420. MR2510024

[24] R. Lyons, R. Pemantle, and Y. Peres, *Ergodic theory on Galton-Watson trees: speed of random walk and dimension of harmonic measure*, Ergodic Theory Dynam. Systems **15** (1995), no. 3, 593–619, DOI 10.1017/S0143385700008543. MR1336708

[25] R. Lyons, R. Pemantle, and Y. Peres, *Conceptual proofs of $L \log L$ criteria for mean behavior of branching processes*, Ann. Probab. **23** (1995), no. 3, 1125–1138. MR1349164

[26] J. Neveu, *Processus ponctuels* (French), École d'Été de Probabilités de Saint-Flour, VI—1976, Springer-Verlag, Berlin, 1977, pp. 249–445. Lecture Notes in Math., Vol. 598. MR0474493

[27] B. G. Nguyen, *Percolation of coalescing random walks*, J. Appl. Probab. **27** (1990), no. 2, 269–277. MR1052299

[28] J. Pach and G. Tardos, *The range of a random walk on a comb*, Electron. J. Combin. **20** (2013), no. 3, Paper 59, 7. MR3118967

[29] I. Rodriguez-Iturbe and A. Rinaldo (2001). *Fractal river basins: chance and self-organization.* Cambridge University Press.

[30] W. Woess, *Random walks on infinite graphs and groups*, Cambridge Tracts in Mathematics, vol. 138, Cambridge University Press, Cambridge, 2000. MR1743100

Department of Mathematics, The University of Texas at Austin, Austin, Texas 78712

Email address: baccelli@math.utexas.edu

Department of Mathematical Sciences, Sharif University of Technology, 11155-9415 Tehran, Iran

Email address: mirsadeghi@sharif.ir

Department of Mathematical Sciences, Tarbiat Modares University, 14115-134 Tehran, Iran

Email address: khezeli@modares.ac.ir

Contemporary Mathematics
Volume **719**, 2018
https://doi.org/10.1090/conm/719/14472

Circular slider graphs: de Bruijn, Kautz, Rauzy, lamplighters and spiders

Vadim A. Kaimanovich

ABSTRACT. We suggest a new point of view on de Bruijn graphs and their subgraphs based on using circular words rather than linear ones.

Introduction

A. De Bruijn graphs. De Bruijn graphs represent overlaps between consecutive subwords of the same length in a longer word. Under various names and in various guises they and their subgraphs currently enjoy a lot of popularity in mathematics (dynamical systems and combinatorics) as well as in the applications to computer science (data networks) and bioinformatics (DNA sequencing).

A very succinct description of these graphs can be found in the following two-line rhyme from the title of the 1975 de Bruijn's historical note [**dB75**]. The

Circular arrangements of 2^n *zeros and ones*
That show each n-letter word exactly once

he is talking about are precisely the Hamiltonian cycles (currently known as de Bruijn sequences) [1] in the de Bruijn graph

$$\overrightarrow{\mathcal{B}}^n_{\mathcal{A}} = \overrightarrow{\mathcal{B}}^n_{|\mathcal{A}|}$$

of span n. This is the directed graph (digraph) whose vertices are all n-letter words $\boldsymbol{a} = \alpha_1\alpha_2\dots\alpha_n$ in a given finite alphabet $\mathcal{A}$ (quite often $\mathcal{A}$ is just the binary alphabet $\{0,1\}$, like in de Bruijn's formulation above), and whose arrows (directed edges)

$$\boldsymbol{a} = \alpha_1\alpha_2\dots\alpha_n \rightsquigarrow \boldsymbol{a}' = \alpha_2\dots\alpha_n\alpha_{n+1} \tag{0.1}$$

are the pairs of n-words with a length $n-1$ overlap, so that the associated transitions (we call them de Bruijn transitions) consist in removing the initial letter α_1 of an n-word and adding instead a new letter α_{n+1} at the end of the word.

2010 *Mathematics Subject Classification.* Primary 05C20, 20E22; Secondary 05C80, 37B10, 68R15.

Key words and phrases. De Bruijn graphs, lamplighter groups.

[1] Throughout the paper we use sans serif when giving a definition or introducing a notation, whereas *italic* is used for emphasizing and in quotes (as usual) or when mentioning a certain term for the first time without defining it.

B. Circular vs linear. The purpose of this note is to suggest a new point of view on de Bruijn graphs and their subgraphs based on using *circular words* rather than *linear* ones. Although the idea of circularity (of de Bruijn sequences) has been present in the subject area ever since the very first known formulation of this setup in 1894 (see Section 1 for more historical details), and the notions of *necklaces* and *Lyndon words* play a pivotal role in various algorithms for generating de Bruijn sequences (e.g., see Fredricksen – Maiorana [**FM78**], Perrin – Restivo [**PR15**], or the latest Sawada – Williams – Wong [**SWW16, SWW17**], and the references therein), it has never been applied to the vertex n-letter words themselves. They have always been treated as linear words with the "giving" and "receiving" ends (the one that loses a letter and the one that acquires a new letter in the process of a de Bruijn transition, respectively) being n symbols apart. [2]

Our approach is based on the totally obvious observation that there is a one-to-one correspondence between linear words and pointed circular words of the same length (the pointer separates the initial and the final letter of the linear word written clockwise). In a more formal language we replace the $\mathbb{Z}$-valued indices $i = 1, 2, \dots, n$ which parameterize the letters α_i of a word $\boldsymbol{a} = \alpha_1\alpha_2\dots\alpha_n$ with the $\mathbb{Z}_n$-valued indices $\iota = i \pmod n$, so that the pointer is positioned between the letters with the indices $\iota = n \equiv 0 \pmod n$ and $\iota = 1 \pmod n$. In this interpretation de Bruijn transitions (0.1) consist in moving the pointer one position clockwise and (possibly) changing the letter located between the old and the new positions of the pointer. Equivalently, moving the pointer along a fixed word is the same as moving the whole word in the opposite direction with respect to a fixed pointer, so that in terms of the circular shift (anticlockwise rotation)

$$(\mathbf{S}\boldsymbol{a})_\iota = (\boldsymbol{a})_{\iota+1 \pmod n}\,, \qquad \boldsymbol{a} \in \mathcal{A}^n \cong \mathcal{A}^{\mathbb{Z}_n}\,, \tag{0.2}$$

de Bruijn transitions are

$$\boldsymbol{a} \rightsquigarrow \boldsymbol{a}' \iff (\boldsymbol{a}')_\iota = (\mathbf{S}\boldsymbol{a})_\iota \quad \forall \iota \neq 0 \iff \alpha'_\iota = \alpha_{\iota+1 \pmod n} \quad \forall \iota \neq 0\,. \tag{0.3}$$

C. Slider graphs. Actually, we find it more convenient to think, instead of a pointer, about a sliding window (slider) of width 2 which covers one letter on either side of the pointer (we call a circular word endowed with a window like this slider pointed). From this point of view the usual de Bruin graph $\overrightarrow{\mathcal{B}}^n_{\mathcal{A}} = \overrightarrow{\mathcal{B}}^n_{|\mathcal{A}|}$ can be identified with the full circular slider graph of span n over an alphabet $\mathcal{A}$. This is the digraph whose vertices are all slider pointed circular words of length n in alphabet $\mathcal{A}$, and whose arrows are the slider transitions which consist in moving the slider one position clockwise and (possibly) replacing the letter at the intersection of the old and the new slider windows, see Figure 0.4.

A general (directed) circular slider graph is then defined as a subgraph of the full circular slider graph. In particular, this class includes the induced circular slider graphs $\overrightarrow{\mathscr{S}}[\mathcal{V}]$ obtained by restricting the full circular slider graph $\overrightarrow{\mathcal{B}}^n_{\mathcal{A}}$ to a subset of vertices $\mathcal{V} \subset \mathcal{A}^n$. One can also further modify circular slider graphs by introducing various *decorations* or *labellings* of their vertices and arrows.

---

[2] The only exception we are aware of is a recent article by Böhmová – Dalfó – Huemer [**BDH15**] and the ensuing papers by Dalfò [**Dal17a, Dal17b**] where *cyclic Kautz graphs* were introduced and studied.

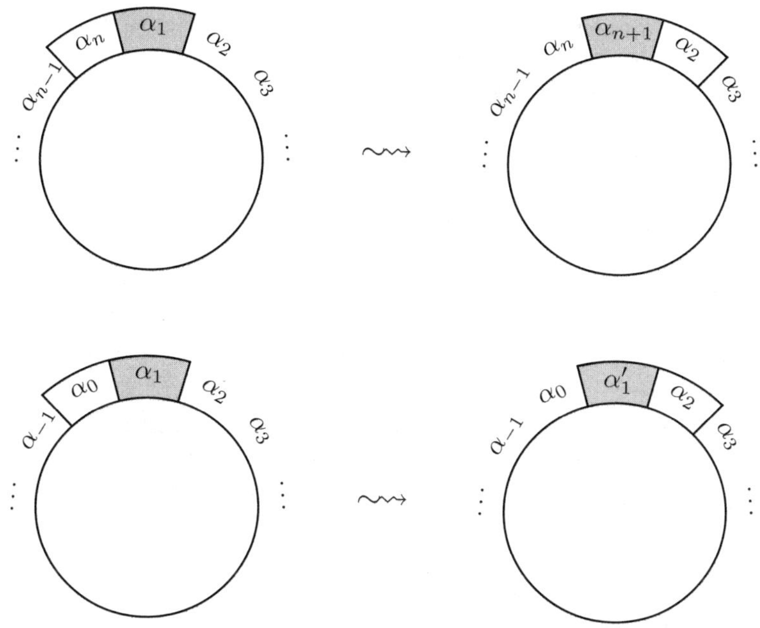

FIGURE 0.4. Slider transitions between circular n-words in the "conventional" parameterization with letter indices $i \in \mathbb{Z}$ and in the cyclic parameterization with letter indices $\iota \in \mathbb{Z}_n$.

D. Motivation. The motivation for our work comes from two sources. The first one is the isomorphism of the Cayley graphs $\mathfrak{Cay}(\mathbb{Z} \wr \mathbb{Z}_m, Q)$ of the *lamplighter groups*

$$\mathbb{Z} \wr \mathbb{Z}_m = \mathbb{Z} \leftthreetimes \operatorname{fun}(\mathbb{Z}, \mathbb{Z}_m)$$

(here $\operatorname{fun}(\mathbb{Z}, \mathbb{Z}_m)$ denotes the additive group of finitely supported $\mathbb{Z}_m$-valued configurations on $\mathbb{Z}$) with the corresponding Diestel – Leader graphs ($\equiv$ *horospheric products* $\mathbb{T}_{m+1} \mathbin{\text{⇅}} \mathbb{T}_{m+1}$ of two copies of pointed at infinity *homogeneous trees* $\mathbb{T}_{m+1}$ of degree $m+1$) discovered by Möller, Neumann and Woess (see the historical accounts by Woess [**Woe13**] and Amchislavska – Riley [**AR15**]). The generating set Q of the group $\mathbb{Z} \wr \mathbb{Z}_m$, for which this isomorphism is established, consists of the set Q_+ of the walk-right—switch generators

$$\left(1, \delta_1^b\right) = (1, \varnothing) \cdot \left(0, \delta_0^b\right)$$

and of the set Q_- of their inverses (the switch—walk-left generators)

$$\left(1, \delta_1^b\right)^{-1} = \left(0, \delta_0^{-b}\right) \cdot (-1, \varnothing) = \left(-1, \delta_0^{-b}\right)$$

(where b runs through $\mathbb{Z}_m$, the zero configuration on $\mathbb{Z}$ is denoted by $\varnothing$, and δ_i^b denotes the configuration on $\mathbb{Z}$ with $\delta_i^b(i) = b$ and $\delta_i^b(j) = 0$ for $j \neq i$). In order to make the symmetry in the definition of these generators more explicit, we shall "mark" the integer line $\mathbb{Z}$ with, instead of lamplighter's position $z \in \mathbb{Z}$, the width 2 slider ("lamplighter's window") over the positions z and $z+1$. Then the multiplication by an element $\left(1, \delta_1^b\right) \in Q_+$ (resp., by its inverse $\left(-1, \delta_0^{-b}\right) \in Q_-$) amounts to moving the slider one position to the right (resp., to the left) and adding b (resp., $-b$) to the state at the intersection of the old and the new sliders. These are precisely the slider transitions (and their inverses) like on Figure 0.4 above between configurations ($\equiv$ strings) from $\mathrm{fun}(\mathbb{Z}, \mathbb{Z}_m)$.

The second source is the link between de Bruijn graphs and lamplighter groups recently discovered and studied by Grigorchuk – Leemann – Nagnibeda [**GLN16**] and Leemann [**Lee16**] who, for a fixed alphabet size m, identified the *Benjamini – Schramm limits* of de Bruijn graphs $\overrightarrow{\mathcal{B}}_m^n$ as $n \to \infty$ with the corresponding directed versions of the Diestel – Leader graphs $\mathbb{T}_{m+1} \upharpoonleft\!\downharpoonright \mathbb{T}_{m+1}$. Their approach is algebraic, and it is based on a presentation of the de Bruijn graph $\overrightarrow{\mathcal{B}}_m^n$ as the *Schreier graph* of a natural spherically transitive action of the associated lamplighter group $\mathbb{Z} \wr \mathbb{Z}_m$ on the n-th level of the m-regular rooted tree. Moreover, by using the classification of the subgroups of $\mathbb{Z} \wr \mathbb{Z}_m$ obtained by Grigorchuk – Kravchenko [**GK14**], they also identify the tensor ($\equiv$ direct) products of de Bruijn graphs $\overrightarrow{\mathcal{B}}_m^n$ and directed cycle graphs $\overrightarrow{\mathcal{C}}_k$ (these products are known as spider-web graphs) with the Cayley graphs of appropriate finite groups for a number of combinations of the parameters k, m, n. In particular, for $k = n$ the spider-web graph $\overrightarrow{\mathcal{C}}_n \otimes \overrightarrow{\mathcal{B}}_m^n$ turns out to be isomorphic to the Cayley digraph $\overrightarrow{\mathfrak{Cay}}(\mathbb{Z}_n \wr \mathbb{Z}_m, Q_+)$ of the finite lamplighter group $\mathbb{Z}_n \wr \mathbb{Z}_m$ with respect to the set Q_+ of the walk-right—switch generators.

The immediate origin of the present paper is the project in progress [**KLN**] (joint with Paul-Henry Leemann and Tatiana Nagnibeda) aimed at understanding to what extent the results on Benjamini – Schramm limits from [**GLN16, Lee16**] could be carried over to generalizations of de Bruijn graphs, and in a sense the present paper can be considered as prolegomena to [**KLN**]. The work on this project started during my gratefully acknowledged visit to the University of Geneva in February-March 2017.

E. New perspective. Of course, formally the circular slider graphs are just subgraphs of the corresponding full de Bruijn graphs, and a number of examples of this kind were considered before (we review them in Section 3). However, our point of view brings in a new perspective.

• We make the aforementioned relationship between de Bruijn graphs and finite lamplighter groups completely transparent. Indeed, as we have already explained, the arrows in $\overrightarrow{\mathfrak{Cay}}(\mathbb{Z} \wr \mathbb{Z}_m, Q_+)$ are precisely our slider transitions between infinite words. The same observation holds for the lamplighter groups $\mathbb{Z}_n \wr \mathbb{Z}_m$ over a finite cyclic group $\mathbb{Z}_n$ as well (with, *mutatis mutandis*, Q and $Q_\pm$ now denoting the corresponding subsets of $\mathbb{Z}_n \wr \mathbb{Z}_m$), so that the only difference between the de Bruijn ($\equiv$ slider) graph $\overrightarrow{\mathcal{B}}_m^n$ and the Cayley digraph $\overrightarrow{\mathfrak{Cay}}(\mathbb{Z}_n \wr \mathbb{Z}_m, Q_+)$ is that in the case of slider graphs the length n circular words in the alphabet $\mathbb{Z}_m$ are marked only once (with the position of the slider), whereas in the case of lamplighter groups the circular words are endowed with two pointers (both the position of the identity of the group $\mathbb{Z}_n$ and the position of the lamplighter). Therefore, in order to describe the slider graphs in terms of the lamplighter group one has to eliminate the additional pointer, or, in other words, to pass from the Cayley digraph $\overrightarrow{\mathfrak{Cay}}(\mathbb{Z}_n \wr \mathbb{Z}_m, Q_+)$ to the corresponding *Schreier digraph* $\overrightarrow{\mathfrak{Sch}}\left(\mathbb{Z}_n \backslash (\mathbb{Z}_n \wr \mathbb{Z}_m), Q_+\right)$ on the quotient of the wreath product $\mathbb{Z}_n \wr \mathbb{Z}_m$ by the cyclic subgroup of translations $\{(z, \varnothing) : z \in \mathbb{Z}_n\} \cong \mathbb{Z}_n$ acting on the left. This Schreier digraph is nothing else than the de Bruijn graph $\overrightarrow{\mathcal{B}}_m^n$, and, conversely, the tensor product $\overrightarrow{\mathcal{C}}_n \otimes \overrightarrow{\mathcal{B}}_m^n$ is precisely the Cayley digraph $\overrightarrow{\mathfrak{Cay}}(\mathbb{Z}_n \wr \mathbb{Z}_m, Q_+)$.

• Our approach gives rise to new interesting classes of circular slider graphs (i.e., of subgraphs of full de Bruijn graphs) inherently linked with the presence of a circular structure. In Section 4 we introduce the periodic slider graphs, which are the induced subgraphs on subsets $\mathcal{V} \subset \mathcal{A}^n$ invariant with respect to the circular shift $\mathbf{S}$ (0.2). In Section 5 we introduce the transversally Markov circular slider graphs, for which one imposes an additional admissibility condition on the replacements $\alpha_1 \rightsquigarrow \alpha_{n+1}$ in the slider transitions from Figure 0.4, i.e., $\alpha_1 \rightsquigarrow \alpha_{n+1} = \alpha_1'$ has to be an arrow of a certain digraph ($\equiv$ topological Markov chain) on the alphabet $\mathcal{A}$. In the particular case when $\mathcal{A}$ is endowed with the structure of a Cayley or a Schreier digraph this definition gives rise to what we call Cayley and Schreier circular slider graphs, respectively.

• In spite of the enormous popularity of de Bruijn graphs and their various modifications, there have been very few attempts to extend this notion to infinite graphs. Laarhoven – de Weger [**LdW13**] in the course of a discussion of a link between de Bruijn graphs and the famous $3n+1$ Collatz conjecture (see Example 3.11) introduced the infinite *p-adic de Bruijn graph* with the vertex set $\mathcal{A}^{\mathbb{Z}_+}$ and the arrows $\boldsymbol{w} \rightsquigarrow \mathbf{S}\boldsymbol{w}$, where $p = |\mathcal{A}|$, and $\mathbf{S} : \alpha_0\alpha_1 \cdots \mapsto \alpha_1\alpha_2 \ldots$ is the (unilateral) shift transformation on $\mathcal{A}^{\mathbb{Z}_+}$. However, this graph does not really grasp a number of significant features of finite de Bruijn graphs, as, for instance, the out-degree of any vertex is 1.

On the other hand, as we have noticed when talking about the slider interpretation of the Cayley graph of the lamplighter group $\mathbb{Z} \wr \mathbb{Z}_m$, an advantage of our approach is that the definition of slider transitions is applicable without much difference both to finite circular and to infinite linear words. Therefore, by letting the span n in the definition of circular slider graphs go to infinity, one naturally arrives at the notion of the full linear slider graph $\overrightarrow{\mathscr{S}_{\mathcal{A}}}$ as the digraph whose vertex set is $\mathcal{A}^{\mathbb{Z}}$ and whose arrows are the slider transitions from the bottom half of Figure 0.4 with "$n = \infty$". General linear slider graphs are further defined as the subgraphs of $\overrightarrow{\mathscr{S}_{\mathcal{A}}}$. We shall return to a discussion of linear slider graphs and their relationship with the horospheric products of trees, lamplighter groups over $\mathbb{Z}$ and their Schreier graphs elsewhere.

• Neighbourhoods in circular and in linear slider graphs look precisely the same, which provides a direct approach to the aforementioned result of Grigorchuk – Leemann – Nagnibeda [**GLN16, Lee16**] on the identification of the Benjamini – Schramm limits of de Bruijn graphs with Diestel – Leader graphs. This idea can actually be used for the identification of the Benjamini – Schramm limits for much more general sequences of circular slider graphs [**KLN**].

• Alternatively, one can directly consider the *stochastic homogenization* for linear slider graphs by looking for equivalence relations with an invariant probability measure graphed by linear slider graphs (cf. [**Kai03**]). Let us remind that two strings $\boldsymbol{a}, \boldsymbol{b} \in \mathcal{A}^{\mathbb{Z}}$ are called (asynchronously) asymptotically equivalent if there exist $\Delta_-, \Delta_+ \in \mathbb{Z}$ such that the strings $\mathbf{S}^{\Delta_-}\boldsymbol{a}$ (resp., $\mathbf{S}^{\Delta_+}\boldsymbol{a}$) and $\boldsymbol{b}$ are cofinal at $-\infty$ (resp., at $+\infty$). If, additionally, $\Delta_- = \Delta_+$ (resp., $\Delta_- = \Delta_+ = 0$), then the strings $\boldsymbol{a}$ and $\boldsymbol{b}$ are called semi-synchronously (resp., synchronously) asymptotically equivalent. The synchronous asymptotic equivalence relation is also known as the homoclinic or Gibbs equivalence relation of the shift transformation $\mathbf{S}$, and it preserves the maximal entropy invariant measure for any subshift of finite type on the alphabet $\mathcal{A}$ [**PS97**].

Since the semi-synchronous asymptotic equivalence relation is the common refinement of the synchronous one and of the orbit equivalence relation of the shift $\mathbf{S}$, it also preserves these measures. The connected components of linear slider graphs are clearly contained in the equivalence classes of the semi-synchronous equivalence relation, which provides a natural stochastic homogenization for the linear slider graphs induced on subshifts of finite type in $\mathcal{A}^{\mathbb{Z}}$.

F. Paper overview. We begin with a brief discussion of the rich history of de Bruijn graphs (Section 1). In Section 2 we introduce the notion of a circular slider graph, after which in Section 3 we review the known examples of subgraphs of de Bruijn graphs from this point of view. In Section 4 we define periodic slider graphs and establish their d-connectedness for arbitrary irreducible sofic subshifts (Theorem 4.5). In Section 5 we introduce and discuss transversally Markov circular slider graphs. In Section 6 we remind the basic definitions concerning the lamplighter groups, after which in Section 7 we prove that de Bruijn digraphs and, more generally, Cayley circular slider digraphs can be interpreted as Schreier digraphs of circular lamplighter groups for a suitable choice of generating subsets (Theorem 7.7 and Theorem 7.10). Finally, in Section 8 we introduce spider slider graphs and identify a certain class of them with the appropriate Cayley graphs of circular lamplighter groups (Theorem 8.1).

Acknowledgements. I would like to thank Reinhard Diestel, Rostislav Grigorchuk, Paul-Henry Leemann and Tatiana Nagnibeda for helpful and inspiring discussions. I am also grateful to Paul-Henry Leemann for valuable comments on a preliminary version of this paper. Last but not least, my thanks go to Florian Sobieczky, the editor of this volume, for all his effort and patience.

1. De Bruijn graphs

Let

$$\mathcal{A}^n = \{\alpha_1\alpha_2\dots\alpha_n : \alpha_i \in \mathcal{A}\}$$

denote the set of words

$$\boldsymbol{a} = \alpha_1\alpha_2\dots\alpha_n$$

of length n in a finite alphabet $\mathcal{A}$.

DEFINITION 1.1. Two words $\boldsymbol{a}, \boldsymbol{a}' \in \mathcal{A}^n$ are linked with a de Bruijn transition

$$\boldsymbol{a} \rightsquigarrow \boldsymbol{a}' \tag{1.2}$$

if there exists a word $\widetilde{\boldsymbol{a}} \in \mathcal{A}^{n+1}$ such that $\boldsymbol{a}$ and $\boldsymbol{a}'$ are its initial and final segments, respectively:

$$\widetilde{\boldsymbol{a}} = \overbrace{\alpha_1\, \alpha_2 \dots \alpha_{n-1}\alpha_n}^{\boldsymbol{a}}\,\alpha_{n+1} \quad\text{with}\quad \boldsymbol{a}' = \underbrace{\alpha_2 \dots \alpha_{n-1}\alpha_n\,\alpha_{n+1}} .$$

DEFINITION 1.3 (e.g., see [**AS03, Rig14**]). The span n de Bruijn digraph

$$\overrightarrow{\mathcal{B}}{}_{\mathcal{A}}^{n} = \overrightarrow{\mathcal{B}}{}_{m}^{n}$$

over an alphabet $\mathcal{A}$ of cardinality $|\mathcal{A}| = m$ is the directed graph *(digraph)* whose vertex set is $\mathcal{A}^n$, and whose arrows are de Bruijn transitions (1.2). Sometimes one forgets the orientation on edges and considers undirected de Bruijn graphs $\mathcal{B}_{\mathcal{A}}^{n}$ as well.

These graphs are called after Nicolaas Govert de Bruijn who in 1946 constructed [**dB46**] what is now known as a *de Bruijn binary sequence* (the formal definition of which is contained in the title of [**dB75**]), i.e., a *Hamiltonian path* in $\overrightarrow{\mathcal{B}}_2^n$. Actually, as it was later discovered by Richard Stanley, in the binary case this had already been done as early as in 1894 (the problem was formulated by de Rivière and solved by Flye Sainte-Marie), and the case of an arbitrary alphabet had also been treated before de Bruijn in 1934 by Martin (the general case was also independently considered in 1946 by Good with a follow-up by Rees). Still, the current generally accepted term for these graphs is *de Bruijn graphs*. It appears to be quite fair as it was de Bruijn himself who in 1975 wrote a very detailed historical note [**dB75**], in which all these works and several other related ones were discussed. This note was eloquently and explicitly called

> *Acknowledgement of priority to C. Flye Sainte-Marie on the counting of circular arrangements of* 2^n *zeros and ones that show each* n*-letter word exactly once,*

an *avis rarissima* by today's standards. By the way (the fact which is not mentioned in [**dB75**]), de Bruijn binary sequence was also apparently independently introduced in the famous *The Logic of Scientific Discovery* by Karl Popper [**Pop59**, Appendix iv] under the name of a *shortest random-like sequence*.

De Bruijn graphs are currently quite popular in computer science and bioinformatics, e.g., see [**BJG09**], [**GYZ14**].

2. Circular slider graphs

We shall now somewhat change the viewpoint and pass from the usual "linear" words to the ones written around a circle (and read clockwise). The position of the initial letter of a word will be marked with a pointer separating the final and the initial letters. The circle itself is allowed to rotate freely, so that the whole word is entirely determined just by the mutual positions of the letters and of the pointer, and therefore there is a natural one-to-one correspondence between linear words and pointed circular words of the same length, see Figure 2.1.

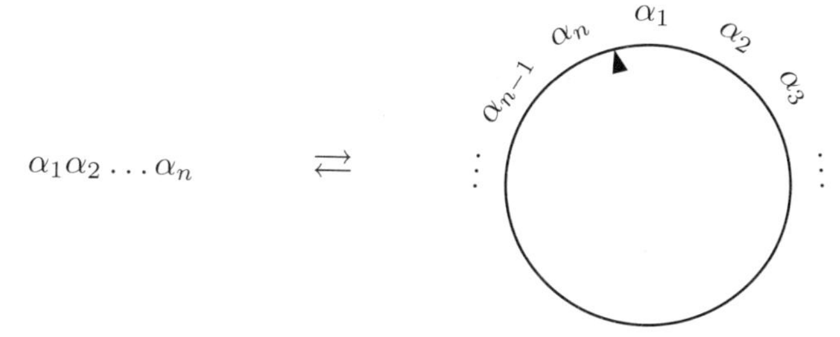

FIGURE 2.1. Linear words and pointed circular words.

This observation is begging to "change the coordinate system" and to pass in the definition of de Bruijn transitions (1.2) from *moving the letters with respect to a fixed pointer* in the "pointer coordinate system", as on Figure 2.2, to *moving the pointer in the opposite direction with respect to fixed letters* in the "letters coordinate system" instead, as on Figure 2.3.

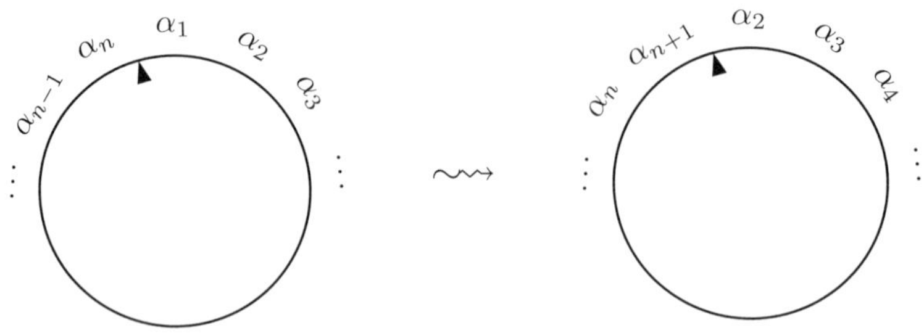

FIGURE 2.2. De Bruijn transitions with respect to a fixed pointer.

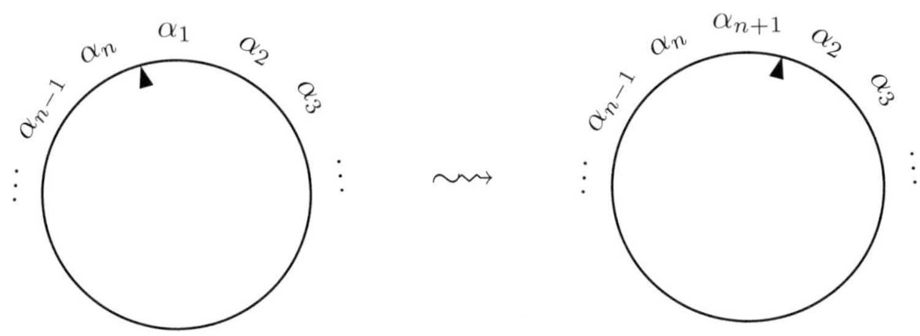

FIGURE 2.3. De Bruijn transitions with respect to a moving pointer.

Then in terms of pointed circular words de Bruijn transitions consist in moving the pointer one position clockwise and (possibly) changing the letter located between the old and the new positions of the pointer. Actually, we find it more convenient to think, instead of a pointer, about a sliding window (slider) of width 2 which covers one letter on either side of the pointer, see Figure 2.4.

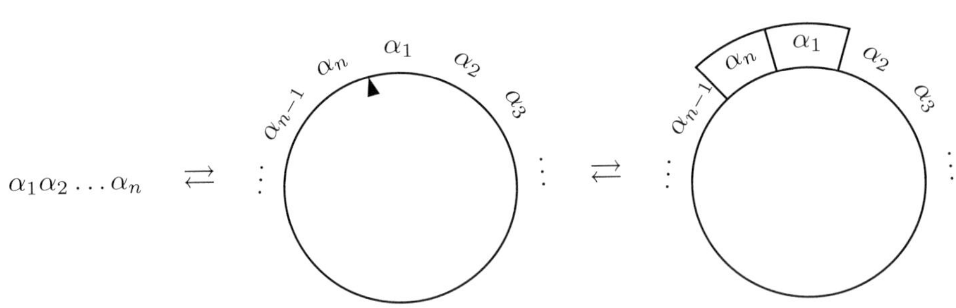

FIGURE 2.4. Slider pointed circular words.

In this terminology de Bruijn transitions (1.2) amount to moving the slider one position clockwise and (possibly) replacing the letter at the intersection of the old and the new slider windows (see Figure 2.6), and Definition 1.3 of de Bruijn graphs

$\overrightarrow{\mathcal{B}}_{\mathcal{A}}^n = \overrightarrow{\mathcal{B}}_{|\mathcal{A}|}^n$ can be then recast as

DEFINITION 2.5. (1) The full circular slider graph of span n over an alphabet $\mathcal{A}$ is the digraph whose vertices are all slider pointed circular words of length n in alphabet $\mathcal{A}$, and whose directed edges (arrows) are the de Bruijn transitions described on Figure 2.6.

(2) A (directed) span n circular slider graph over an alphabet $\mathcal{A}$ is a subgraph of the full circular slider graph, i.e., its set of vertices is a subset of the set $\mathcal{A}^n$ of the words of length n, and its set of directed edges (arrows) is a subset of the set of de Bruijn transitions.

(3) For a subset $\mathcal{V} \subset \mathcal{A}^n$ we shall denote by $\overrightarrow{\mathscr{S}}[\mathcal{V}]$ the corresponding induced circular slider graph, i.e., the associated *induced subgraph* of $\overrightarrow{\mathcal{B}}_{\mathcal{A}}^n$ (its set of vertices is $\mathcal{V}$, and its set of arrows consists of all de Bruijn transitions between these vertices).

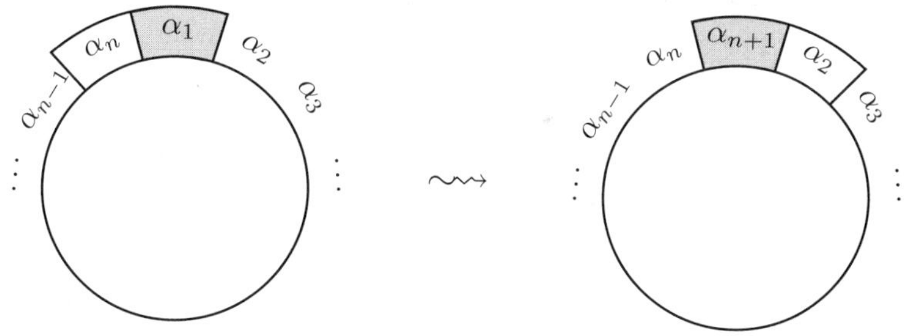

FIGURE 2.6. De Bruijn transitions with respect to a moving slider.

In the same way one can define undirected circular slider graphs as well. Note that *a priori* we impose no *connectedness conditions.* Although in the literature de Bruijn graphs and their subgraphs have so far always been considered over finite alphabets only, we do not impose any *finiteness conditions* on the alphabet $\mathcal{A}$ either. In fact, there are meaningful examples of locally finite circular slider graphs over infinite alphabets as well (e.g., the *Cayley* and *Schreier circular slider graphs* introduced in Definition 5.5).

REMARK 2.7. Of course, formally the circular slider graphs from Definition 2.5 are just subgraphs of the corresponding full de Bruijn graphs $\overrightarrow{\mathcal{B}}_{\mathcal{A}}^n$, and numerous examples of this kind have been considered before (see Section 3 below). However, our point of view puts this notion in a new perspective as outlined in Section **E** of the Introduction.

REMARK 2.8. Our inspiration for choosing the term "slider" comes from the analogy with an old-fashioned mechanical analogue computer known as *slide rule*, which was equipped with a sliding window split into two equal halves by a cursor line, see Figure 2.9 (this window was actually called "runner" though). It is interesting that there existed both linear and circular slide rules (cf. the comparison of the notions of linear and circular slider graphs in Section **E** of the Introduction)!

FIGURE 2.9. Linear and circular slide rules.

3. Examples

General subgraphs of the de Bruijn graph $\overrightarrow{\mathcal{B}}_{\mathcal{A}}^n$ (i.e., circular slider graphs in the sense of Definition 2.5) are parameterized by subsets $\mathcal{E} \subset \mathcal{A}^{n+1}$. The set of vertices of the associated subgraph is the collection of the length n initial and final segments of the words from $\mathcal{E}$, and any $\boldsymbol{w} \in \mathcal{E}$ produces an arrow between its length n initial and final segments [**Mor05**].

Most examples of the subgraphs of de Bruijn graphs considered in the literature are determined by the set

$$\mathcal{L}^{(n+1)} = \mathcal{A}^{n+1} \cap \mathcal{L}$$

of the length $n+1$ words in a certain factorial language $\mathcal{L}$, i.e., such language that any factor ($\equiv$ truncation, both on the left and on the right) of a word from $\mathcal{L}$ is also in $\mathcal{L}$. The set of vertices is then $\mathcal{L}^n$. We shall denote these graphs

$$\overrightarrow{\mathscr{S}}^n(\mathcal{L})$$

and call them factorial slider graphs.

If the language $\mathcal{L}$ is in addition prolongable (i.e., for any $\boldsymbol{w} \in \mathcal{L}$ there are $\alpha, \beta \in \mathcal{A}$ such that $\alpha \boldsymbol{w} \beta \in \mathcal{L}$; the languages with this property are also called extendable), then $\mathcal{L}$ coincides with the language $\mathcal{L}(\Sigma)$ of the subwords of a certain subshift (closed shift invariant subset)

$$\Sigma \subset \mathcal{A}^{\mathbb{Z}}$$

(e.g., see [**LM95**]), and, conversely, the language $\mathcal{L}(\Sigma)$ is obviously factorial and prolongable for any subshift Σ. In this situation

$$\mathcal{L}^{(n)} = \Sigma^{(n)}$$

is the set of the length n subwords appearing in the infinite words from Σ, and we shall use the notation

$$\overrightarrow{\mathscr{S}}^n(\Sigma) = \overrightarrow{\mathscr{S}}^n(\mathcal{L})$$

for the slider graph of span n determined by a subshift Σ.

EXAMPLE 3.1. $\mathcal{L} = \mathcal{L}(\boldsymbol{w})$ is the language of the factors of a *finite word* $\boldsymbol{w}$ in the alphabet $\mathcal{A}$. In bioinformatics one considers the situation when $\boldsymbol{w}$ is a finite (if very long) *DNA "word"* (called reference sequence), and the associated graphs $\overrightarrow{\mathscr{S}}^n(\mathcal{L}) = \overrightarrow{\mathscr{S}}^n(\boldsymbol{w})$ are called de Bruijn graphs (along with the usual full de Bruijn graphs $\overrightarrow{\mathcal{B}}_{\mathcal{A}}^n$), see [**CPT11**], [**LP14**]. One has also introduced the notion of a colored de Bruijn graph [**ICT+12**] to describe the situation when there are several reference sequences.

EXAMPLE 3.2. $\mathcal{L} = \mathcal{L}(\Sigma)$ for the subshift $\Sigma \subset \mathcal{A}^{\mathbb{Z}}$ determined by the condition that *any two consecutive letters in the words from* Σ *are distinct* (equivalently, $\mathcal{L}$ is the language of the irreducible words representing the elements of the *free product* of several copies of the cyclic 2-element group $\mathbb{Z}_2$ indexed by the alphabet $\mathcal{A}$). The associated graphs $\overrightarrow{\mathscr{S}}^n(\Sigma)$ are known as the Kautz graphs [**Kau71**]. Actually, in the case of a 3-letter alphabet $\mathcal{A}$ these graphs appear already in the original de Bruijn's paper [**dB46**].

EXAMPLE 3.3. $\mathcal{L} = \mathcal{L}(\Sigma)$ for the subshift Σ corresponding to a *topological Markov chain* [3] on the alphabet $\mathcal{A}$. The associated graphs $\overrightarrow{\mathscr{S}}^n(\Sigma)$ were introduced in 1983 by Fiol, Yebra and Fàbrega under the name of sequence graphs [**FYF83**], also see [**GFY92**] (however, these papers have remained virtually unknown outside of the Spanish graph theory community). If Σ is the full shift, then, of course, $\overrightarrow{\mathscr{S}}^n(\Sigma)$ is the full de Bruin graph $\overrightarrow{\mathcal{B}}^n_{\mathcal{A}}$.

EXAMPLE 3.4. The graphs $\overrightarrow{\mathscr{S}}^n(\Sigma)$ determined by (the language of) a *general subshift of finite type* $\Sigma \subset \mathcal{A}^{\mathbb{Z}}$ were introduced by Moreno [**Mor03**] who was apparently not aware of [**FYF83, GFY92**].

REMARK 3.5. In Example 3.2 and Example 3.3 the graphs $\overrightarrow{\mathscr{S}}^n(\mathcal{L})$ coincide with the induced subgraphs $\overrightarrow{\mathscr{S}}\left[\mathcal{L}^{(n)}\right]$ determined by the sets of length n words of the language $\mathcal{L}$. Provided n is large enough, this is also the case for Example 3.3.

EXAMPLE 3.6. $\mathcal{L} = \mathcal{L}(\boldsymbol{w})$ is the language of the factors of a *single semi-infinite word* $\boldsymbol{w} \in \mathcal{A}^{\mathbb{Z}_+}$. In this case the associated graphs $\overrightarrow{\mathscr{S}}^n(\mathcal{L})$ were introduced by Rauzy [**Rau83**] and are currently known as Rauzy graphs.

EXAMPLE 3.7. $\mathcal{L} = \mathcal{L}(\Sigma)$ for a *minimal subshift* $\Sigma \subset \mathcal{A}^{\mathbb{Z}}$ (then by minimality $\mathcal{L} = \mathcal{L}(\boldsymbol{w}_+)$ for the positive subword $\boldsymbol{w}_+$ of any $\boldsymbol{w} \in \Sigma$, so that this setup is essentially equivalent to that of Rauzy from Example 3.6). The associated graphs $\overrightarrow{\mathscr{S}}^n(\Sigma)$ were used by Boshernitzan in his analysis of the unique ergodicity for interval exchange maps [**Bos85**, Theorem 6.9]. Their appearance in [**Bos85**] remained unknown outside of a very limited circle of specialists until 2010 when it was emphasized by Ferenczi – Monteil in [**FM10**].

EXAMPLE 3.8. $\mathcal{L}$ is a *general factorial prolongable language* in the alphabet $\mathcal{A}$ (as we have already explained, this is the same as saying that $\mathcal{L} = \mathcal{L}(\Sigma)$ is the language of a *subshift* $\Sigma \subset \mathcal{A}^{\mathbb{Z}}$). The associated graphs $\overrightarrow{\mathscr{S}}^n(\Sigma) = \overrightarrow{\mathscr{S}}^n(\mathcal{L})$, considered as a generalization of the ones from Example 3.6 above, were called Rauzy graphs by Cassaigne [**Cas96**] who used them for studying linear complexity languages.

REMARK 3.9. Example 3.2, Example 3.3, Example 3.4, Example 3.7 are particular cases of the situation when Σ is a *general subshift over an alphabet* $\mathcal{A}$ which is described in Example 3.8. As is typical for this area (cf. de Bruijn's acknowledgement [**dB75**] and its discussion in Section 1), the authors of [**FYF83**], [**Rau83**] and [**Bos85**] were apparently unaware of each other's work. One should note, however, that although formally all these papers deal with essentially the

[3] We distinguish general subshifts of finite type determined by a finite collection of *forbidden words* and topological Markov chains, for which all forbidden words have length 2 (or, equivalently, all two-letter subwords must belong to a fixed set of *admissible transitions*), e.g., see [**LM95**].

same setup of a subshift $\Sigma \subset \mathcal{A}^{\mathbb{Z}}$, the focus in [**FYF83**], [**Mor03**] (Example 3.2, Example 3.3, Example 3.4), on one hand, and in [**Rau83**], [**Bos85**], [**Cas96**] (Example 3.6, Example 3.7, Example 3.8), on the other hand, is entirely different. It is well-known (e.g., see [**LM95**]) that the subshifts of finite type considered in [**FYF83**], [**Mor03**] are in a sense completely opposite to the low complexity subshifts treated in [**Rau83**], [**Bos85**], [**Cas96**]. For instance, the former ones have a lot of periodic words, whereas the latter ones have none (cf. Section 4 below).

EXAMPLE 3.10. Let α be a *finite partition* of a state space X. Denote by $\mathcal{A}$ the set of elements of α, and let $x \mapsto \alpha(x) \in \mathcal{A}$ be the map which assigns to any point $x \in X$ its α-name, i.e., the element of the partition α which contains x. Then for any transformation $T : X \to X$ the symbolic encoding map

$$\mathfrak{S} : x \mapsto \big(\alpha(x), \alpha(Tx), \alpha(T^2x), \dots\big) \in \mathcal{A}^{\mathbb{Z}_+}$$

is a semi-conjugacy between the original map T and the shift transformation

$$\mathbf{S} : \mathcal{A}^{\mathbb{Z}_+} \to \mathcal{A}^{\mathbb{Z}_+}$$

(note that *a priori*, without imposing any additional conditions, the image set $\mathfrak{S}(X) \subset \mathcal{A}^{\mathbb{Z}_+}$ need not be closed or shift invariant). The map $\mathfrak{S}$ is a standard tool in the theory of dynamical systems both in the measurable and in the topological setups (for instance, see [**Kea91**] or [**LM95**]).

As usual, let $T^{-n}\alpha$ denote the partition of X into the T^n-preimages of the elements of the partition α, so that $x, y \in X$ belong to the same element of $T^{-n}\alpha$ iff $T^n x$ and $T^n y$ belong to the same element of α, i.e., $\alpha(T^n x) = \alpha(T^n y)$, and let

$$\alpha_n = \alpha \vee T^{-1}\alpha \vee \dots \vee T^{-n+1}\alpha$$

be the common refinement of the partitions $T^{-k}\alpha$, $0 \le k \le n-1$. The α_n-names of points $x \in X$ are length n words in the alphabet $\mathcal{A}$, so that the triple (X, T, α) determines a subgraph $\overrightarrow{\mathscr{S}}^n_{\alpha,T}$ of $\overrightarrow{\mathcal{B}}^n_{\mathcal{A}}$ with the vertex set

$$\{\alpha_n(x) : x \in X\} \subset \mathcal{A}^n$$

and the set of arrows

$$\{\alpha_n(x) \rightsquigarrow \alpha_n(Tx) : x \in X\}$$

which can be identified with

$$\{\alpha_{n+1}(x) : x \in X\} \subset \mathcal{A}^{n+1}$$

(because $\alpha_n(x)$ and $\alpha_n(Tx)$ are the initial and the final segments of $\alpha_{n+1}(x)$, respectively).

Clearly, $\overrightarrow{\mathscr{S}}^n_{\alpha,T}$ is completely determined just by the action of the shift transformation $\mathbf{S}$ on the image set $\mathfrak{S}(X) \subset \mathcal{A}^{\mathbb{Z}_+}$ with respect to the time 0 coordinate partition of $\mathcal{A}^{\mathbb{Z}_+}$, so that Example 3.8 is a particular case of this construction.

EXAMPLE 3.11. Let

$$\mathbf{C}(x) = \begin{cases} x/2\,, & x \text{ is even}\,, \\ (3x+1)/2\,, & x \text{ is odd}\,, \end{cases}$$

be the Collatz function on the set of positive integers $\mathbb{N}$. The famous (and still very much open) Collatz conjecture claims that the $\mathbf{C}$-orbit of any starting point will eventually reach the number 1 (e.g., see [**Lag10**]). Let α be the partition of $\mathbb{N}$ into even and odd numbers. It is easy to see that if $x \equiv y \pmod{2^n}$, then $\alpha_n(x) = \alpha_n(y)$.

Moreover, as it has been independently established by Terras [**Ter76**, Theorem 1.2] and Everett [**Eve77**, Theorem 1] (also see Lagarias [**Lag85**, Theorem B]), the map $x \mapsto \alpha_n(x)$ is a surjection, which implies that for any $n > 0$ the associated graph $\overrightarrow{\mathscr{S}}^n_{\alpha,\mathbf{C}}$ is the full de Bruijn graph $\overrightarrow{\mathcal{B}}^n_2$ [**LdW13**].

4. Periodic slider graphs: connectedness and step d induced graphs

All examples of factorial slider graphs described in Section 3 are based on using the linear word structure. However, our approach makes natural to consider circular words as well.

DEFINITION 4.1. For a subshift $\Sigma \subset \mathcal{A}^{\mathbb{Z}}$, let

$$\mathcal{P}_n(\Sigma) \subset \Sigma$$

be the set of its n-periodic words, and let

$$\Sigma^{(n)}_{\circlearrowleft} \subset \mathcal{A}^n$$

be the set of all their length n factors ($\equiv$ Σ-admissible circular words of length n). The associated induced circular slider graph $\overrightarrow{\mathscr{S}}\left[\Sigma^{(n)}_{\circlearrowleft}\right]$ is called the n-periodic slider graph determined by the subshift Σ.

EXAMPLE 4.2. If Σ is a topological Markov chain on $\mathcal{A}$, then $\Sigma^{(n)}_{\circlearrowleft}$ consists of all words $\boldsymbol{a} = \alpha_1\alpha_2\dots\alpha_n$ such that, in addition to all transitions $\alpha_i\alpha_{i+1}$, $1 \le i \le n-1$, the transition $\alpha_n\alpha_1$ is also admissible.

The periodic slider graphs corresponding to the Kautz topological Markov chain (no double letters, see Example 3.2 above) have been recently introduced by Böhmová, Dalfó and Huemer [**BDH15**] and further studied by Dalfó [**Dal17a**, **Dal17b**]. We are not aware of any considerations of the periodic slider graphs for any other topological Markov chains or subshifts of finite type. All these subshifts have a lot of periodic words (as well as more general *sofic subshifts* [**LM95**] or *subshifts of quasi-finite type* [**Buz05**]), and the class of the associated periodic slider graphs should be quite interesting and promising for a future study.

The question about *connectedness* does not really arise for the factorial slider graphs considered in Section 3, as under the standard assumptions all these graphs are strongly connected in the sense that for any two vertices x, y there exists a directed path from x to y (except for, possibly, Example 3.1 and Example 3.6 of the languages $\mathcal{L}(\boldsymbol{w})$ generated by, respectively, a single finite or a single semi-infinite word $\boldsymbol{w}$). Indeed, in the case of *sofic subshifts* (in particular, of topological Markov chains or subshifts of finite type) it follows from the usual irreducibility assumption (which consists in requiring that for any two words $\boldsymbol{u}, \boldsymbol{v}$ from the subshift language $\mathcal{L}$ there is a word $\boldsymbol{w}$ such that $\boldsymbol{uwv} \in \mathcal{L}$, which is precisely what is needed for the strong connectivity of the associated slider graph). For the *low complexity shifts* irreducibility follows from minimality, which is also a standard condition in this setup.

However, already for the periodic slider graphs (let alone the induced slider graphs determined by a general subset $\mathcal{V} \subset \mathcal{A}^n$) the situation is different, and these graphs need not be even weakly connected (i.e., as undirected graphs) in the simplest situations, for instance, for irreducible aperiodic topological Markov chains.

EXAMPLE 4.3. Let Σ be the topological Markov chain on the 3-letter alphabet

$$\mathcal{A} = \{\alpha, \beta, \gamma\}$$

with the admissible transitions

$$\alpha \rightsquigarrow \alpha \rightsquigarrow \beta \rightsquigarrow \gamma \rightsquigarrow \alpha .$$

This chain has the property that its position at any moment of time t is uniquely determined by the positions at times $t-1$ and $t+1$, i.e., for any two letters $\zeta_1, \zeta_2 \in \mathcal{A}$ there exists at most one letter $\zeta \in \mathcal{A}$ which can be inserted between ζ_1 and ζ_2 in such a way that the word $\zeta_1\zeta\zeta_2$ is admissible. Therefore, in this situation the admissible de Bruijn transitions (Figure 2.6) on $\Sigma^{(n)}_{\circlearrowleft}$ consist just in moving the slider along circular words without replacing any letters, so that the connected components of the undirected n-periodic slider graph $\overrightarrow{\mathscr{S}}\left[\Sigma^{(n)}_{\circlearrowleft}\right]$ are just the rotation orbits (consisting of cyclic permutations) of n-periodic words. For instance, for $n = 3$ there are just two periodic words $\alpha\alpha\alpha$ and $\alpha\beta\gamma$ (up to a rotation $\equiv$ cyclic permutation), so that the 3-periodic slider graph has 4 vertices

$$\alpha\alpha\alpha\,, \quad \alpha\beta\gamma\,, \quad \beta\gamma\alpha\,, \quad \gamma\alpha\beta$$

with the arrows

$$\alpha\beta\gamma \rightsquigarrow \beta\gamma\alpha \rightsquigarrow \gamma\alpha\beta \rightsquigarrow \alpha\beta\gamma\,, \qquad \alpha\alpha\alpha \rightsquigarrow \alpha\alpha\alpha\,,$$

and it has two (weakly) connected components $\{\alpha\alpha\alpha\}$ and $\{\alpha\beta\gamma, \beta\gamma\alpha, \gamma\alpha\beta\}$.

However, one can easily modify Definition 2.5 to make the periodic slider graphs connected by allowing slider transitions of uniformly bounded "step length". In spite of its naturalness, we could not find the following definition in the literature.

DEFINITION 4.4. Let Γ be a directed graph with the vertex set V, and let $X \subset V$. Fix a positive integer d. Let us connect two vertices $x, y \in X$ with a directed arrow if there exists a path of length $\le d$ joining x and y in the ambient graph Γ. The resulting digraph with the vertex set X will be called the step d graph on X induced from Γ (or, step d induced graph in short). In the same way one defines step d undirected induced graphs in the situation when the ambient graph is undirected. One can also talk about strictly step d induced graphs when only the paths of length precisely d in the ambient graph are considered. If the step d induced graph on X is connected (strongly or weakly), then we shall say that the set X is step d connected (resp., strongly or weakly) in the ambient graph Γ.

THEOREM 4.5. *If $\Sigma \subset \mathcal{A}^{\mathbb{Z}}$ is an irreducible sofic subshift on a finite alphabet $\mathcal{A}$, then there exists an integer d such that all periodic slider graphs $\overrightarrow{\mathscr{S}}\left[\Sigma^{(n)}_{\circlearrowleft}\right]$ are step d connected in the corresponding ambient graphs $\overrightarrow{\mathscr{S}}\left[\Sigma^{(n)}\right]$.*

PROOF. Since sofic subshifts are factors of finite type subshifts, it is sufficient to establish the claim just for the latter ones, so that by replacing, if necessary, the alphabet $\mathcal{A}$ with the alphabet $\mathcal{A}' = \mathcal{A}^N$ of the length N words for a certain integer N, the general case reduces in the usual way to the situation when Σ is an irreducible topological Markov chain on $\mathcal{A}$ (cf. [**LM95**]). For simplicity we shall also assume that this chain is aperiodic (otherwise one would have to add a couple of usual pretty obvious technicalities). Then the entries of a certain power of the admissibility matrix of Σ are all positive, i.e., there exists a positive integer κ with the property that for any $\alpha, \beta \in \mathcal{A}$ there is $\boldsymbol{u} \in \mathcal{A}^\kappa$ such that $\alpha\boldsymbol{u}\beta \in \mathcal{L}(\Sigma)$.

By the above, for any admissible circular word $\boldsymbol{a} \in \Sigma_{\circlearrowleft}^{(n)}$ and any letter $\alpha \in \mathcal{A}$, any contiguous length $2\kappa + 1$ segment in $\boldsymbol{a}$ can be replaced with a same length segment $\boldsymbol{u}\alpha\boldsymbol{v}$, $\boldsymbol{u}, \boldsymbol{v} \in \mathcal{A}^{\kappa}$, with the letter $\alpha \in \mathcal{A}$ in the middle, in such a way that the resulting new circular word $\boldsymbol{a}'$ is also admissible. Then the set $\{\boldsymbol{a}, \boldsymbol{a}'\}$ is obviously step $(2\kappa + 2)$ connected in $\overrightarrow{\mathscr{S}}\left[\Sigma^{(n)}\right]$.

If $\boldsymbol{b} \in \Sigma_{\circlearrowleft}^{(n)}$ is another admissible circular word, then the above argument shows that $\boldsymbol{a}$ is step $(2\kappa + 2)$ connected with an admissible circular word $\boldsymbol{c} \in \Sigma_{\circlearrowleft}^{(n)}$ which coincides with $\boldsymbol{b}$ on a certain subset of positions Z of the cycle $\mathbb{Z}_n$ such that the distance between any two neighbours $z, z' \in Z$ does not exceed $2\kappa + 1$. Since $\boldsymbol{c}$ and $\boldsymbol{b}$ are step $(2\kappa + 2)$ connected in $\overrightarrow{\mathscr{S}}\left[\Sigma^{(n)}\right]$, the claim follows with $d = 2\kappa + 2$. □

5. Missing links and transversally Markov circular slider graphs

In spite of a number of studies of the *fault tolerance* of de Bruijn and Kautz graphs to edge failures (e.g., see [**RNT12**], [**LZL17**] and the references therein), circular slider graphs other than the ones determined by various factorial languages (see Section 3) have not attracted much attention *per se*.

We shall give here the definition of a natural class of circular slider graphs, which is in an essential way based on using the circular word structure and can not be described by any factorial language. As far as we know, this notion has not appeared in the literature so far.

DEFINITION 5.1. Let Σ be a topological Markov chain over a finite alphabet $\mathcal{A}$ determined by its set of admissible transitions $A = \Sigma^{(2)} \subset \mathcal{A}^2$. The associated transversally Markov circular slider graph

$$\overrightarrow{\mathscr{S}}_{\Sigma}(\mathcal{A}^n)$$

of span n is the circular slider graph with the vertex set $\mathcal{A}^n$, whose arrows are the de Bruijn transitions described on Figure 2.6, with the additional condition that $\alpha_1\alpha_{n+1} \in A$. In other words, one retains only the transversally Markov de Bruijn transitions, i.e., those for which the replacement $\alpha_1 \rightsquigarrow \alpha_{n+1}$ is Σ-admissible. Moreover, given a subset $\mathcal{V} \subset \mathcal{A}^n$, one can further consider the induced transversally Markov circular slider graph $\overrightarrow{\mathscr{S}}_{\Sigma}[\mathcal{V}]$, whose vertex set is $\mathcal{V}$, and whose arrows are the transversally Markov de Bruijn transitions between the words from $\mathcal{V}$.

More generally, one can also impose more complicated rules for admissibility of de Bruijn transitions, for instance, it can depend not just on the values of α_1 and α_{n+1}, but also on the letters in a certain fixed neighbourhood of the slider. Ultimately, when this neighbourhood becomes the whole circular word, one arrives at the definition of a general circular slider graph.

EXAMPLE 5.2. The golden mean subshift $\Sigma = \Sigma(A)$ is the topological Markov chain over the binary alphabet $\mathcal{A} = \{0, 1\}$ with the forbidden transition $1 \rightsquigarrow 1$, i.e., with the set of admissible transitions $A = \{00, 01, 10\}$ (e.g., see [**LM95**]). On the left of Figure 5.3 is the full slider graph $\overrightarrow{\mathcal{B}}_2^3$ of span 3 over the alphabet $\mathcal{A} = \{0, 1\}$. Its vertices are all 3-letter words $\alpha_1\alpha_2\alpha_3$ in the alphabet $\mathcal{A}$, and its arrows are all de Bruijn transitions $\alpha_1\alpha_2\alpha_3 \rightsquigarrow \alpha_2\alpha_3\alpha_4$ labelled with the corresponding replacement letters α_4. On the right of Figure 5.3 is the transversally Markov slider graph $\overrightarrow{\mathscr{S}}_{\Sigma}\left(\mathcal{A}^3\right)$ obtained by removing the de Bruijn transitions $\alpha_1\alpha_2\alpha_3 \rightsquigarrow \alpha_2\alpha_3\alpha_4$ with $\alpha_1 = \alpha_4 = 1$.

The induced subgraphs $\vec{\mathscr{S}}_\Sigma\left[\Sigma^{(3)}\right]$ and $\vec{\mathscr{S}}_\Sigma\left[\Sigma^{(3)}_{\circlearrowleft}\right]$ of $\vec{\mathscr{S}}_\Sigma\left(\mathcal{A}^3\right)$ determined by the subsets

$$\Sigma^{(3)} = \{000, 001, 010, 100, 101\}$$

and

$$\Sigma^{(3)}_{\circlearrowleft} = \{000, 001, 010, 100\}$$

of all Σ-admissible and of Σ-admissible circular words, respectively, are presented on Figure 5.4 [4].

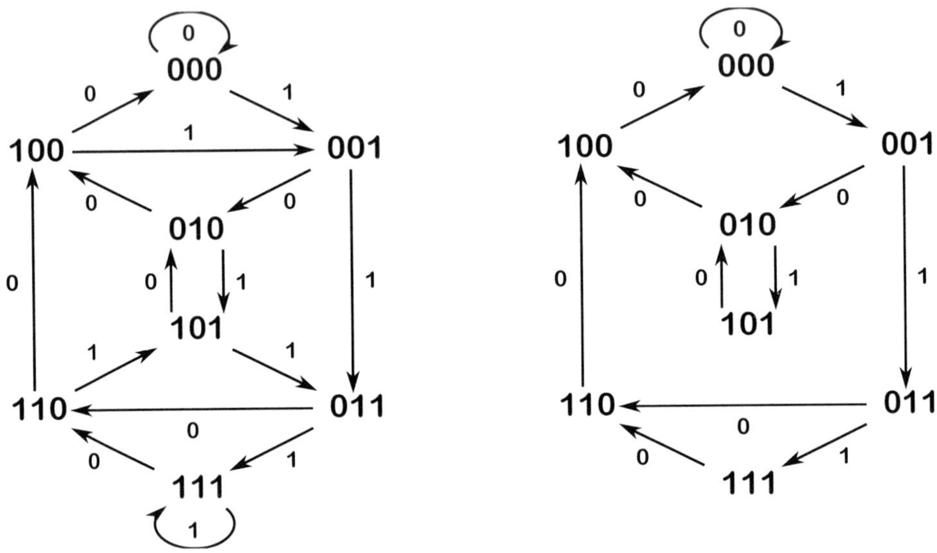

FIGURE 5.3. The full slider graph and the corresponding transversally Markov slider graph determined by the golden mean topological Markov chain.

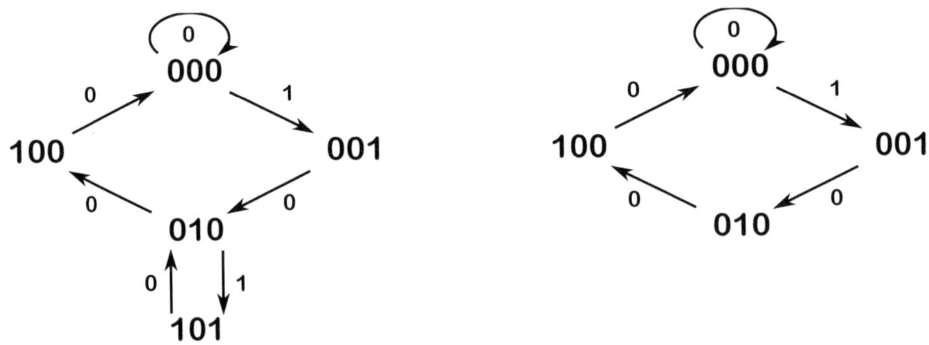

FIGURE 5.4. Induced subgraphs of the transversally Markov slider graph determined by the golden mean topological Markov chain.

[4] Of course, in general there is no need to use the same topological Markov chain both for defining a transversally Markov circular slider graph and for defining its induced subgraphs. However, there is not much choice in the case of a two letter alphabet.

The following class of examples of transversally Markov circular slider graphs is inspired by the *lamplighters* (see below Section 6) and is based on the notion of the *Cayley graph of a group* (the use of which is currently becoming popular in the theory of interconnection networks, see [**RNT12**], [**CPFV14**]). Let us first remind that the (directed) Cayley graph ($\equiv$ Cayley topological Markov chain)

$$\overrightarrow{\mathfrak{Cay}}(G, K)$$

on a group G determined by a subset $K \subset G$ has the vertex set G and the arrows

$$g \rightsquigarrow gk\ , \qquad g \in G, k \in K\ .$$

More generally, given a (right) action of a group G on a set X, and a subset $K \subset G$, one defines the associated (directed) Schreier graph ($\equiv$ Schreier topological Markov chain)

$$\overrightarrow{\mathfrak{Sch}}(X, K)$$

with the vertex set X and the arrows

$$x \rightsquigarrow xk\ , \qquad x \in X, k \in K\ .$$

DEFINITION 5.5. Let K be a subset of a group G. The associated Cayley circular slider graph

$$\overrightarrow{\mathscr{S}}_K(G^n) = \overrightarrow{\mathscr{S}}_{\overrightarrow{\mathfrak{Cay}}(G,K)}(G^n)$$

is the transversally Markov circular slider of span n over the alphabet G determined by the Cayley topological Markov chain $\overrightarrow{\mathfrak{Cay}}(G, K)$, i.e., its vertex set is G^n, and the arrows are

$$(g_1, g_2, \ldots, g_n) \rightsquigarrow (g_2, g_3, \ldots, g_n, g_1 h)\ , \qquad g_i \in G,\ h \in K\ .$$

More generally, given a (right) action of a group G on an action space X and a subset $K \subset G$, the associated Schreier circular slider graph

$$\overrightarrow{\mathscr{S}}_K(X^n) = \overrightarrow{\mathscr{S}}_{\overrightarrow{\mathfrak{Sch}}(X,K)}(X^n)$$

is the transversally Markov circular slider of span n over the alphabet X determined by the Schreier topological Markov chain $\overrightarrow{\mathfrak{Sch}}(X, K)$, i.e., its vertex set is X^n, and the arrows are

$$(x_1, x_2, \ldots, x_n) \rightsquigarrow (x_2, x_3, \ldots, x_n, x_1 h)\ , \qquad x_i \in X,\ h \in K\ .$$

Following Definition 5.1, one can also consider the induced subgraphs $\overrightarrow{\mathscr{S}}_K[\mathcal{V}]$ determined by various subsets $\mathcal{V} \subset G^n$ or $\mathcal{V} \subset X^n$ (cf. Section 3).

REMARK 5.6. In the case $K = G$ the graph $\overrightarrow{\mathscr{S}}_K(G^n)$ coincides with the full slider graph $\overrightarrow{\mathscr{B}}^n_G$, whereas for $K \subsetneq G$ the graphs $\overrightarrow{\mathscr{S}}_K(G^n)$ have the same vertex set G^n as $\overrightarrow{\mathscr{B}}^n_G$, but fewer arrows. As we shall see below (Theorem 7.10), the graphs $\overrightarrow{\mathscr{S}}_K(G^n)$ are actually *Schreier graphs of circular lamplighter groups*.

6. Lamplighters over cyclic groups

Before discussing the relationship between slider graphs and lamplighter groups let us first remind the basic definitions concerning wreath products and lamplighters.

DEFINITION 6.1. The (restricted) wreath product

$$G = A \wr B$$

with the active (or, base) group A and the passive group (or, group of states) B is the semi-direct product

$$G = A \leftthreetimes \mathrm{fun}(A, B)$$

of the group A and the group

$$\mathrm{fun}(A, B) \cong \bigoplus_{a \in A} B$$

of finitely supported B-valued configurations (i.e., those that take values different from the identity of B at finitely many points only) on A with the operation of pointwise multiplication (i.e., the direct sum of the copies of the group B indexed by A), on which the group A acts by translations.

In our notation for the wreath and semi-direct products the active group is always on the left (in accordance with the syntactic structure of the English language). However, quite often one also uses the notation in which A and B are switched. In this paper we are interested just in the situation when the *active group A is cyclic* (in particular, abelian, so that we shall use the additive notation for the group operation in A).

In the context of functional and stochastic analysis the groups $\mathbb{Z}^d \wr \mathbb{Z}_2$ were first introduced by Vershik and the author [**KV83**] under the name of the groups of dynamical configurations. Nonetheless, this term did not stick, and the current generally accepted standard is to call them lamplighter groups (apparently, this usage goes back to [**LPP96**]). More general wreath products are also sometimes called lamplighter groups.

Below, if the cyclic group A is finite (resp., infinite), we shall call $A \wr B$ a circular (resp., linear) lamplighter group.

As a set, the group $G = A \wr B$ is the usual product of A and the group of configurations $\mathrm{fun}(A, B)$. For a group element $(a, \Phi) \in G$ its A component a and its $\mathrm{fun}(A, B)$ component Φ are usually referred to as the lamlighter position and the lamp configuration, respectively. The group operation in G is "skewed" by using the left action of A on $\mathrm{fun}(A, B)$ by the group automorphisms

$$\mathbf{T}^a \Phi(x) = \Phi(x - a) ,$$

so that the group multiplication in G is

$$(a_1, \Phi_1) \cdot (a_2, \Phi_2) = (a_1 + a_2, \Phi_1 \cdot \mathbf{T}^{a_1} \Phi_2) . \tag{6.2}$$

The identity of G is the pair $(0, \varnothing)$, where $\varnothing$ (the identity of the group of configurations $\mathrm{fun}(A, B)$) is the empty configuration

$$\varnothing(a) = e \qquad \forall\, a \in A ,$$

and e is the identity of the group of states B.

The **standard generators** of G are

$$(\pm 1, \varnothing) \quad \text{and} \quad \{(0, \delta_0^b)\}_{b \in K} , \tag{6.3}$$

where K is a fixed (symmetric) generating set of B, and $\delta_a^b \in \operatorname{fun}(A, B)$ denotes the configuration defined as

$$\delta_a^b(x) = \begin{cases} b , & x = a , \\ e , & x \neq a \end{cases}$$

(in other words, δ_a^b is the generator b in the a-indexed copy of the group B from the direct sum $\operatorname{fun}(A, B) \cong \bigoplus_{a \in A} B$).

Then, by (6.2), for any $(a, \Phi) \in G$

$$(a, \Phi) \cdot (\pm 1, \varnothing) = (a \pm 1, \Phi) ,$$

so that the right multiplication by the **walk generators** $(\pm 1, \varnothing)$ means that the "lamplighter" moves along A one step to the left or to the right, whereas the lamp configuration Φ remains intact. In the same way,

$$(a, \Phi) \cdot (0, \delta_0^b) = (a, \Phi \cdot \delta_a^b) ,$$

so that the right multiplication by the **switch generators** $(0, \delta_0^b)$ means that the position a of the lamplighter in the group A remains the same, whereas the state of the lamp at a changes from $\Phi(a)$ to $\Phi(a) \cdot b$.

7. Lamplighters and circular slider graphs

We shall now introduce another generating set for the group $G = A \wr B$. Let

$$\begin{aligned} \widetilde{B}_+ &= \{(1, \delta_1^b)\}_{b \in B} = \{(1, \varnothing) \cdot (0, \delta_0^b)\}_{b \in B} , \\ \widetilde{B}_- &= \left(\widetilde{B}_+\right)^{-1} = \{(-1, \delta_0^b)\}_{b \in B} = \{(0, \delta_0^b) \cdot (-1, \varnothing)\}_{b \in B} , \\ \widetilde{B} &= \widetilde{B}_- \cup \widetilde{B}_+ . \end{aligned} \tag{7.1}$$

The Cayley graphs $\overrightarrow{\mathfrak{Cay}}\left(G, \widetilde{B}_+\right)$ and $\overrightarrow{\mathfrak{Cay}}\left(G, \widetilde{B}_-\right)$ of the group G determined by the sets $\widetilde{B}_+$ and $\widetilde{B}_-$, respectively, are the digraphs obtained from the undirected Cayley graph $\mathfrak{Cay}\left(G, \widetilde{B}\right)$ with respect to the symmetric generating set $\widetilde{B}$ by taking two opposite orientations of its edges.

By (6.2), the result of the right multiplication of an element (a, Φ) by an increment $(1, \delta_1^b) \in \widetilde{B}_+$ is

$$(a, \Phi) \cdot (1, \delta_1^b) = (a + 1, \Phi \cdot \delta_{a+1}^b) ,$$

i.e., the lamplighter moves from the position a to the position $a + 1$, and the value of the configuration at the arrival point $a + 1$ changes from $\Phi(a + 1)$ to $\Phi(a + 1) \cdot b$. In the same way, the right multiplication by $(-1, \delta_0^b) \in \widetilde{B}_-$ amounts to changing the value of the configuration at the departure point a from $\Phi(a)$ to $\Phi(a) \cdot b$ and moving from a to $a - 1$. Therefore, the generators from $\widetilde{B}_+$ (resp., from $\widetilde{B}_-$) can be called **walk-right—switch** (resp., **switch—walk-left**) **generators** (this nomenclature for various kinds of elements of lamplighter groups was, in the context of random walks, coined by Wolfgang Woess, and apparently first appeared in print in [**LNW08**]).

Now, in order to make the picture more symmetric, it is convenient to "mark" the active group A with the width 2 **slider** (sliding window) over the positions a and

$a+1$ (or, equivalently, to distinguish the edge between a and $a+1$) rather than to point at lamplighter's position a, see Figure 7.2.

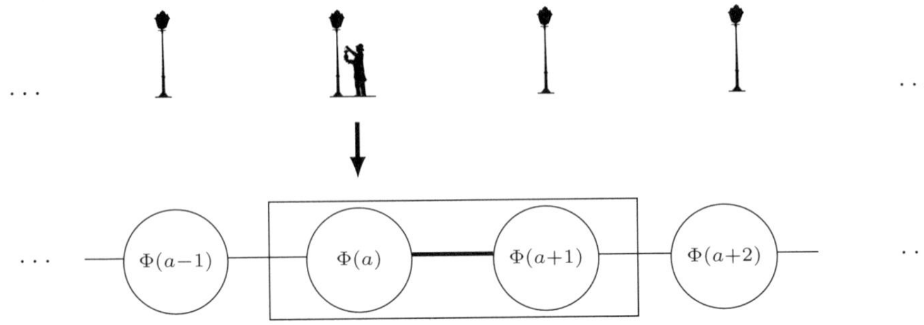

FIGURE 7.2. Lamplighter's window (slider).

Then the multiplication by an element $(1,\delta_1^b)\in\widetilde{B}_+$ amounts to shifting the slider one position to the right and multiplying the state at the intersection of the old and the new sliders by b, see Figure 7.3.

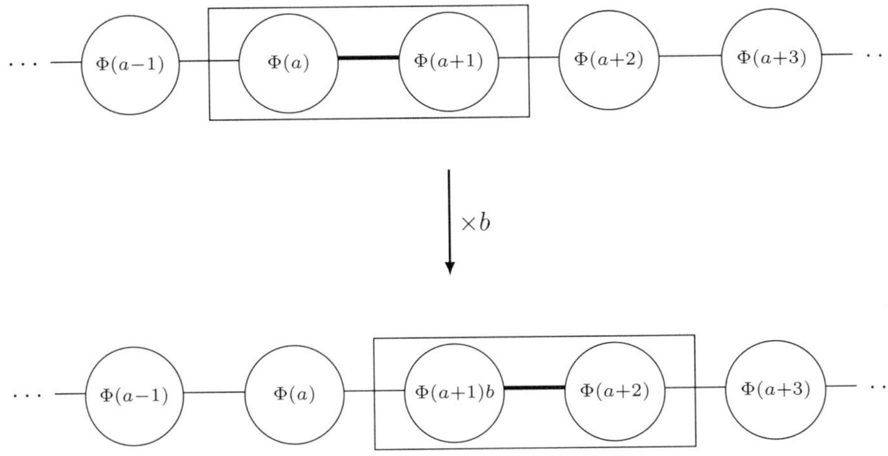

FIGURE 7.3. Walk-right—switch generators.

In a perfectly symmetrical way the multiplication by

$$(-1,\delta_0^b)=(1,\delta_1^{b^{-1}})^{-1}\in\widetilde{B}_-$$

amounts to shifting the slider one position to the left and multiplying the state at the intersection of the old and the new sliders by b, see Figure 7.4.

Thus, the arrows of the Cayley graph $\overrightarrow{\mathfrak{Cay}}\left(G,\widetilde{B}_+\right)$ on Figure 7.3 look precisely like the de Bruijn transitions from Figure 2.6. An important difference, however, is that in the case of slider graphs the circular words in the alphabet B are marked

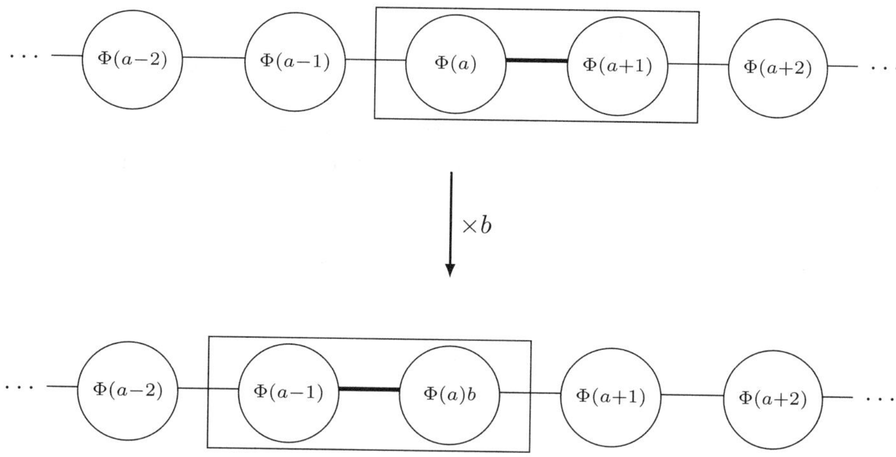

FIGURE 7.4. Switch—walk-left generators.

only once (with the position of the slider), whereas in the case of lamplighter groups the circular words are endowed with two pointers (both the position of the identity of the group A and the position of the lamplighter). Therefore, in order to describe the slider graphs in terms of the lamplighter group one just has to eliminate the additional pointer, or, in other words, to eliminate the A component of the elements $(a, \Phi) \in G = A \wr B$.

At the formal level, the group A embeds into G as the subgroup

$$\{(a, \varnothing) : a \in A\} \cong A .$$

By formula (6.2),

$$(a', \varnothing)(a, \Phi) = (a' + a, \mathbf{T}^{a'}\Phi) \qquad \forall\, a' \in A, (a, \Phi) \in G ,$$

which means that the left multiplication by $(a', \varnothing)$ amounts to shifting both lamplighter's position a and the lamp configuration Φ by a'. Thus, the map

$$(a, \Phi) \mapsto \mathbf{T}^{-a}\Phi \tag{7.5}$$

is constant on the cosets $Ag \subset G$, and therefore it allows one to identify the homogeneous space $A\backslash G$ with the space of configurations $\mathrm{fun}(A, B)$, or, if the group A is finite, just with the set B^n of the words of length $n = |A|$ in the alphabet B. Namely, the word $\boldsymbol{b} \cong \mathbf{T}^{-a}\Phi$ is the sequence of the values of Φ read clockwise starting from the position adjacent to lamplighter's position a, i.e.,

$$\boldsymbol{b} = \beta_1\beta_2\ldots\beta_n , \quad \text{where} \quad \beta_i = \Phi(a+i) \quad (\text{addition mod } n) .$$

In these terms the right action of the group G on the space $A\backslash G \cong B^n$ takes the form

$$\boldsymbol{b}.(a, \Phi) = \boldsymbol{b}.(0, \Phi) \cdot (a, \varnothing) = \mathbf{T}^{-a}(\boldsymbol{b} \cdot \Phi) . \tag{7.6}$$

The reason for the appearance of the minus sign in formula (7.6) is that the map (7.5) essentially consists in passing from the "fixed coordinate system" to the "lamplighter coordinate system" (it is the position of the lamplighter that becomes the reference point), and therefore, if the lamplighter moves in the clockwise direction (say, if $a = 1$) in the fixed coordinate system, then the lamp configuration moves in the opposite anticlockwise direction with respect to the lamplighter (cf. Figure 2.2 and Figure 2.3).

Then, as it follows from comparing Figure 7.3 and Figure 2.6, the right action of the elements from $\widetilde{B}_+$ on $A\backslash G \cong B^{|A|}$ consists precisely in de Bruijn transitions. Thus, we have proved

THEOREM 7.7. *Let A be a finite cyclic group. Then the Schreier digraph*

$$\mathfrak{Sch}\left(A\backslash G, \widetilde{B}_+\right) \cong \mathfrak{Sch}\left(B^{|A|}, \widetilde{B}_+\right)$$

of the action (7.6) *of the wreath product $G = A \wr B$ on the homogeneous space $A\backslash G \cong B^{|A|}$ with respect to the generating set $\widetilde{B}_+$* (7.1) *is isomorphic to the de Bruijn digraph $\overrightarrow{\mathcal{B}}_B^{|A|}$ of span $|A|$ over the alphabet B.*

REMARK 7.8. A realization of de Bruijn graphs as Schreier graphs of the *infinite* lamplighter groups $\mathbb{Z} \wr \mathbb{Z}_d$ was recently obtained by Grigorchuk, Leemann and Nagnibeda [**GLN16**, Theorem 4.4.1]. In a sense, their Theorem 6.1.3 implicitly contains our result as it identifies the Cayley graph of the finite lamplighter group $\mathbb{Z}_n \wr \mathbb{Z}_d$ with the *spider-web graph* built from the de Bruijn graph $\overrightarrow{\mathcal{B}}_d^n$ by taking its direct (or tensor) product with $\mathbb{Z}_n$ (see below Section 8). However, our argument is much more direct.

If one considers the generating set (7.1), then the arising Cayley and Schreier graphs only depend on the size of the group B. Of course, this is no longer the case if the set $\widetilde{B}_+$ (7.1) is replaced with

$$\widetilde{K}_+ = \{(1, \delta_1^b)\}_{b \in K} , \tag{7.9}$$

for a proper subset $K \subsetneq B$. However, the same argument as above still yields

THEOREM 7.10. *Let A be a finite cyclic group, and $K \subset B$. Then the Schreier digraph*

$$\mathfrak{Sch}\left(A\backslash G, \widetilde{K}_+\right) \cong \mathfrak{Sch}\left(B^{|A|}, \widetilde{K}_+\right)$$

of the action (7.6) *of the wreath product $G = A \wr B$ on the homogeneous space $A\backslash G \cong B^{|A|}$ with respect to the generating set $\widetilde{K}_+$* (7.9) *is isomorphic to the Cayley circular slider digraph $\overrightarrow{\mathscr{S}}_K\left(B^{|A|}\right)$ from Definition* 5.5.

One can also obtain a similar description of the Schreier circular slider graphs from Definition 5.5 in terms of Schreier graphs of lamplighter groups.

8. Spider slider graphs

We shall denote by $\overrightarrow{\mathcal{C}}$ the Cayley digraph of a cyclic group C determined by the generator 1, so that the arrows of $\overrightarrow{\mathcal{C}}$ are $i \rightsquigarrow i+1$ (addition mod $|C|$). Let us recall that the tensor (or, direct) product of digraphs Γ_1, Γ_2 is the digraph $\Gamma_1 \otimes \Gamma_2$ such that both its vertex set and its edge set are the products, respectively, of the vertex sets and of the edge sets of Γ_1 and Γ_2, with the natural incidence relations (e.g., see [**HIK11**]). As we have already mentioned in Remark 7.8, the tensor products $\overrightarrow{\mathcal{C}} \otimes \overrightarrow{\mathcal{B}}_m^n$ are known as spider-web graphs (see Grigorchuk – Leemann – Nagnibeda [**GLN16**] and the references therein). They are parameterized by the alphabet size m, the span n of the de Bruijn graph, and by the size $|C|$ of the group C.

Theorem 6.1.3 of Grigorchuk – Leemann – Nagnibeda [**GLN16**] identifies, by using an algebraic approach based on a classification of all subgroups of the infinite lamplighter groups $\mathbb{Z} \wr \mathbb{Z}_m$ earlier obtained by Grigorchuk – Kravchenko [**GK14**], the spider-web graphs $\overrightarrow{\mathcal{C}} \otimes \overrightarrow{\mathcal{B}}_m^n$ with the Cayley graphs of certain finite groups for a number of combinations of the parameters $m, n, |C|$. In particular, if $|C| = n$, they establish an isomorphism of $\overrightarrow{\mathcal{C}}_n \otimes \overrightarrow{\mathcal{B}}_m^n$ with the Cayley graph of the finite lamplighter group $\mathbb{Z}_n \wr \mathbb{Z}_m$ endowed with the generating set (7.1).

As we have just explained in Theorem 7.7 and Theorem 7.10, the Cayley circular slider graphs $\overrightarrow{\mathscr{S}}_K\left(B^{|A|}\right)$ (in particular, the de Bruijn graphs $\overrightarrow{\mathcal{B}}_B^{|A|}$) can be obtained from the Cayley graphs of circular lamplighter groups by removing an extra pointer (which amounts to passing to the corresponding Schreier graph). Conversely, in order to recover the Cayley graph from the corresponding de Bruijn graph one just has to add a parameter $a \in A$ describing the relative position of the slider and of the identity of the group A, which yields

THEOREM 8.1. *Let A be a finite cyclic group, and $K \subset B$. Then the Cayley graph* $\mathfrak{Cay}\left(A \wr B, \widetilde{K}_+\right)$ *is isomorphic to the tensor product* $\overrightarrow{A} \otimes \overrightarrow{\mathscr{S}}_K\left(B^{|A|}\right)$.

In particular, for $K = B = \mathbb{Z}_m$ we recover (in a much more direct way) the aforementioned result from [**GLN16**, Theorem 6.1.3].

The construction of spider-web graphs obviously carries over to arbitrary circular slider graphs:

DEFINITION 8.2. We shall call the tensor product $\overrightarrow{A} \otimes \overrightarrow{\mathscr{S}}$ of the Cayley digraph $\overrightarrow{A}$ of a cyclic group A and of a circular slider graph $\overrightarrow{\mathscr{S}}$ a spider slider graph.

Numerous interesting properties of the classical spider-web graphs discussed in [**GLN16**] and, in more detail, in [**Lee16**] suggest that it would be interesting to study more general spider slider graphs associated with the slider graphs considered in Section 3, Section 4 and Section 5.

References

[AR15] Margarita Amchislavska and Timothy Riley, *Lamplighters, metabelian groups, and horocyclic products*, Enseign. Math. **61** (2015), no. 3-4, 373–415, DOI 10.4171/LEM/61-3/4-5. MR3539843

[AS03] Jean-Paul Allouche and Jeffrey Shallit, *Automatic sequences*, Cambridge University Press, Cambridge, 2003. Theory, applications, generalizations. MR1997038

[BDH15] Katerina Böhmová, Cristina Dalfó, and Clemens Huemer, *The diameter of cyclic Kautz digraphs*, Filomat **31** (2017), no. 20, 6551–6560. MR3746889

[BJG09] Jørgen Bang-Jensen and Gregory Gutin, *Digraphs*, 2nd ed., Springer Monographs in Mathematics, Springer-Verlag London, Ltd., London, 2009. Theory, algorithms and applications. MR2472389

[Bos85] Michael Boshernitzan, *A condition for minimal interval exchange maps to be uniquely ergodic*, Duke Math. J. **52** (1985), no. 3, 723–752, DOI 10.1215/S0012-7094-85-05238-X. MR808101

[Buz05] Jérôme Buzzi, *Subshifts of quasi-finite type*, Invent. Math. **159** (2005), no. 2, 369–406, DOI 10.1007/s00222-004-0392-1. MR2116278

[Cas96] Julien Cassaigne, *Special factors of sequences with linear subword complexity*, Developments in language theory, II (Magdeburg, 1995), World Sci. Publ., River Edge, NJ, 1996, pp. 25–34. MR1466182

[CPFV14] Miguel Camelo, Dimitri Papadimitriou, Lluís Fàbrega, and Pere Vilà, *Efficient routing in data center with underlying Cayley graph*, Complex Networks V: Proceedings of the 5th Workshop on Complex Networks CompleNet 2014 (Pierluigi Contucci, Ronaldo Menezes, Andrea Omicini, and Julia Poncela-Casanovas, eds.), Springer International Publishing, 2014, pp. 189–197.

[CPT11] Phillip E. C. Compeau, Pavel A. Pevzner, and Glenn Tesler, *How to apply de Bruijn graphs to genome assembly*, Nat. Biotechnol. **29** (2011), no. 11, 987–991.

[Dal17a] C. Dalfó, *From subKautz digraphs to cyclic Kautz digraphs*, arXiv:1709.01882, 2017.

[Dal17b] C. Dalfó, *The spectra of subKautz and cyclic Kautz digraphs*, Linear Algebra Appl. **531** (2017), 210–219, DOI 10.1016/j.laa.2017.05.046. MR3682699

[dB46] N. G. de Bruijn, *A combinatorial problem*, Nederl. Akad. Wetensch., Proc. **49** (1946), 758–764 = Indagationes Math. 8, 461–467 (1946). MR0018142

[dB75] N. G. de Bruijn, *Acknowledgement of priority to C. Flye Sainte-Marie on the counting of circular arrangements of 2^n zeros and ones that show each n-letter word exactly once*, TH report, Technische Hogeschool Eindhoven Nederland, 1975.

[Eve77] C. J. Everett, *Iteration of the number-theoretic function $f(2n) = n$, $f(2n+1) = 3n+2$*, Adv. Math. **25** (1977), no. 1, 42–45, DOI 10.1016/0001-8708(77)90087-1. MR0457344

[FM78] Harold Fredricksen and James Maiorana, *Necklaces of beads in k colors and k-ary de Bruijn sequences*, Discrete Math. **23** (1978), no. 3, 207–210, DOI 10.1016/0012-365X(78)90002-X. MR523071

[FM10] Sébastien Ferenczi and Thierry Monteil, *Infinite words with uniform frequencies, and invariant measures*, Combinatorics, automata and number theory, Encyclopedia Math. Appl., vol. 135, Cambridge Univ. Press, Cambridge, 2010, pp. 373–409. MR2759110

[FYF83] M. A. Fiol, J. L. A. Yebra, and J. Fàbrega, *Sequence graphs and interconnection networks*, Ars Combin. **16** (1983), no. A, 7–14. MR737074

[GFY92] J. Gómez, M. A. Fiol, and J. L. A. Yebra, *Graphs on alphabets as models for large interconnection networks*, Discrete Appl. Math. **37/38** (1992), 227–243, DOI 10.1016/0166-218X(92)90135-W. MR1176855

[GK14] R. Grigorchuk and R. Kravchenko, *On the lattice of subgroups of the lamplighter group*, Internat. J. Algebra Comput. **24** (2014), no. 6, 837–877, DOI 10.1142/S0218196714500374. MR3278388

[GLN16] R. Grigorchuk, P.-H. Leemann, and T. Nagnibeda, *Lamplighter groups, de Brujin graphs, spider-web graphs and their spectra*, J. Phys. A **49** (2016), no. 20, 205004, 35, DOI 10.1088/1751-8113/49/20/205004. MR3499181

[GYZ14] Jonathan L. Gross and Jay Yellen, *Graph theory and its applications*, 2nd ed., Discrete Mathematics and its Applications (Boca Raton), Chapman & Hall/CRC, Boca Raton, FL, 2006. MR2181153

[HIK11] Richard Hammack, Wilfried Imrich, and Sandi Klavžar, *Handbook of product graphs*, 2nd ed., Discrete Mathematics and its Applications (Boca Raton), CRC Press, Boca Raton, FL, 2011. With a foreword by Peter Winkler. MR2817074

[ICT+12] Zamin Iqbal, Mario Caccamo, Isaac Turner, Paul Flicek, and Gil McVean, De novo *assembly and genotyping of variants using colored de Bruijn graphs*, Nature Genetics **44** (2012), 226–232.

[Kai03] Vadim A. Kaimanovich, *Random walks on Sierpiński graphs: hyperbolicity and stochastic homogenization*, Fractals in Graz 2001, Trends Math., Birkhäuser, Basel, 2003, pp. 145–183. MR2091703

[Kau71] W. H. Kautz, *Design of optimal interconnection networks for multiprocessors*, Structure et conception des ordinateurs. Architecture and design of digital computers (NATO Advanced Summer Institute, 1969) (Guy G. Boulaye, ed.), Dunod, Paris, 1971, pp. 249–272.

[Kea91] Michael S. Keane, *Ergodic theory and subshifts of finite type*, Ergodic theory, symbolic dynamics, and hyperbolic spaces (Trieste, 1989), Oxford Sci. Publ., Oxford Univ. Press, New York, 1991, pp. 35–70. MR1130172

[KLN] Vadim A. Kaimanovich, Paul-Henry Leemann, and Tatiana Nagnibeda, *Limits of Rauzy digraphs*, in preparation.

[KV83] V. A. Kaĭmanovich and A. M. Vershik, *Random walks on discrete groups: boundary and entropy*, Ann. Probab. **11** (1983), no. 3, 457–490. MR704539

[Lag85] Jeffrey C. Lagarias, *The $3x+1$ problem and its generalizations*, Amer. Math. Monthly **92** (1985), no. 1, 3–23, DOI 10.2307/2322189. MR777565

[Lag10] Jeffrey C. Lagarias, *The $3x+1$ problem: an annotated bibliography (1963–1999)*, The ultimate challenge: the $3x+1$ problem, Amer. Math. Soc., Providence, RI, 2010, pp. 267–341. MR2560718

[LdW13] Thijs Laarhoven and Benne de Weger, *The Collatz conjecture and de Bruijn graphs*, Indag. Math. (N.S.) **24** (2013), no. 4, 971–983, DOI 10.1016/j.indag.2013.03.003. MR3124811

[Lee16] Paul-Henry Leemann, *On subgroups and Schreier graphs of finitely generated groups*, Thèse de doctorat, Genève, 2016.

[LM95] Douglas Lind and Brian Marcus, *An introduction to symbolic dynamics and coding*, Cambridge University Press, Cambridge, 1995. MR1369092

[LNW08] Franz Lehner, Markus Neuhauser, and Wolfgang Woess, *On the spectrum of lamplighter groups and percolation clusters*, Math. Ann. **342** (2008), no. 1, 69–89, DOI 10.1007/s00208-008-0222-7. MR2415315

[LP14] Yu Lin and Pavel A. Pevzner, *Manifold de Bruijn graphs*, Algorithms in bioinformatics, Lecture Notes in Comput. Sci., vol. 8701, Springer, Heidelberg, 2014, pp. 296–310, DOI 10.1007/978-3-662-44753-6_22. MR3254497

[LPP96] Russell Lyons, Robin Pemantle, and Yuval Peres, *Random walks on the lamplighter group*, Ann. Probab. **24** (1996), no. 4, 1993–2006, DOI 10.1214/aop/1041903214. MR1415237

[LZL17] Shangwei Lin, Chanchan Zhou, and Chunfang Li, *Arc fault tolerance of Kautz digraphs*, Theoret. Comput. Sci. **687** (2017), 1–10, DOI 10.1016/j.tcs.2017.04.011. MR3659427

[Mor03] Eduardo Moreno, *Lyndon words and de Bruijn sequences in a subshift of finite type*, Proceedings of WORDS'03, TUCS Gen. Publ., vol. 27, Turku Cent. Comput. Sci., Turku, 2003, pp. 400–410. MR2081372

[Mor05] Eduardo Moreno, *De Bruijn sequences and de Bruijn graphs for a general language*, Inform. Process. Lett. **96** (2005), no. 6, 214–219, DOI 10.1016/j.ipl.2005.05.028. MR2179213

[Pop59] Karl R. Popper, *The logic of scientific discovery*, Hutchinson and Co., Ltd., London, 1959. MR0107593

[PR15] Dominique Perrin and Antonio Restivo, *Words*, Handbook of enumerative combinatorics, Discrete Math. Appl. (Boca Raton), CRC Press, Boca Raton, FL, 2015, pp. 485–539. MR3409349

[PS97] Karl Petersen and Klaus Schmidt, *Symmetric Gibbs measures*, Trans. Amer. Math. Soc. **349** (1997), no. 7, 2775–2811, DOI 10.1090/S0002-9947-97-01934-X. MR1422906

[Rau83] Gérard Rauzy, *Suites à termes dans un alphabet fini* (French), Seminar on number theory, 1982–1983 (Talence, 1982/1983), Univ. Bordeaux I, Talence, 1983, pp. Exp. No. 25, 16. MR750326

[Rig14] Michel Rigo, *Formal languages, automata and numeration systems. 2*, Networks and Telecommunications Series, ISTE, London; John Wiley & Sons, Inc., Hoboken, NJ, 2014. Applications to recognizability and decidability; With a foreword by Valérie Berthé. MR3526118

[RNT12] Junghun Ryu, Eric Noel, and K. Wendy Tang, *Distributed and fault-tolerant routing for Borel Cayley graphs*, International Journal of Distributed Sensor Networks **8** (2012), no. 10, 124245.

[SWW16] Joe Sawada, Aaron Williams, and Dennis Wong, *Generalizing the classic greedy and necklace constructions of de Bruijn sequences and universal cycles*, Electron. J. Combin. **23** (2016), no. 1, Paper 1.24, 20. MR3484729

[SWW17] Joe Sawada, Aaron Williams, and Dennis Wong, *A simple shift rule for k-ary de Bruijn sequences*, Discrete Math. **340** (2017), no. 3, 524–531, DOI 10.1016/j.disc.2016.09.008. MR3584839

[Ter76] Riho Terras, *A stopping time problem on the positive integers*, Acta Arith. **30** (1976), no. 3, 241–252, DOI 10.4064/aa-30-3-241-252. MR0568274

[Woe13] Wolfgang Woess, *What is a horocyclic product, and how is it related to lamplighters?*, Internat. Math. Nachrichten **67** (2013), no. 224, 1–27.

Department of Mathematics and Statistics, University of Ottawa, 585 King Edward, Ottawa, Ontario K1N 6N5, Canada

Email address: vkaimano@uottawa.ca, vadim.kaimanovich@gmail.com

Contemporary Mathematics
Volume **719**, 2018
https://doi.org/10.1090/conm/719/14473

All properly ergodic Markov chains over a free group are orbit equivalent

Lewis Bowen

Abstract. Previous work showed that all Bernoulli shifts over a free group are orbit-equivalent. This result is strengthened here by replacing Bernoulli shifts with the wider class of properly ergodic countable state Markov chains over a free group. A list of related open problems is provided.

Contents

1. Introduction

Consider countable groups Γ, Λ, standard probability spaces $(X,\mu), (Y,\nu)$ and probability-measure-preserving (pmp) actions

$$\Gamma \curvearrowright (X,\mu), \quad \Lambda \curvearrowright (Y,\nu).$$

These actions are **orbit-equivalent** (OE) if there exists a measure-space isomorphism $\Phi : (X,\mu) \to (Y,\nu)$ such that $\Phi(\Gamma x) = \Lambda x$ for a.e. x (so Φ takes orbits to orbits). More generally, these actions are **stably orbit-equivalent** (SOE) if there exist positive measure sets $X' \subset X$, $Y' \subset Y$ and a measurable isomorphism $\Phi : X' \to Y'$ such that $\Phi_*(\mu \upharpoonright X')$ is a scalar multiple of $\nu \upharpoonright Y'$ and $\Phi(\Gamma x \cap X') = \Lambda\Phi(x) \cap Y'$ for a.e. x.

Dye proved that any two essentially free ergodic pmp actions of the integers are OE [**Dye59, Dye63**]. More generally, if Γ, Λ are countably infinite amenable groups, then any two essentially free ergodic pmp actions of Γ, Λ are OE by a theorem of Ornstein-Weiss [**OW80**] (see also [**CFW81**] for the non-singular case).

2010 *Mathematics Subject Classification.* Primary 37A20.
Key words and phrases. Tree-indexed Markov chains, orbit equivalence, Bernoulli shifts.
Supported in part by NSF grant DMS-1500389, NSF CAREER Award DMS-0954606.

On the other hand, when Γ is non-amenable, then Epstein showed that there exist uncountably many pairwise non-OE essentially free ergodic pmp actions of Γ [**Eps08, IKT09**]. This followed the work of many authors on various special cases (see [**Hjo05, GP05, Ioa11**] for example).

Here we are motivated by the problem of classifying a special class of actions, called Bernoulli shifts, up to OE. Given a standard probability space (K, κ), let K^Γ be the set of all functions $x : \Gamma \to K$. We denote such a function by $x = (x_g)_{g\in\Gamma} \in K^\Gamma$. Then Γ acts on K^Γ by shifting $(g \cdot x)_f := x_{fg}$. This action preserves the product measure κ^Γ. The system $\Gamma \curvearrowright (K, \kappa)^\Gamma$ is called the **Bernoulli shift over Γ with base space** (K, κ). Bernoulli shifts play a central role in the classification theory of measure-preserving actions [**Bow17**].

It is a consequence of Popa's cocycle-super-rigidity Theorems [**Pop08, Pop06b**] and Kida's OE-rigidity Theorems [**Kid10, Kid08**] together with sofic entropy theory [**Bow10a**] that there are many groups Γ with the property that if two Bernoulli shifts over Γ are OE then their base spaces have the same Shannon entropy. This is explained in more detail in [**Bow10a**]. For such groups there is a continuum of pairwise non-OE Bernoulli shifts.

On the other hand, free groups appear to be remarkably flexible. I showed in [**Bow11a**] that all Bernoulli shifts over a non-abelian free group $\mathbb{F}$ are OE. Moreover, Bernoulli shifts over non-abelian free groups of different finite rank are stably-orbit-equivalent [**Bow11b**] (see also [**MRV13**] for a nice exposition and further results). In [**Bow15**] I showed that the OE-class of any essentially free action of $\mathbb{F}$ is weakly dense in the space of actions.

The main result of this paper is:

THEOREM 1.1. *Let $\mathbb{F}$ be a non-abelian free group of finite rank. Then all properly ergodic countable-state Markov chains over $\mathbb{F}$ are OE. In particular, they are all OE to a Bernoulli shift over $\mathbb{F}$.*

REMARK 1. An action is **properly ergodic** if it is ergodic and there does not exist a co-null set on which the group acts transitively. Markov chains over free groups are carefully defined in §2.2.

1.1. Aside on randomly generated networks. Can unimodular random networks be classified? (see [**AL07**] for definitions) Of course, this depends on what one means by 'classify' but in any case anti-classification results of ergodic theory should give one pause. To make this a more reasonable endeavor, we specialize to those random networks whose underlying graphs are trees, especially those that arise naturally (Galton-Watson trees, uniform spanning trees, etc). The notion of classification considered here is stable-orbit-equivalence.

For further simplicity, let us say that a network is **asymmetric** if its automorphism group is trivial. Suppose μ_1, μ_2 are two unimodular measures on the space of rooted networks that are concentrated on asymmetric networks. We will say they are **stably orbit-equivalent** (OE) if there are measurably sets A_1, A_2 with $\mu_1(A_1) > 0, \mu_2(A_2) > 0$ and a measure-space isomorphism

$$\Phi : \left(A_1, \frac{\mu_1 \upharpoonright A_1}{\mu_1(A_1)}\right) \to \left(A_2, \frac{\mu_2 \upharpoonright A_2}{\mu_2(A_2)}\right)$$

such that if $\mathcal{N}_1, \mathcal{N}_2 \in A_1$ are isomorphic networks then $\Phi(\mathcal{N}_1), \Phi(\mathcal{N}_2)$ are also isomorphic networks (forgetting the roots). They are **orbit-equivalent** (OE) if A_1, A_2 can be chosen to be μ_1-conull, μ_2-conull respectively.

For example, suppose $\mathbb{F}$ is a non-abelian free group and $\mathbb{F}\curvearrowright(A^{\mathbb{F}},\mu)$ is a Markov process (as defined in §2.2). Let G be a Cayley tree of $\mathbb{F}$. Any $x \in A^{\mathbb{F}}$ determines a marking of the vertices of G. By sampling x from μ we obtain a unimodular random network (where the randomness is in the marking since the graph is deterministic).

This leads to a general question: given a unimodular random network whose underlying graph is a tree a.s., is it SOE to a Bernoulli process over a free group? In Question 1 of the next section, explicit examples where this problem is open, are given.

1.2. Questions and comments.

(1) Is there a nice characterization of the measured equivalence relations that are SOE to a Bernoulli shift over $\mathbb{F}$? Such a characterization should help determine which of the following actions are SOE to a Bernoulli shift over $\mathbb{F}$: unimodular Galton-Watson trees [**AL07**, Example 1.1], Bernoulli shifts over surface groups or other treeable groups, Poisson point processes in the hyperbolic plane conditioned on the origin being contained in the point process, actions of the form $\Gamma\curvearrowright \mathrm{Aut}(T_d)/\Lambda$ where Γ,Λ are lattices in $\mathrm{Aut}(T_d)$ and $\mathrm{Aut}(T_d)$ is the automorphism group of the $d>2$ regular tree, actions of $\mathbb{F}$ with completely positive Rokhlin entropy (for the definition of this see [**Bow17**]), factors of Bernoulli shifts, non-weakly-compact Gaussian actions of Bernoulli shifts, non-hyperfinite ergodic subequivalence relations of Bernoulli shifts over $\mathbb{F}$, inverse limits of Bernoulli shifts, direct products of arbitrary ergodic actions with Bernoulli shifts, Markovian planar stochastic hyperbolic infinite triangulations as in [**Cur16**], the free spanning forest FSF of the Cayley graph of a surface group [**BLPS01**], and the cluster relation of Bernoulli percolation in the non-uniqueness phase of non-amenable Cayley graph [**BS96**].

(2) Weak compactness of actions was defined in [**OP10**]. It is an SOE invariant and was shown in [**Bow17**] to imply zero Rokhlin entropy. Therefore no weakly compact ergodic action of any group can be SOE to a Bernoulli shift over $\mathbb{F}$. For example, if $\mathbb{F}$ is embedded densely into a compact group K then the translation action $\mathbb{F}\curvearrowright K$ is compact and therefore, weakly compact.

(3) Rigid actions were defined in [**Pop06a**] (see also [**Ioa10**] for an ergodic-theoretic formulation). Rigidity is also an SOE invariant and is not a property of Bernoulli shift actions. For example, the usual action of $SL(2,\mathbb{Z})$ on the 2-torus is rigid [**Ioa10**].

(4) There exist properly ergodic Markov chains whose Koopman representation is not contained in the countable sum of left-regular representations. Such Markov chains necessarily factor onto zero entropy actions [**Bow17**]. Since Bernoulli shifts do not have this property, this cannot be an OE invariant.

(5) Bernoulli actions are **solidly ergodic** in the sense that every subequivalence relation of the orbit relation decomposes into a hyperfinite piece and at most countably many strongly ergodic pieces [**CI10**]. Because this property is an OE invariant, Theorem 1.1 implies properly ergodic Markov chains over $\mathbb{F}$ are solidly ergodic.

(6) If $\mathbb{F}\curvearrowright(X,\mu)$ is solidly ergodic, essentially free, not weakly compact and not rigid then is it OE to a Bernoulli shift?

1.3. Remarks on the proof. There are two main parts: first we show that any properly ergodic Markov chain is OE to a Markov chain that is "generator-ergodic" in the sense that its symbolic restriction to any generator subgroup is ergodic and essentially free (Proposition 5.1). Second we show that every generator-ergodic Markov chain is OE to a Bernoulli shift.

The first step is by explicit construction, involving some "edge-sliding" arguments. In fact, the orbit-equivalences are continuous. The second step uses Dye's Theorem for actions of $\mathbb{Z}$ (as a black box) and so is considerably less constructive.

Acknowledgements. Thanks to Brandon Seward, Robin Tucker-Drob and Peter Burton for discussing this problem with me. The picture greatly clarified from these discussions.

2. Preliminaries

2.1. Notation. Let S be a finite set and $\mathbb{F} = \langle S \rangle$ be the free group generated by S. Let A be a finite or countable set called the **alphabet**. Then $A^{\mathbb{F}}$ is the set of all functions from $\mathbb{F}$ to A. We denote such a function by $x = (x_g)_{g \in \mathbb{F}}$. Let $\mathbb{F}$ act on $A^{\mathbb{F}}$ by

$$(g \cdot x)_f = x_{fg}.$$

This is called the **shift action** and is denoted by $\mathbb{F} \curvearrowright A^{\mathbb{F}}$. Let $\mathrm{Prob}_{\mathbb{F}}(A^{\mathbb{F}})$ denote the set of all shift-invariant Borel probability measures on $A^{\mathbb{F}}$.

Similarly, define

$$T : A^{\mathbb{Z}} \to A^{\mathbb{Z}}, \quad (Tx)_n = x_{n+1}$$

and let $\mathrm{Prob}_{\mathbb{Z}}(A^{\mathbb{Z}})$ denote the set of all T-invariant Borel probability measures on $A^{\mathbb{Z}}$.

If (X, μ) is a measure space, Y is a Borel space and $\phi : X \to Y$ is measurable then the **pushforward measure** $\phi_*\mu$ on Y is defined by $\phi_*\mu(E) = \mu(\phi^{-1}(E))$ for measurable $E \subset Y$.

For $g \in \mathbb{F}$, let $\pi_g(x) = x_g$. Similarly, for $n \in \mathbb{Z}$ and $x \in A^{\mathbb{Z}}$, let $\pi_n(x) = x_n$. We will frequently abuse notation by writing π for either π_e or π_0 (depending on whether the argument is in $A^{\mathbb{F}}$ or $A^{\mathbb{Z}}$).

2.2. Markov Chains. Here we define Markov chains over free groups. Let $|\cdot|$ denote the word length on $\mathbb{F}$. So for $g \in \mathbb{F}$, $|g|$ is the smallest integer n such that g is a product of n elements in $S \cup S^{-1}$.

DEFINITION 1. For $s \in S \cup S^{-1}$, let

$$\mathrm{past}(s) = \{g \in \mathbb{F} : \ |gs^{-1}| = |g| - 1\}.$$

So $\mathrm{past}(s)$ consists of all reduced words that end in s. Note that $\mathbb{F}$ is the disjoint union of $\{e\}$ and $\mathrm{past}(s)$ for $s \in S \cup S^{-1}$.

DEFINITION 2. Let $\mu \in \mathrm{Prob}_{\mathbb{F}}(A^{\mathbb{F}})$ be a shift-invariant measure. For $s \in S$, μ is s-**Markov** if the following is true. Let $\mathcal{F}_s$ be the sigma-algebra of Borel subsets of $A^{\mathbb{F}}$ generated by the functions

$$x \mapsto x_g$$

for $g \in \mathrm{past}(s)$. Also let $\hat{\mathcal{F}}_s$ be the sigma-algebra of Borel subsets of $A^{\mathbb{F}}$ generated by the functions

$$x \mapsto x_g$$

for $g \notin \text{past}(s)$. Recall that $\pi : A^{\mathbb{F}} \to A$ is the time 0 map $\pi(x) = x_e$. Then μ is **Markov** if $\mathfrak{F}_s$ is independent of $\hat{\mathfrak{F}}_s$ conditioned on π with respect to μ. Equivalently, if for every $E \subset \mathfrak{F}_s, \widehat{E} \subset \hat{\mathfrak{F}}_s$ and $a \in A$,

$$\mu(E \cap \widehat{E} | \pi = a) = \mu(E | \pi = a)\mu(\widehat{E} | \pi = a).$$

We say that μ is **Markov with respect to** S if it is s-Markov for every $s \in S$. Usually we will simply say that μ is Markov if S is understood. For example, Bernoulli shifts of the form $\mathbb{F} \curvearrowright (K, \kappa)^{\mathbb{F}}$ in which K is a countable or finite set are Markov.

REMARK 2. The entropy theory of Markov chains over free groups is studied in [**Bow10b**].

2.3. Aside on tree-indexed Markov chains. Under an additional symmetry condition and conditioning on the root, the Markov chains studied here are tree-indexed Markov chains in the sense of [**BP94a**, **BP94b**]. The tree under consideration is the Cayley graph of $\mathbb{F}$. More precisely, it is the left Cayley graph with vertex set $\mathbb{F}$ and edges (g, sg) for $g \in \mathbb{F}, s \in S$ (where $\mathbb{F} = \langle S \rangle$). We can consider the identity element to be the root vertex.

Now suppose $\mu \in \text{Prob}_{\mathbb{F}}(A^{\mathbb{F}})$ is Markov with respect to S. For each $g \in \mathbb{F}$, let $X_g : A^{\mathbb{F}} \to A$ be the coordinate projection, regarded as a random variable (with respect to μ). The Markov measure μ is **symmetric** if the transition probabilities $P(X_g = a | X_{sg} = b)$ do not depend on $s \in S \cup S^{-1}$ (they never depend on g because μ is translation-invariant). Now suppose that we condition on the value of the root variable X_e and assume that μ is symmetric. Then the family of conditioned random variables $(X_g | X_e)_{g \in \mathbb{F}}$ is a tree-indexed Markov process in the sense of [**BP94a**, **BP94b**].

3. General results regarding Markov chains

This section establishes some general results on Markov chains. It also establishes the very useful Lemma 3.3 showing that if an action is Markov with respect to $|S| - 1$ generators, and the restriction to the last generator is Markov, then the action itself is Markov. This will be used in both parts of the proof of the main theorem.

3.1. Cylinder sets. To begin, we obtain a formula for Markov measures of cylinder sets.

DEFINITION 3. The **left-Cayley graph** of $\mathbb{F}$ has vertex set $\mathbb{F}$ and edge set (g, sg) for $g \in \mathbb{F}, s \in S$. Because of the way we define the action $\mathbb{F} \curvearrowright A^{\mathbb{F}}$, the left-Cayley graph is more relevant to our concerns than the more usual right-Cayley graph. A subset $W \subset \mathbb{F}$ is **left-connected** if its induced subgraph is connected, equivalently if for every $w_1, w_2 \in W$ there exist $s_1, \ldots, s_n \in S \cup S^{-1}$ such that $w_2 = s_n \cdots s_1 w_1$ and $s_i \cdots s_1 w_1 \in W$ for all $1 \le i \le n$.

DEFINITION 4. Given $D \subset \mathbb{F}$ and $\phi : D \to A$, let

$$\text{Cyl}(\phi) = \{x \in A^{\mathbb{F}} : \ x_g = \phi(g) \ \forall g \in D\}$$

be the **cylinder set** of ϕ. A similar definition applies to $\mathbb{Z}$ in place of $\mathbb{F}$.

LEMMA 3.1. *Let $\mu \in \mathrm{Prob}_{\mathbb{F}}(A^{\mathbb{F}})$ be a shift-invariant measure. Then μ is Markov if and only if for every left-connected finite set $D \subset \mathbb{F}$ such that $e \in D$ and every $\phi : D \to A$,*

$$\mu(\mathrm{Cyl}(\phi)) = \mu(x_e = \phi(e)) \prod_{g \in D \setminus \{e\}} \mu\Big(x_g = \phi(g) | x_{\sigma(g)} = \phi(\sigma(g))\Big)$$

where $\sigma(g) \in \mathbb{F}$ is the unique element satisfying $|\sigma(g)| = |g| - 1$ and $g\sigma(g)^{-1} \in S \cup S^{-1}$. In other words, $\sigma(g)$ is on the unique path from g to e and $|\sigma(g)| = |g| - 1$.

PROOF. If μ satisfies the condition above then it is clearly Markov. So suppose that μ is Markov. We prove the formula above by induction on $|D|$. If $|D| = 1$ then the statement is trivial. So suppose $|D| > 1$. Then there exists $g \in D$ such that $D' := D \setminus \{g\}$ is left-connected.

We will reduce to the special case in which $g \in S \cup S^{-1}$. To do this, define $h \in \mathbb{F}$ by: if $g = e$ then let $h \in (S \cup S^{-1}) \cap W$. Otherwise, set $h = \sigma(g)$. Because μ is shift-invariant,

$$\mu(h\,\mathrm{Cyl}(\phi)) = \mu(\mathrm{Cyl}(\phi)).$$

However,

$$h\,\mathrm{Cyl}(\phi) = \mathrm{Cyl}(\psi)$$

where $\psi : Dh^{-1} \to A$ is defined by $\psi(dh^{-1}) = \phi(d)$. By choice of h, note that $gh^{-1} \in S \cup S^{-1}$ and therefore $\sigma(gh^{-1}) = e$. After replacing ϕ with ψ and g with gh^{-1}, we see that it suffices to prove the claim when $g \in S \cup S^{-1}$ which we now assume.

Let $D' = D \setminus \{g\}$ and ϕ' be the restriction of ϕ to D'. By induction,

$$\begin{aligned}\mu(\mathrm{Cyl}(\phi)) &= \mu(\mathrm{Cyl}(\phi'))\mu(x_g = \phi(g) | x \in \mathrm{Cyl}(\phi')) \\ &= \mu(x_e = \phi(e)) \prod_{f \in D \setminus \{e,g\}} \mu\Big(x_f = \phi(f) | x_{\sigma(f)} = \phi(\sigma(f))\Big)\mu(x_g = \phi(g) | x \in \mathrm{Cyl}(\phi')).\end{aligned}$$

Because $D' \setminus \{e\} \subset \mathbb{F} \setminus \mathrm{past}(g)$, the Markov property implies

$$\mu(x_g = \phi(g) | x \in \mathrm{Cyl}(\phi')) = \mu(x_g = \phi(g) | x_e = \phi(e)) = \mu\Big(x_g = \phi(g) | x_{\sigma(g)} = \phi(\sigma(g))\Big).$$

Combined with the previous equation, this completes the induction step. □

3.2. Ergodicity and freeness. In this subsection, we establish criteria for ergodicity and essential freeness of Markov chains.

DEFINITION 5. Let $\mu \in \mathrm{Prob}_{\mathbb{F}}(A^{\mathbb{F}})$ be a shift-invariant measure. For $s \in S$, define the **symbolic restriction map**

$$R^s : A^{\mathbb{F}} \to A^{\mathbb{Z}} \quad R^s(x)_n = x_{s^n}$$

and $\mu_s \in \mathrm{Prob}_{\mathbb{Z}}(A^{\mathbb{Z}})$ by $\mu_s = R^s_*\mu$. The measure μ_s is called the **symbolic restriction of μ to the subgroup generated by s**.

DEFINITION 6. Given $\mu \in \mathrm{Prob}_{\mathbb{F}}(A^{\mathbb{F}})$ and $s \in S \cup S^{-1}$, let E^μ_s be the set of all $(a, b) \in A \times A$ such that

$$\mu_s(\{x \in A^{\mathbb{Z}} : (x_0, x_1) = (a, b)\}) > 0.$$

Let $E^\mu = \cup_{s \in S \cup S^{-1}} E^\mu_s$. Let $\mathcal{R}^\mu_s, \mathcal{R}^\mu \subset A \times A$ be the equivalence relation generated by E^μ_s, E^μ respectively. A $\mathcal{R}^\mu_s$-class $A' \subset A$ is called **periodic** if every $a' \in A'$ has in-degree and out-degree 1 in the directed graph (A, E^μ_s). Otherwise A' is called **aperiodic**.

LEMMA 3.2. *Suppose* $\mu \in \mathrm{Prob}_{\mathbb{F}}(A^{\mathbb{F}})$ *is Markov and* $\pi_*\mu$ *has full support on* A. *For any* $s \in S$, μ_s *is ergodic if and only if* $\mathcal{R}^{\mu}_s = A \times A$. *Also* μ_s *is essentially free if and only if every* $\mathcal{R}^{\mu}_s$*-class is aperiodic. Similarly,* μ *is ergodic if and only if* $\mathcal{R}^{\mu} = A \times A$. *If* μ *is ergodic then* μ *is properly ergodic if and only if there exists some* $s \in S$ *such that some* $\mathcal{R}^{\mu}_s$*-class is aperiodic.*

PROOF. The first statement is well-known. It follows, for example, from [**Dur96**, Section 6.1, page 338 in the 2nd edition]. The second statement is a trivial exercise. The last two statements are similar. □

3.3. A sufficient condition for a measure to be Markov.

LEMMA 3.3. *For each* $s \in S$, *let* $\nu_s \in \mathrm{Prob}_{\mathbb{Z}}(A^{\mathbb{Z}})$ *be a shift-invariant measure. Let* $t \in S$. *Suppose that* ν_s *is Markov for every* $s \in S \setminus \{t\}$ *and* $\pi_*\nu_s = \pi_*\nu_r$ *for every* $r, s \in S$ *where* $\pi : A^{\mathbb{F}} \to A$ *is the time 0 map* $\pi(x) = x_e$.

Then there exists a unique shift-invariant measure $\rho \in \mathrm{Prob}_{\mathbb{F}}(A^{\mathbb{F}})$ *such that*

- $\rho_s = \nu_s$ *for all* $s \in S$,
- ρ *is* s*-Markov for every* $s \in S \setminus \{t\}$.

Moreover, if ν_t *is also Markov then* ρ *is Markov.*

PROOF. Let $D \subset \mathbb{F}$ be finite, left-connected and satisfying $e \in D$. Let $\phi : D \to A$. For $g \in \mathbb{F} \setminus \{e\}$, define $\sigma(g) \in \mathbb{F}$ is as in Lemma 3.1. Set

$$\widehat{D} := \{g \in D \setminus \{e\} : \ g\sigma(g)^{-1} \notin \{t, t^{-1}\}\}.$$

For $g \in D$, let

$$D_g = \{i \in \mathbb{Z} : \ t^i g \in D\}, \quad \phi_g : D_g \to A, \ \phi_g(i) = \phi(t^i g).$$

If ρ exists then since it is s-Markov for every $s \in S \setminus \{t\}$,

$$\begin{aligned}&\rho(\mathrm{Cyl}(\phi))\\ &= \nu_t\big(\mathrm{Cyl}(\phi_e)\big) \prod_{g \in \widehat{D}} \nu_{g\sigma(g)^{-1}}\Big(y_1 = \phi(g) | x_0 = \phi\big(\sigma(g)\big)\Big) \nu_t\big(\mathrm{Cyl}(\phi_g) | x_0 = \phi(g)\big).\end{aligned}$$

This proves uniqueness. It also implies existence because we can define ρ by the above equation since it satisfies the hypotheses of the Kolmogorov Extension Theorem. If ν_t is Markov, then Lemma 3.1 implies ρ is Markov. □

4. General constructions of orbit-equivalences

To prove the main theorem we will construct orbit-equivalences by first constructing alternative actions of $\mathbb{F}$ on $A^{\mathbb{F}}$ with the same orbits as the usual action. General facts regarding this construction are presented here.

LEMMA 4.1. *Suppose* $\tau : (S \cup S^{-1}) \times A^{\mathbb{F}} \to \mathbb{F}$ *is a function satisfying*

$$\tau(s^{-1}, \tau(s,x) \cdot x) = \tau(s,x)^{-1} \quad \forall s \in S \cup S^{-1}, x \in A^{\mathbb{F}}.$$

There there exists an action $* : \mathbb{F} \times A^{\mathbb{F}} \to A^{\mathbb{F}}$ *of* $\mathbb{F}$ *satisfying*

$$s * x = \tau(s) \cdot x \quad \forall s \in S \cup S^{-1}, x \in A^{\mathbb{F}}$$

and a function $\omega : \mathbb{F} \times A^{\mathbb{F}} \to \mathbb{F}$ *extending* τ *and satisfying the cocycle equation*

$$\omega(gh, x) = \omega(g, h * x)\omega(h, x).$$

Moreover, $g * x = \omega(g,x)\cdot x$. *Also, if* $\Omega : A^{\mathbb{F}} \to A^{\mathbb{F}}$ *is defined by*

$$\Omega(x)_h = x_{\omega(h,x)}$$

then Ω *is* $(*,\cdot)$*-equivariant in the sense that*

$$g\cdot(\Omega x) = \Omega(g*x).$$

PROOF. The existence of $*$ is immediate since $\mathbb{F}$ is freely generated as a semi-group by $S\cup S^{-1}$ and

$$s^{-1}*(s*x) = \tau(s^{-1},\tau(s,x)\cdot x)\cdot(\tau(s,x)\cdot x) = x.$$

Similarly the existence of ω is immediate and the equation $g*x=\omega(g,x)\cdot x$ follows from the cocycle equation by inducting on $|g|$.

To see that Ω is $(*,\cdot)$-equivariant, let $g,h\in\mathbb{F}$. Then

$$\begin{aligned}\Omega(g*x)_h &= (g*x)_{\omega(h,g*x)} = (\omega(g,x)\cdot x)_{\omega(h,g*x)} = \\ &= x_{\omega(h,g*x)\omega(g,x)} = x_{\omega(hg,x)} = (\Omega x)_{hg} = (g\cdot\Omega x)_h.\end{aligned}$$

□

LEMMA 4.2. *Let* $*,\omega,\Omega$ *be as in Lemma 4.1. Also let* $s\in S$. *If for every* $x\in A^{\mathbb{F}}$

$$g\in \operatorname{past}(s) \Leftrightarrow \omega(g,x)\in\operatorname{past}(s)$$

and $\mu\in \operatorname{Prob}_{\mathbb{F}}(A^{\mathbb{F}})$ *is* s*-Markov then* $\Omega_*\mu$ *is* s*-Markov.*

PROOF. Define $\mathcal{F}_s, \widehat{\mathcal{F}}_s$ as in Definition 2.

Claim 1. $\Omega^{-1}(\mathcal{F}_s)\subset\mathcal{F}_s$.

PROOF. For $g\in\mathbb{F}$, let $\pi_g : A^{\mathbb{F}}\to A$ be the coordinate function $\pi_g(x)=x_g$. Because $\mathcal{F}_s$ is generated by sets of the form $\pi_g^{-1}(B)$ for $B\subset A$ and $g\in\operatorname{past}(s)$, $\Omega^{-1}(\mathcal{F}_s)$ is generated by sets of the form $\Omega^{-1}\pi_g^{-1}(B) = (\pi_g\circ\Omega)^{-1}(B)$. So it suffices to show that if $g\in\operatorname{past}(s)$ then $\pi_g\circ\Omega$ is $\mathcal{F}_s$-measurable. Since $\pi_g\Omega x = (\Omega x)_g = x_{\omega(g,x)}$ and $\omega(g,x)\in\operatorname{past}(s)$, it follows that $\pi_g\Omega$ is $\mathcal{F}_s$-measurable. □

A similar argument shows that $\Omega^{-1}(\widehat{\mathcal{F}}_s)\subset\widehat{\mathcal{F}}_s$. To prove $\Omega_*\mu$ is s-Markov, let $E_1\in\mathcal{F}_s$ and $E_2\in\widehat{\mathcal{F}}_s$. Claim 1 implies $\Omega^{-1}(E_1)\in\mathcal{F}_s$ and $\Omega^{-1}(E_2)\in\widehat{\mathcal{F}}_s$. Since μ is s-Markov and $\pi_e\Omega=\pi_e$, for any $a\in A$,

$$\begin{aligned}\Omega_*\mu(E_1\cap E_2|\pi_e=a) &= \mu(\Omega^{-1}(E_1)\cap\Omega^{-1}(E_2)|\pi_e=a)\\ &= \mu(\Omega^{-1}(E_1)|\pi_e=a)\mu(\Omega^{-1}(E_2)|\pi_e=a)\\ &= \Omega_*\mu(E_1|\pi_e=a)\Omega_*\mu(E_2|\pi_e=a).\end{aligned}$$

Since E_1,E_2,a are arbitrary this shows $\Omega_*\mu$ is s-Markov. □

LEMMA 4.3. *Let* $*,\omega,\Omega$ *be as in Lemma 4.1. Suppose there is another function* $\hat\tau : (S\cup S^{-1})\times A^{\mathbb{F}}\to\mathbb{F}$ *satisfying*

$$\hat\tau(s^{-1},\hat\tau(s,x)\cdot x) = \hat\tau(s,x)^{-1}$$

for every $s\in S\cup S^{-1}$. *Let* $\star : \mathbb{F}\times A^{\mathbb{F}}\to\mathbb{F}$, $\widehat\omega : \mathbb{F}\times A^{\mathbb{F}}\to\mathbb{F}, \hat\Omega : A^{\mathbb{F}}\to A^{\mathbb{F}}$ *denote the associated action, cocycle and map as in Lemma 4.1. Suppose as well that* $\omega(\widehat\omega(s,\Omega x),x)=s$ *for every* $s\in S\cup S^{-1}$. *Then*

$$\omega(\widehat\omega(g,\Omega x),x) = g\ \forall g\in\mathbb{F} \tag{1}$$

and $\hat\Omega\Omega x = x$ *for all* x.

PROOF. To prove the first claim, it suffices to prove: if $g_1, g_2 \in \mathbb{F}$ satisfy (1) for all $x \in A^{\mathbb{F}}$ then the product $g_1 g_2$ also satisfies (1). This follows from the cocycle equations

$$\begin{aligned} \omega(\widehat{\omega}(g_1 g_2, \Omega x), x) &= \omega(\widehat{\omega}(g_1, g_2 \star \Omega x)\widehat{\omega}(g_2, \Omega x), x) \\ &= \omega(\widehat{\omega}(g_1, g_2 \star \Omega x), \widehat{\omega}(g_2, \Omega x) * x)\omega(\widehat{\omega}(g_2, \Omega x), x) \end{aligned}$$

Since $\omega(\widehat{\omega}(g_2, \Omega x), x) = g_2$ it suffices to show $\omega(\widehat{\omega}(g_1, g_2 \star \Omega x), \widehat{\omega}(g_2, \Omega x) * x) = g_1$. This will follow from the assumption that g_1 satisfies (1) once we show that

$$g_2 \star \Omega x = \Omega(\widehat{\omega}(g_2, \Omega x) * x).$$

This follows from Lemma 4.1 and since Ω is $(*, \cdot)$-equivariant:

$$g_2 \star \Omega x = \widehat{\omega}(g_2, \Omega x) \cdot (\Omega x) = \Omega(\widehat{\omega}(g_2, \Omega x) * x).$$

The proves the first claim. To prove the last, let $x \in A^{\mathbb{F}}$ and $h \in \mathbb{F}$. Then

$$(\hat{\Omega}\Omega x)_h = (\Omega x)_{\widehat{\omega}(h, \Omega x)} = x_{\omega(\widehat{\omega}(h, \Omega x), x)} = x_h.$$

□

5. From properly ergodic to generator-ergodic

The main result of this section is:

PROPOSITION 5.1. *Let $\mu \in \mathrm{Prob}_{\mathbb{F}}(A^{\mathbb{F}})$ be properly ergodic and Markov. Then there exists a countable set B and a shift-invariant measure $\rho \in \mathrm{Prob}_{\mathbb{F}}(B^{\mathbb{F}})$ such that $\mathbb{F} \curvearrowright (A^{\mathbb{F}}, \mu)$ is OE to $\mathbb{F} \curvearrowright (B^{\mathbb{F}}, \rho)$, ρ is Markov and for every $s \in S$, ρ_s is essentially free and ergodic.*

To prove this result, we will construct a very specific kind of orbit-equivalence which we then apply multiple times with slightly varying hypotheses. The orbit-equivalence we build depends on a choice of a subset $\mathcal{E} \subset E^{\mu}_u$ (where $u \in S$ and E^{μ}_u is as in Definition 6) satisfying some technical conditions described next.

DEFINITION 7. Let $\mu \in \mathrm{Prob}_{\mathbb{F}}(A^{\mathbb{F}})$ and $s \in S$. A subset $\mathcal{E} \subset E^{\mu}_s$ is said to be μ_s-**special** if

(1) for every $a \in A$ there does not exist $b, c \in A$ such that both $(a, b) \in \mathcal{E}$ and $(c, a) \in \mathcal{E}$, and
(2) if $(a, b) \in \mathcal{E}$ then the $\mathcal{R}^{\mu}_s$ classes of a and b are aperiodic.

The next result is the key lemma towards proving Proposition 5.1. We will apply it multiple times to obtain Proposition 5.1. The reader who is only interested in the special case in which the alphabet A is finite can assume that $\mathcal{E} = \{(a, b)\}$ is a singleton.

LEMMA 5.2. *Let $\mu \in \mathrm{Prob}_{\mathbb{F}}(A^{\mathbb{F}})$ be Markov, properly ergodic such that $\pi_* \mu \in \mathrm{Prob}(A)$ is fully supported. Let $u, t \in S$ be distinct and let $\mathcal{E}$ be μ_u-special. Then there exists a Markov measure $\rho \in \mathrm{Prob}_{\mathbb{F}}(A^{\mathbb{F}})$ such that*

- *$\mathbb{F} \curvearrowright (A^{\mathbb{F}}, \mu)$ is OE to $\mathbb{F} \curvearrowright (A^{\mathbb{F}}, \rho)$,*
- *$\mu_s = \rho_s$ for all $s \in S \setminus \{t\}$,*
- *$\mathcal{R}^{\rho}_t \supset \mathcal{E} \cup \mathcal{R}^{\mu}_t$,*
- *for every $(a, b) \in \mathcal{E}$, the $\mathcal{R}^{\rho}_t$-classes of a and b are aperiodic.*

Proof of Lemma 5.2. Let $(a,b) \in \mathcal{E}$. Because the $\mathcal{R}_u^\mu$-class of b is aperiodic, there exist a smallest number $n = n(b) > 0$, elements $b_0, \ldots, b_n \in A$ and $\eta(b) \in A$ such that

- $b = b_0$,
- $(b_i, b_{i+1}) \in E_u^\mu$ for all $0 \leq i < n$
- $b_n \neq \eta(b)$ and $(b_{n-1}, \eta(b)) \in E_u^\mu$.

Choose a function η satisfing the above. Let

$$\mathcal{F} = \big\{(a, b, \eta(b)) : \ (a,b) \in \mathcal{E}\big\}.$$

Let

$$F(x) := (x_{u^{-1}}, x_e, x_{u^{n(x_e)}})$$

whenever $(x_{u^{-1}}, x_e) \in \mathcal{E}$.

Define $\tau : (S \cup S^{-1}) \times A^{\mathbb{F}} \to \mathbb{F}$ by

$$\tau(s,x) = s \quad \forall s \in S \cup S^{-1} \setminus \{t, t^{-1}\},$$

$$\tau(t,x) = \begin{cases} ut & \text{if } F(ut \cdot x) \in \mathcal{F} \\ u^{-1}t & \text{if } F(t \cdot x) \in \mathcal{F} \\ t & \text{otherwise} \end{cases}$$

$$\tau(t^{-1},x) = \begin{cases} (ut)^{-1} & \text{if } F(x) \in \mathcal{F} \\ (u^{-1}t)^{-1} & \text{if } F(u \cdot x) \in \mathcal{F} \\ t^{-1} & \text{otherwise} \end{cases}$$

For example, $F(ut \cdot x) \in \mathcal{F}$ means that $F(ut \cdot x)$ is well-defined and $F(ut \cdot x) \in \mathcal{F}$.

Because $\mathcal{E}$ is μ_u-special, τ is well-defined (for example, it cannot be that $F(ut \cdot x) \in \mathcal{F}$ and $F(t \cdot x) \in \mathcal{F}$). Also τ satisfies the hypotheses of Lemma 4.1. Let $*, \omega, \Omega$ be as in Lemma 4.1.

We will show that $\mathbb{F} \curvearrowright^* (A^{\mathbb{F}}, \mu)$ has the same orbits as $\mathbb{F} \curvearrowright (A^{\mathbb{F}}, \mu)$ (modulo μ null sets) and $\mathbb{F} \curvearrowright^* (A^{\mathbb{F}}, \mu)$ is measurably-conjugate to an action of the form $\mathbb{F} \curvearrowright (A^{\mathbb{F}}, \rho)$ where ρ satisfies the conclusion.

Claim 1. For μ-a.e. x, $\mathbb{F} \cdot x = \mathbb{F} * x$.

Proof. It is immediate that $\mathbb{F} \cdot x \supset \mathbb{F} * x$ for a.e. x. To show the opposite inclusion, it suffices to show that for a.e. $x \in A^{\mathbb{F}}$ and every $g \in \mathbb{F}$ there exists $h \in \mathbb{F}$ such that $\omega(h,x) = g$. Because of the cocycle equation, it suffices to prove this for $g \in S \cup S^{-1}$. The special case of $g \in S \cup S^{-1} \setminus \{t, t^{-1}\}$ is clear since in that case $\omega(g,x) = g$.

We claim:

$$\begin{aligned} F(ut \cdot x) \in \mathcal{F} \quad &\Rightarrow \quad \omega(u^{-1}t, x) = t \\ F(t \cdot x) \in \mathcal{F} \quad &\Rightarrow \quad \omega(ut, x) = t \\ F(ut \cdot x) \notin \mathcal{F} \wedge F(t \cdot x) \notin \mathcal{F} \quad &\Rightarrow \quad \omega(t,x) = t. \end{aligned}$$

To see the first equation, assume $F(ut \cdot x) \in \mathcal{F}$. By the cocycle equation

$$\omega(ut, x) = \omega(u^{-1}, t * x)\omega(t,x) = u^{-1}(ut) = t.$$

The other cases are similar. This shows for a.e. $x \in A^{\mathbb{F}}$, there exists h such that $\omega(h,x) = t$. The statement with t^{-1} in place of t is similar. □

Claim 2. For any $x \in A^{\mathbb{F}}$ and $g \in \mathbb{F}$,

$$\omega(\omega(g, \Omega x), x) = g$$

and $\Omega(\Omega x) = x$.

PROOF. The first claim is immediate for $g \in S \cup S^{-1} \setminus \{t, t^{-1}\}$. To handle the case $g = t$, suppose that $F(ut \cdot x) \in \mathcal{F}$. Then $\omega(t, x) = ut$ and for any $m \in \mathbb{Z}$,

$$\omega(u^m t, x) = \omega(u^m, t * x)\omega(t, x) = u^m(ut) = u^{m+1}t.$$

So

$$(t \cdot \Omega x)_{u^m} = (\Omega x)_{u^m t} = x_{\omega(u^m t, x)} = x_{u^{m+1}t}.$$

Let $n = n(x_{ut}) = n((t \cdot \Omega x)_e)$. Then

$$F(t \cdot \Omega x) = ((t \cdot \Omega x)_{u^{-1}}, (t \cdot \Omega x)_e, (t \cdot \Omega x)_{u^n}) = (x_t, x_{ut}, x_{u^{n+1}t}) = F(ut \cdot x) \in \mathcal{F}.$$

So $\omega(t, \Omega x) = u^{-1}t$ and

$$\omega(\omega(t, \Omega x), x) = \omega(u^{-1}t, x) = t.$$

The other cases are similar. Claim 2 now follows from Lemma 4.3. □

It now suffices to prove that if $\rho := \Omega_* \mu$ then ρ satisfies the conclusions of this lemma. Claims 1 and 2 show that $\mathbb{F} \curvearrowright (A^{\mathbb{F}}, \mu)$ is OE to $\mathbb{F} \curvearrowright (A^{\mathbb{F}}, \rho)$. The next three claims show that ρ is Markov.

Claim 3. If $s \in S \cup S^{-1} \setminus \{t^{-1}, u, u^{-1}\}$ and $g \in \text{past}(s)$ then $\omega(g, x) \in \text{past}(s)$ for all $x \in A^{\mathbb{F}}$.

PROOF. We prove this by induction on $|g|$. If $|g| = 1$ then $g = s$ and either $s \neq t$ in which case $\omega(g, x) = g \in \text{past}(s)$, or $s = t$ and $\omega(g, x) \in \{t, ut, u^{-1}t\} \subset \text{past}(s)$.

So assume $|g| > 1$. Then we can write $g = hk$ for some $h \in S \cup S^{-1}$, $k \in \text{past}(s)$ such that $|k| < |g|$. By induction we can assume $\omega(k, x) \in \text{past}(s)$. Then

$$\omega(hk, x) = \omega(h, k * x)\omega(k, x).$$

To obtain a contradiction, suppose that $\omega(hk, x) \notin \text{past}(s)$.

Since $h \in S \cup S^{-1}$, $|\omega(h, k * x)| \in \{1, 2\}$. If $|\omega(h, k * x)| = 1$ then since $f\,\text{past}(s) \subset \text{past}(s)$ for all $f \in S \cup S^{-1} \setminus \{s^{-1}\}$ it must be that $\omega(h, k * x) = s^{-1}$. Since $s^{-1}\,\text{past}(s) = \{e\} \cup \text{past}(s)$, it must be that $\omega(hk, x) = e$ (since $\omega(\cdot, x)$ is injective) which implies $hk = g = e$, contradicting that $g \in \text{past}(s)$.

So suppose $|\omega(h, k * x)| = 2$. Since $|h| = 1$ this implies $h \in \{t, t^{-1}\}$ and $\omega(h, k * x) \in \{ut, u^{-1}t, (ut)^{-1}, (u^{-1}t)^{-1}\}$. If $f \in \mathbb{F}$ is any element with $|f| = 2$ then

$$f\,\text{past}(s) \subset \text{past}(s) \cup \{e\} \cup S \cup S^{-1}.$$

If $\omega(h, k * x)\omega(k, x) = e$ then $hk = g = e$, contradicting that $g \in \text{past}(s)$. So we may assume that

$$\omega(h, k * x)\omega(k, x) \in S \cup S^{-1} \setminus \text{past}(s). \tag{2}$$

Since $\omega(h, k * x) \in \{ut, u^{-1}t, (ut)^{-1}, (u^{-1}t)^{-1}\}$ and $\omega(k, x) \in \text{past}(s)$, this implies that $s \in \{u, u^{-1}, t, t^{-1}\}$. By assumption $s \notin \{t^{-1}, u, u^{-1}\}$. So $s = t$ and $\omega(k, x) \in \{t, ut, u^{-1}t\}$. Thus

$$\omega(g, x) = \omega(h, k * x)\omega(k, x) \in \{ut, u^{-1}t, (ut)^{-1}, (u^{-1}t)^{-1}\}\{t, ut, u^{-1}t\} \subset \text{past}(t) \cup \{e\}.$$

Since we are assuming $\omega(g, x) \notin \text{past}(t)$, this implies $\omega(g, x) = e$. Since $\omega(\cdot, x)$ is injective, this implies $g \in \{u, u^{-1}\}$, contradicting that $g \in \text{past}(t)$. □

Claim 4. If $s \in S \cup S^{-1} \setminus \{t^{-1}, u, u^{-1}\}$ and $g \notin \text{past}(s)$ then $\omega(g, x) \notin \text{past}(s)$ for all $x \in A^{\mathbb{F}}$.

PROOF. To obtain a contradiction, suppose there exists $g \notin \text{past}(s)$ and $y \in A^{\mathbb{F}}$ such that $\omega(g, y) \in \text{past}(s)$. Let $\Omega y = x$. By Claim 2, $\Omega x = y$ and

$$\omega(\omega(g, y), x) = \omega(\omega(g, \Omega x), x) = g.$$

Since $\omega(g, y) \in \text{past}(s)$, Claim 3 implies $\omega(\omega(g, y), x) = g \in \text{past}(s)$. This contradiction proves Claim 4. □

Claim 5. ρ is Markov.

PROOF. It follows from Claims 3 and 4 that if $s \in S \cup S^{-1} \setminus \{t^{-1}, u, u^{-1}\}$ and $x \in A^{\mathbb{F}}$ then

$$g \in \text{past}(s) \Leftrightarrow \omega(g, x) \in \text{past}(s).$$

By Lemma 4.2, ρ is s-Markov for every $s \in S \setminus \{u\}$. Since $\rho_u = \mu_u$ is also Markov, Lemma 3.3 implies ρ is Markov. □

Let $x \in A^{\mathbb{F}}$, $s \in S \setminus \{t\}$ and $n \in \mathbb{Z}$. Then

$$(\Omega x)_{s^n} = x_{\omega(s^n, x)} = x_{s^n}.$$

Thus $R^s \Omega x = R^s x$ which implies $\rho_s = \mu_s$ for all $s \in S \setminus \{t\}$.

Claim 6. $\mathcal{R}_t^\mu \subset \mathcal{R}_t^\rho$.

PROOF. Let $(\alpha, \beta) \in E_t^\mu$. We will show that $(\alpha, \beta) \in E_t^\rho$. Indeed,

$$\mu(\{x \in A^{\mathbb{F}} : x_e = \alpha, x_t = \beta, F(ut \cdot x) \notin \mathcal{F} \text{ and } F(t \cdot x) \notin \mathcal{F}\}) > 0.$$

This is because the event $ut \cdot x \in \mathcal{F}$ depends only on x_e, x_t and $x_{u^n t}$ where $n = n(x_{ut})$. Since μ is Markov, the event that $ut \cdot x \in \mathcal{F}$ given that $x_t = \beta$ does not depend on x_e. A similar statement hold for the event $F(t \cdot x) \in \mathcal{F}$. Moreover, because $\mathcal{E}$ is μ_u-special, depending only on β, one of the events $F(ut \cdot x) \notin \mathcal{F}$, $F(t \cdot x) \notin \mathcal{F}$ must occur.

Now suppose that $x \in A^{\mathbb{F}}$ satisfies $x_e = \alpha, x_t = \beta, F(ut \cdot x) \notin \mathcal{F}$, and $F(t \cdot x) \notin \mathcal{F}$. Then $\omega(t, x) = t$, $(\Omega x)_e = \alpha$ and

$$(\Omega x)_t = x_{\omega(t,x)} = x_t = \beta.$$

Thus shows

$$\begin{aligned} &\Omega^{-1}(\{y \in A^{\mathbb{F}} : y_e = \alpha, y_t = \beta\}) \\ &\quad \supset \{x \in A^{\mathbb{F}} : x_e = \alpha, x_t = \beta, F(ut \cdot x) \notin \mathcal{F}, \text{ and } F(t \cdot x) \notin \mathcal{F}\}. \end{aligned}$$

Therefore,

$$\Omega_* \mu(\{y \in A^{\mathbb{F}} : y_e = \alpha, y_t = \beta\}) > 0.$$

Since $\rho = \Omega_* \mu$, this implies $(\alpha, \beta) \in E_t^\rho$. Since (α, β) is arbitrary, $E_t^\mu \subset E_t^\rho$. Since $\mathcal{R}_t^\mu$ is generated by E_t^μ, it follows that $\mathcal{R}_t^\mu \subset \mathcal{R}_t^\rho$. □

Claim 7. Let $(a, b) \in \mathcal{E}$ and let $\alpha \in A$ be such that $(\alpha, a) \in E_t^\mu$. Then $(\alpha, b) \in E_t^\rho$.

PROOF. Since $(a, b) \in \mathcal{E} \subset E_u^\mu$, the Markov property (via Lemma 3.1) implies

$$\mu(\{x \in A^{\mathbb{F}} : x_e = \alpha, x_t = a, x_{ut} = b\}) > 0.$$

In fact,

$$\mu(\{x \in A^{\mathbb{F}} : x_e = \alpha, x_t = a, x_{ut} = b, F(ut \cdot x) \in \mathcal{F}\}) > 0.$$

This is because the event $F(ut\cdot x) \in \mathcal{F}$ given $x_t = a, x_{ut} = b$ depends only on $x_{u^{n+1}t}$ where $n = n(b)$.

If $x \in A^{\mathbb{F}}$ is such that $x_e = \alpha, x_t = a, x_{ut} = b, F(ut \cdot x) \in \mathcal{F}$ then $\omega(t, x) = ut$. So

$$(\Omega x)_e = x_e = \alpha, \quad (\Omega x)_t = x_{\omega(t,x)} = x_{ut} = b.$$

So

$$\Omega^{-1}(\{y \in A^{\mathbb{F}} : y_e = \alpha, y_t = b\}) \supset \{x \in A^{\mathbb{F}} : x_e = \alpha, x_t = a, x_{ut} = b, F(ut\cdot x) \in \mathcal{F}\}.$$

Therefore

$$\Omega_*\mu(\{y \in A^{\mathbb{F}} : y_e = \alpha, y_t = b\}) > 0.$$

This shows $(\alpha, b) \in E^\rho_t$. □

It follows from Claim 6 and 7 that $(a, b) \in \mathcal{R}^\rho_t$ for every $(a, b) \in \mathcal{E}$. Therefore $\mathcal{E} \subset \mathcal{R}^\rho_t$. By Claim 6, $\mathcal{E} \cup \mathcal{R}^\mu_t \subset \mathcal{R}^\rho_t$.

Let a, b, α be as in Claim 7 and observe that $(\alpha, a), (\alpha, b) \in E^\rho_t$ by Claims 6 and 7. Since $a \neq b$ (because $\mathcal{E}$ is special) the out-degree of α in the directed graph (A, E^ρ_t) is at least 2. So the $\mathcal{R}^\rho_t$-class of α is aperiodic. Since a and b are $\mathcal{R}^\rho_t$-equivalent to α, the $\mathcal{R}^\rho_t$-classes of a and b are aperiodic. Since $(a, b) \in \mathcal{E}$ is arbitrary, this finishes the lemma. □

Proof of Proposition 5.1. Without loss of generality we may assume $\pi_*\mu$ is a fully supported measure on A. Because μ is properly ergodic, Lemma 3.2 implies there exist $a \in A$ and $u \in S$ such that the $\mathcal{R}^\mu_u$ class of a is aperiodic. Let $[a]^\mu_u$ denote the $\mathcal{R}^\mu_u$-class of a. Let $\mathcal{T}_u \subset E^\mu_u$ be a spanning tree of the induced subgraph of $[a]^\mu_u$ in (A, E^μ_u). Because trees are bi-partitite, there exists a partition $A_0 \sqcup A_1$ of $[a]^\mu_u$ such that

$$\mathcal{T}_u \subset (A_0 \times A_1) \cup (A_1 \times A_0).$$

Let

$$\mathcal{E}_1 = \mathcal{T}_u \cap (A_0 \times A_1), \quad \mathcal{E}_2 = \mathcal{T}_u \cap (A_1 \times A_0).$$

Then each $\mathcal{E}_i$ is μ_u-special. After applying Lemma 5.2 successively using $\mathcal{E}_1, \mathcal{E}_2$ and letting t vary over $S \setminus \{u\}$, we obtain the existence of a Markov measure $\rho \in \mathrm{Prob}_{\mathbb{F}}(A^{\mathbb{F}})$ such that

- $\mathbb{F}\curvearrowright(A^{\mathbb{F}}, \mu)$ is OE to $\mathbb{F}\curvearrowright(A^{\mathbb{F}}, \rho)$,
- for every $s \in S$, $\mathcal{R}^\rho_s \supset \mathcal{R}^\mu_s \cup \mathcal{T}_u$,
- for every $s \in S$, the $\mathcal{R}^\rho_s$-class of a is aperiodic.

Thus after replacing μ with ρ if necessary, we may assume that the $\mathcal{R}^\mu_s$-class of a is aperiodic for every s.

We can now apply the same argument as above for any $s \in S$ in place of u. Thus we obtain the existence of a Markov measure $\rho \in \mathrm{Prob}_{\mathbb{F}}(A^{\mathbb{F}})$ such that

(1) $\mathbb{F}\curvearrowright(A^{\mathbb{F}}, \mu)$ is OE to $\mathbb{F}\curvearrowright(A^{\mathbb{F}}, \rho)$,
(2) for every $s, u \in S$, $\mathcal{R}^\rho_s \supset \mathcal{R}^\mu_s \cup \mathcal{T}_u$,
(3) for every $s \in S$, the $\mathcal{R}^\rho_s$-class of a is aperiodic.

It follows from item (2) that $\mathcal{R}^\rho_s \ni (a, b)$ for every b such that there exists some $u \in S$ with $(a, b) \in \mathcal{R}^\mu_u$. This is because $\mathcal{T}_u$ generates the $\mathcal{R}^\mu_u$-class of a. However, since μ is properly ergodic and $\pi_*\mu$ is fully supported, this implies $\mathcal{R}^\rho_s \ni (a, b)$ for every $b \in A$. So $\mathcal{R}^\rho_s = A \times A$. Thus ρ_s is ergodic and by (3) essentially free. □

6. Proof of the main theorem

The main theorem is obtained by applying a specific kind of orbit-equivalence to a given Markov system multiple times. This kind of orbit-equivalence does not change the 1-dimensional marginal $\pi_*\mu$ and preserves the Markov property. At the same time, it replaces one of the symbolic restrictions μ_t with a Bernoulli measure.

To build this orbit equivalence, we will first need some well-known facts about full groups of measured equivalence relations (Definition 8, Lemma 6.1). We then apply these facts to obtain a slightly enhanced version of Dye's Theorem (Lemma 6.2). Then Lemma 6.3 establishes the specific kind of orbit-equivalence we need to prove the main theorem.

DEFINITION 8. Recall that $T : A^{\mathbb{Z}} \to A^{\mathbb{Z}}$ is defined by $(Tx)_n = x_{n+1}$ and $\mathrm{Prob}_{\mathbb{Z}}(A^{\mathbb{Z}})$ is the space of T-invariant Borel probability measures on $A^{\mathbb{Z}}$. For $\mu \in \mathrm{Prob}_{\mathbb{Z}}(A^{\mathbb{Z}})$ let $[T,\mu]$ denote the **full group** of the orbit-equivalence relation of T modulo μ. To be precise, $[T,\mu]$ consists of all measurable automorphisms $S : X \to X$ (where $X \subset A^{\mathbb{Z}}$ is μ-conull and T-invariant) such that for every $x \in X$ there exists $n \in \mathbb{Z}$ with $Sx = T^n x$. Two such automorphisms are identified if they agree on a μ-conull set.

LEMMA 6.1. *Let $\mu \in \mathrm{Prob}_{\mathbb{Z}}(A^{\mathbb{Z}})$ be ergodic and essentially free. Let B be a finite or countable set and let $\phi : A^{\mathbb{Z}} \to B, \psi : A^{\mathbb{Z}} \to B$ be measurable maps with the same pushforward measures (so $\phi_*\mu = \psi_*\mu$). Then there exists $S \in [T,\mu]$ such that $\psi = \phi \circ S$.*

PROOF. This result is well-known but I did not find a suitable reference (it partially generalizes a lemma in [**HIK74**]).

Let $\{n_i\}_{i=1}^{\infty} = \mathbb{Z}$ be an enumeration of the integers. Let

$$X_1 = \{x \in A^{\mathbb{F}} : \psi(x) = \phi(T^{n_1}x)\}.$$

Define $S_1 : X_1 \to A^{\mathbb{F}}$ by $S_1(x) = T^{n_1}x$. If X_k and S_k have been defined, let X_{k+1} be the set of all $x \in A^{\mathbb{F}} \setminus \bigcup_{i=1}^k X_i$ such that

$$\psi(x) = \phi(T^{n_{k+1}}x) \text{ and } T^{n_{k+1}}x \notin \bigcup_{i=1}^{k} S_i(X_i).$$

Define $X = \cup_k X_k$ and $S : X \to A^{\mathbb{F}}$ by $Sx = S_k x$ for $x \in X_k$. Because the X_k's are pairwise disjoint, S is well-defined.

By ergodicity, for each $b \in B$, $\phi^{-1}(b) \subset \cup_{n\in\mathbb{Z}} T^n\psi^{-1}(b)$ modulo μ-null sets. Therefore, X is μ-conull. By design, $\psi = \phi \circ S$. Moreover, since the $S_k(X_k)$'s are pairwise disjoint and each S_k is injective, S is invertible. This shows S is in the full group $[T,\mu]$. □

LEMMA 6.2. *Let $\mu, \nu \in \mathrm{Prob}_{\mathbb{Z}}(A^{\mathbb{Z}})$ be shift-invariant, ergodic, essentially free measures. Let $\pi : A^{\mathbb{Z}} \to A$ be the time 0 map. Suppose $\pi_*\mu = \pi_*\nu$ (in other words, for every $a \in A$,*

$$\mu(\{x \in A^{\mathbb{Z}} : x_e = a\}) = \nu(\{x \in A^{\mathbb{Z}} : x_e = a\}).$$

Then there exists an orbit equivalence $\Psi : A^{\mathbb{Z}} \to A^{\mathbb{Z}}$ from the shift action $\mathbb{Z} \curvearrowright^T (A^{\mathbb{Z}},\mu)$ to $\mathbb{Z} \curvearrowright^T (A^{\mathbb{Z}},\nu)$ such that $\pi = \pi\Psi$.

PROOF. By Dye's Theorem, there exists an orbit-equivalence $\Psi' : A^{\mathbb{Z}} \to A^{\mathbb{Z}}$ from the shift action $\mathbb{Z} \curvearrowright^T (A^{\mathbb{Z}}, \mu)$ to $\mathbb{Z} \curvearrowright^T (A^{\mathbb{Z}}, \nu)$. By Lemma 6.1 there exists $S \in [T, \mu]$ such that $\pi = \pi\Psi' S$.

Let $\Psi = \Psi' S$. Then Ψ is an orbit-equivalence from $\mathbb{Z} \curvearrowright^T (A^{\mathbb{Z}}, \mu)$ to $\mathbb{Z} \curvearrowright^T (A^{\mathbb{Z}}, \nu)$ since pre-composing with an element of the full group does not change orbits. Also $\pi = \pi\Psi$. □

LEMMA 6.3. *Let $\mu \in \mathrm{Prob}_{\mathbb{F}}(A^{\mathbb{F}})$ be Markov. Let $t \in S$ and suppose that the symbolic restriction $\mu_t \in \mathrm{Prob}_{\mathbb{Z}}(A^{\mathbb{Z}})$ is such that $\mathbb{Z} \curvearrowright^T (A^{\mathbb{Z}}, \mu_t)$ is essentially free and ergodic. Also let $\nu \in \mathrm{Prob}_{\mathbb{Z}}(A^{\mathbb{Z}})$ be an ergodic, essentially free, shift invariant measure. By abuse of notation, let $\pi : A^{\mathbb{F}} \to A$ denote the map $\pi(x) = x_e$ and let $\pi : A^{\mathbb{Z}} \to A$ denote the map $\pi(x) = x_0$. Suppose that $\pi_*\mu = \pi_*\nu$.*

Then the action $\mathbb{F} \curvearrowright (A^{\mathbb{F}}, \mu)$ is OE to $\mathbb{F} \curvearrowright (A^{\mathbb{F}}, \rho)$ where ρ is a shift-invariant measure uniquely determined by the following.

- $\rho_t = \nu$,
- $\rho_s = \mu_s$ *for all* $s \in S \setminus \{t\}$,
- ρ *is Markov along* s *for every* $s \in S \setminus \{t\}$.

PROOF. By Lemma 6.2 there exists an orbit equivalence

$$\Psi : A^{\mathbb{Z}} \to A^{\mathbb{Z}}$$

from the shift action $\mathbb{Z} \curvearrowright^T (A^{\mathbb{Z}}, \mu_t)$ to $\mathbb{Z} \curvearrowright^T (A^{\mathbb{Z}}, \nu)$ such that $\pi = \pi\Psi$.

Define

$$\begin{array}{lll} \tilde{T} : A^{\mathbb{Z}} \to A^{\mathbb{Z}} & \text{by} & \tilde{T} = \Psi^{-1} T \Psi \\ \alpha : \mathbb{Z} \times A^{\mathbb{Z}} \to \mathbb{Z} & \text{by} & \tilde{T}^{\alpha(n,x)} x = T^n x \\ \beta : \mathbb{Z} \times A^{\mathbb{Z}} \to \mathbb{Z} & \text{by} & T^{\beta(n,x)} x = \tilde{T}^n x \end{array}$$

Because Ψ is an OE, α, β are well-defined, satisfy the cocycle equations below and the inverse equation:

$$\alpha(n+m, x) = \alpha(n, T^m x) + \alpha(m, x) \tag{3}$$

$$\beta(n+m, x) = \beta(n, \tilde{T}^m x) + \beta(m, x) \tag{4}$$

$$\beta(\alpha(n, x), x) = \alpha(\beta(n, x), x) = n \tag{5}$$

for all $x \in A^{\mathbb{Z}}, n, m \in \mathbb{Z}$. To ease notation, let $R = R^t : A^{\mathbb{F}} \to A^{\mathbb{Z}}$ denote the restriction map as in Definition 5. Define $\tau : (S \cup S^{-1}) \times A^{\mathbb{F}} \to \mathbb{F}$ by

$$\tau(s, x) = s \text{ for } s \in S \cup S^{-1} \setminus \{t, t^{-1}\},$$

$$\tau(t^n, x) = t^{\beta(n, Rx)} \text{ for } n \in \{-1, +1\}.$$

Then

$$\begin{aligned} \tau(t^{-1}, \tau(t, x) \cdot x) &= \tau(t^{-1}, t^{\beta(1,Rx)} \cdot x) = t^{\beta(-1, R(t^{\beta(1,Rx)} \cdot x))} \\ &= t^{\beta(-1, T^{\beta(1,Rx)} Rx)} = t^{\beta(-1, \tilde{T} Rx)} = t^{-\beta(1,Rx)} = \tau(t, x)^{-1} \end{aligned}$$

where the second-to-last equality follows from the cocycle equation for β. Similarly, $\tau(t, \tau(t^{-1}, x) \cdot x) = \tau(t^{-1}, x)^{-1}$. So τ satisfies the conditions of Lemma 4.1. Let $*, \omega, \Omega$ be as in Lemma 4.1.

We claim that

$$t^n * x = t^{\beta(n, Rx)} \cdot x \text{ for } n \in \mathbb{Z}. \tag{6}$$

By definition this statement is true if $n \in \{-1, 1\}$. So it suffices to prove that if $n, m \in \mathbb{Z}$ satisfy (6) then $n+m$ also satisfies (6). We claim that $R(t^m * x) = \tilde{T}^m Rx$. This follows from:

$$\begin{aligned} R(t^m * x)_n &= (t^m * x)_{t^n} = (t^{\beta(m,Rx)} \cdot x)_{t^n} = x_{t^{n+\beta(m,Rx)}} \\ &= (Rx)_{n+\beta(m,Rx)} = (T^{\beta(m,Rx)} Rx)_n = (\tilde{T}^m Rx)_n. \end{aligned}$$

Therefore,

$$\begin{aligned} t^{n+m} * x &= t^n * (t^m * x) = t^{\beta(n,Rt^m * x)} \cdot (t^m * x) = t^{\beta(n,Rt^m * x)} \cdot (t^{\beta(m,Rx)} \cdot x) \\ &= t^{\beta(n,Rt^m * x)+\beta(m,Rx)} \cdot x = t^{\beta(n,\tilde{T}^m Rx)+\beta(m,Rx)} \cdot x = t^{\beta(n+m,Rx)} \cdot x. \end{aligned}$$

This proves (6).

To finish the lemma, we will show that $*$ has the same orbits as $\cdot$ (modulo μ null sets) and afterwards that $\mathbb{F} \curvearrowright^* (A^{\mathbb{F}}, \mu)$ is measure-conjugate to $\mathbb{F} \curvearrowright (A^{\mathbb{F}}, \rho)$.

Claim 1. $\mathbb{F} \curvearrowright^* (A^{\mathbb{F}}, \mu)$ has the same orbits as $\mathbb{F} \curvearrowright (A^{\mathbb{F}}, \mu)$.

PROOF. Let $\kappa : \mathbb{F} \times A^{\mathbb{F}} \to \mathbb{F}$ be the unique function satisfying the following:

$$\begin{aligned} \kappa(s, x) &= s \text{ for } s \in S \cup S^{-1} - \{t, t^{-1}\} \\ \kappa(t^n, x) &= t^{\alpha(n,Rx)} \text{ for } n \in \mathbb{Z} \\ \kappa(gh, x) &= \kappa(g, h \cdot x)\kappa(h, x). \end{aligned}$$

This is well-defined because α satisfies the cocycle equation and $R(t^n \cdot x) = T^n Rx$. The following inverse equation also holds:

$$\kappa(\omega(h, x), x) = \omega(\kappa(h, x), x) = h. \tag{7}$$

To see this, first note that it is obvious when $h \in S \cup S^{-1} \setminus \{t, t^{-1}\}$. Then observe that the inverse equations (5) imply the statement for h in the subgroup generated by t. If the statement holds for elements h_1, h_2 then it must hold for their product because of:

$$\begin{aligned} \kappa(\omega(h_1h_2, x), x) &= \kappa(\omega(h_1, h_2 * x)\omega(h_2, x), x) \\ &= \kappa(\omega(h_1, h_2 * x), \omega(h_2, x) \cdot x)\kappa(\omega(h_2, x), x) \\ &= \kappa(\omega(h_1, h_2 * x), h_2 * x)\kappa(\omega(h_2, x), x) = h_1h_2 \end{aligned}$$

and the related equation with the orders of ω, κ reversed. By induction (7) is true for all $h \in \mathbb{F}$. Equation (7) implies that the actions $*$ and $\cdot$ have the same orbits (modulo μ null sets). □

Claim 2. Ω is a measure-conjugacy between $\mathbb{F} \curvearrowright^* (A^{\mathbb{F}}, \mu)$ and $\mathbb{F} \curvearrowright (A^{\mathbb{F}}, \Omega_* \mu)$.

PROOF. By Lemma 4.1, Ω is $(*, \cdot)$-equivariant. In order to show that Ω is invertible, define

$$\begin{aligned} \widehat{T} : A^{\mathbb{Z}} \to A^{\mathbb{Z}} &\quad \text{by} \quad \widehat{T} = \Psi T \Psi^{-1} \\ \hat{\beta} : \mathbb{Z} \times A^{\mathbb{Z}} \to \mathbb{Z} &\quad \text{by} \quad T^{\hat{\beta}(n,x)} x = \widehat{T}^n x. \end{aligned}$$

Because Ψ is an OE, $\hat{\beta}$ satisfies the cocycle equation

$$\hat{\beta}(n + m, x) = \hat{\beta}(n, \widehat{T}^m x) + \hat{\beta}(m, x). \tag{8}$$

It also satisfies the inverse equations:

$$\hat{\beta}(\beta(n, \Psi^{-1} x), x) = \beta(\hat{\beta}(n, \Psi x), x) = n. \tag{9}$$

To see this observe that

$$\begin{aligned} T^{\hat\beta(\beta(n,\Psi^{-1}x),x)}x &= \widehat{T}^{\beta(n,\Psi^{-1}x)}x = \Psi T^{\beta(n,\Psi^{-1}x)}\Psi^{-1}x \\ &= \Psi\tilde{T}^n\Psi^{-1}x = T^n x. \end{aligned}$$

This shows $\hat\beta(\beta(n,\Psi^{-1}x),x) = n$. The other equality is similar.

Define $\hat\tau : (S\cup S^{-1})\times A^{\mathbb{F}} \to \mathbb{F}$ by

$$\hat\tau(s,x) = s \text{ for } s \in S\cup S^{-1}\setminus\{t,t^{-1}\},$$

$$\hat\tau(t^n,x) = t^{\hat\beta(n,Rx)} \text{ for } n\in\{-1,1\}.$$

As in the case of τ, $\hat\tau$ satisfies the hypotheses of Lemma 4.1. Let $\star, \widehat{\omega}, \hat\Omega$ be the action, cocycle and map defined by Lemma 4.1.

The following restriction equations hold:

$$(10) \qquad R\hat\Omega = \Psi^{-1}R, \quad R\Omega = \Psi R.$$

The first equation above is proven by:

$$\begin{aligned} (R\hat\Omega x)_n &= (\hat\Omega x)_{t^n} = x_{\widehat{\omega}(t^n,x)} = x_{t^{\hat\beta(n,Rx)}} \\ &= (T^{\hat\beta(n,Rx)}Rx)_0 = (\widehat{T}^n Rx)_0 = (\Psi T^n\Psi^{-1}Rx)_0 = \pi(\Psi T^n \Psi^{-1}Rx) \\ &= \pi(T^n\Psi^{-1}Rx) = (\Psi^{-1}Rx)_n. \end{aligned}$$

The second equation is similar. We now claim the following inverse equations:

$$(11) \qquad \widehat{\omega}(\omega(g,\hat\Omega x),x) = \omega(\widehat{\omega}(g,\Omega x),x) = g.$$

This is immediate if $g\in S\cup S^{-1}\setminus\{t,t^{-1}\}$. The case $g = t^n$ follows from (10) and (9):

$$\widehat{\omega}(\omega(t^n,\hat\Omega x),x) = \widehat{\omega}(t^{\beta(n,R\hat\Omega x)},x) = \widehat{\omega}(t^{\beta(n,\Psi^{-1}Rx)},x) = t^{\hat\beta(\beta(n,\Psi^{-1}Rx),Rx)} = t^n.$$

The general case follows from Lemma 4.3 which also shows $\hat\Omega\Omega x = \Omega\hat\Omega x = x$. Therefore, Ω is invertible with inverse equal to $\hat\Omega$. □

The measure ρ is well-defined by Lemma 3.3. It now suffices to show $\Omega_*\mu = \rho$. This is obtained by verifying that $\Omega_*\mu$ satisfies the same conditions defining ρ.

Claim 3. For every $s\in S$, $(\Omega_*\mu)_s = \rho_s$.

PROOF. By (10),

$$(\Omega_*\mu)_t = (R\Omega)_*\mu = (\Psi R)_*\mu = \Psi_*\mu_t = \nu = \rho_t.$$

Fix $s\in S\setminus\{t\}$. Then $\omega(s^n,x) = s^n$ for all n. We claim that $R^s\Omega = R^s$. This follows from

$$(R^s\Omega x)_n = (\Omega x)_{s^n} = x_{\omega(s^n,x)} = x_{s^n} = (R^s x)_n.$$

So

$$(\Omega_*\mu)_s = (R^s\Omega)_*\mu = R^s_*\mu = \mu_s = \rho_s.$$

□

It now suffices to show that $\Omega_*\mu$ is s-Markov for every $s\in S\setminus\{t\}$. We will use Lemma 4.2 and the next two claims.

Claim 4. For any $s\in S\cup S^{-1}\setminus\{t,t^{-1}\}$,

$$g\in \mathrm{past}(s) \Rightarrow \omega(g,x)\in\mathrm{past}(s)\ \forall x\in A^{\mathbb{F}}.$$

PROOF. The proof of the Claim is by induction on $|g|$. If $|g| \le 1$ then $g = s$ and $\omega(g,x) = s \in \mathrm{past}(s)$. So assume $|g| > 1$. Then $g = hk$ for some $h \in S \cup S^{-1}$ and $k \in \mathrm{past}(s)$ with $|k| < |g|$. So

$$\omega(g,x) = \omega(hk,x) = \omega(h,k*x)\omega(k,x).$$

By induction, we may assume $\omega(k,x) \in \mathrm{past}(s)$. To obtain a contradiction, suppose $\omega(g,x) \notin \mathrm{past}(s)$.

If $h \in S \cup S^{-1} \setminus \{t,t^{-1}\}$ then $\omega(h,k*x) = h$ has length 1. If $f \in \mathbb{F}$ is any element with length 1 then $f\,\mathrm{past}(s) \subset \{e\} \cup \mathrm{past}(s)$. So if $\omega(g,x) = \omega(h,k*x)\omega(k,x) \notin \mathrm{past}(s)$ then $\omega(g,x) = e$. But ω is injective by (11). This implies $g = e$ contradicting that $g \in \mathrm{past}(s)$.

On the other hand, if $h = t^m$ for some m then $\omega(h,k*x) = t^n$ for some n. Since $t^n\,\mathrm{past}(s) \subset \mathrm{past}(s)$ (since $s \notin \{t,t^{-1}\}$), this shows $\omega(g,x) \in \mathrm{past}(s)$. □

Claim 5. If $g \in \mathrm{past}(t) \cup \mathrm{past}(t^{-1})$ then $\omega(g,x) \in \mathrm{past}(t) \cup \mathrm{past}(t^{-1})$ for μ-a.e. $x \in A^{\mathbb{F}}$.

PROOF. If $|g| \le 1$ then $g \in \{t,t^{-1}\}$ and

$$\omega(g,x) = t^n \in \mathrm{past}(t) \cup \mathrm{past}(t^{-1})$$

for some n by (6) and the definition of ω from Lemma 4.1. We are using here that ω is injective by (11) and therefore $n \neq 0$.

So we may assume $|g| > 1$. Then $g = hk$ for some $h \in S \cup S^{-1}$ and $k \in \mathrm{past}(t) \cup \mathrm{past}(t^{-1})$ with $|k| < |g|$. So

$$\omega(g,x) = \omega(hk,x) = \omega(h,k*x)\omega(k,x).$$

By induction we may assume $\omega(k,x) \in \mathrm{past}(t) \cup \mathrm{past}(t^{-1})$. To obtain a contradiction, assume $\omega(g,x) \notin \mathrm{past}(t) \cup \mathrm{past}(t^{-1})$.

If $h \in S \cup S^{-1} \setminus \{t,t^{-1}\}$ then $\omega(h,k*x) = h$. Since

$$(S \cup S^{-1} \setminus \{t,t^{-1}\})[\mathrm{past}(t) \cup \mathrm{past}(t^{-1})] \subset \mathrm{past}(t) \cup \mathrm{past}(t^{-1}),$$

this shows $\omega(g,x) \in \mathrm{past}(t) \cup \mathrm{past}(t^{-1})$.

So assume $h = t^n$ for some $n \in \{-1,+1\}$. Then $\omega(h,k*x) = t^m$ for some m. Since

$$t^m[\mathrm{past}(t) \cup \mathrm{past}(t^{-1})] \subset \mathrm{past}(t) \cup \mathrm{past}(t^{-1}) \cup \{e\}$$

it follows that $\omega(k,x) = t^{-m}$ so $\omega(g,x) = e$. But this implies $g = e$ since ω is injective (11), a contradiction. □

Lemma 4.2 implies $\Omega_*\mu$ is s-Markov for every $s \in S \setminus \{t\}$. So $\Omega_*\mu$ satisfies the same defining properties as ρ. Thus $\Omega_*\mu = \rho$. □

PROOF OF THEOREM 1.1. Let $\mu \in \mathrm{Prob}_{\mathbb{F}}(A^{\mathbb{F}})$ be Markov and properly ergodic. Since all Bernoulli shifts over $\mathbb{F}$ are OE (by [**Bow11a**]) it suffices to show that $\mathbb{F} \curvearrowright (A^{\mathbb{F}},\mu)$ is OE to a Bernoulli shift. By Proposition 5.1 we may assume that μ_s is essentially free and ergodic for every $s \in S$.

Let $S = \{s_1,\dots,s_r\}$. Let ν be the Bernoulli product measure $\nu = (\pi_*\mu)^{\mathbb{Z}} \in \mathrm{Prob}_{\mathbb{Z}}(A^{\mathbb{Z}})$. Define shift-invariant measures $\mu^{(0)},\mu^{(1)},\dots,\mu^{(r)} \in \mathrm{Prob}_{\mathbb{F}}(A^{\mathbb{F}})$ as follows. First, $\mu^{(0)} = \mu$. For $i > 0$, $\mu^{(i)}$ is characterized by:

- $\mu^{(i)}_{s_j} = \nu$ for all $j \le i$
- $\mu^{(i)}_{s_j} = \mu_{s_k}$ for all $i < j \le r$,
- $\mu^{(i)}$ is s_j-Markov for all $j \neq i$.

By Lemma 3.3, $\mu^{(i)}$ is Markov. By Lemma 6.3, $\mathbb{F}\curvearrowright(A^{\mathbb{F}}, \mu^{(i)})$ is OE to $\mathbb{F}\curvearrowright(A^{\mathbb{F}}, \mu^{(i+1)})$ for all $i < r$. Since $\mu^{(r)} = (\pi_*\mu)^{\mathbb{F}}$ is Bernoulli this completes the proof. □

References

[AL07] David Aldous and Russell Lyons, *Processes on unimodular random networks*, Electron. J. Probab. **12** (2007), no. 54, 1454–1508, DOI 10.1214/EJP.v12-463. MR2354165

[BLPS01] Itai Benjamini, Russell Lyons, Yuval Peres, and Oded Schramm, *Uniform spanning forests*, Ann. Probab. **29** (2001), no. 1, 1–65, DOI 10.1214/aop/1008956321. MR1825141

[Bow10a] Lewis Bowen, *Measure conjugacy invariants for actions of countable sofic groups*, J. Amer. Math. Soc. **23** (2010), no. 1, 217–245, DOI 10.1090/S0894-0347-09-00637-7. MR2552252

[Bow10b] Lewis Bowen, *Non-abelian free group actions: Markov processes, the Abramov-Rohlin formula and Yuzvinskii's formula*, Ergodic Theory Dynam. Systems **30** (2010), no. 6, 1629–1663, DOI 10.1017/S0143385709000844. MR2736889

[Bow11a] Lewis Bowen, *Orbit equivalence, coinduced actions and free products*, Groups Geom. Dyn. **5** (2011), no. 1, 1–15, DOI 10.4171/GGD/114. MR2763777

[Bow11b] Lewis Bowen, *Stable orbit equivalence of Bernoulli shifts over free groups*, Groups Geom. Dyn. **5** (2011), no. 1, 17–38, DOI 10.4171/GGD/115. MR2763778

[Bow15] Lewis Bowen, *Weak density of orbit equivalence classes of free group actions*, Groups Geom. Dyn. **9** (2015), no. 3, 811–830, DOI 10.4171/GGD/329. MR3420545

[Bow17] Lewis Bowen. Examples in the entropy theory of countable group actions. *arXiv:1704.06349*, 2017.

[BP94a] Itai Benjamini and Yuval Peres, *Markov chains indexed by trees*, Ann. Probab. **22** (1994), no. 1, 219–243. MR1258875

[BP94b] Itai Benjamini and Yuval Peres, *Tree-indexed random walks on groups and first passage percolation*, Probab. Theory Related Fields **98** (1994), no. 1, 91–112, DOI 10.1007/BF01311350. MR1254826

[BS96] Itai Benjamini and Oded Schramm, *Percolation beyond* $\mathbf{Z}^d$*, many questions and a few answers*, Electron. Comm. Probab. **1** (1996), no. 8, 71–82, DOI 10.1214/ECP.v1-978. MR1423907

[CFW81] A. Connes, J. Feldman, and B. Weiss, *An amenable equivalence relation is generated by a single transformation*, Ergodic Theory Dynamical Systems **1** (1981), no. 4, 431–450 (1982). MR662736

[CI10] Ionut Chifan and Adrian Ioana, *Ergodic subequivalence relations induced by a Bernoulli action*, Geom. Funct. Anal. **20** (2010), no. 1, 53–67, DOI 10.1007/s00039-010-0058-7. MR2647134

[Cur16] Nicolas Curien, *Planar stochastic hyperbolic triangulations*, Probab. Theory Related Fields **165** (2016), no. 3-4, 509–540, DOI 10.1007/s00440-015-0638-4. MR3520011

[Dur96] Richard Durrett, *Probability: theory and examples*, 2nd ed., Duxbury Press, Belmont, CA, 1996. MR1609153

[Dye59] H. A. Dye, *On groups of measure preserving transformation. I*, Amer. J. Math. **81** (1959), 119–159, DOI 10.2307/2372852. MR0131516

[Dye63] H. A. Dye, *On groups of measure preserving transformations. II*, Amer. J. Math. **85** (1963), 551–576, DOI 10.2307/2373108. MR0158048

[Eps08] Inessa Epstein, *Some results on orbit inequivalent actions of non-amenable groups*, ProQuest LLC, Ann Arbor, MI, 2008. Thesis (Ph.D.)–University of California, Los Angeles. MR2711968

[GP05] Damien Gaboriau and Sorin Popa, *An uncountable family of nonorbit equivalent actions of* $\mathbb{F}_n$, J. Amer. Math. Soc. **18** (2005), no. 3, 547–559, DOI 10.1090/S0894-0347-05-00480-7. MR2138136

[HIK74] Arshag Hajian, Yuji Ito, and Shizuo Kakutani, *Orbits, sections, and induced transformations*, Israel J. Math. **18** (1974), 97–115, DOI 10.1007/BF02756865. MR0369658

[Hjo05] Greg Hjorth, *A converse to Dye's theorem*, Trans. Amer. Math. Soc. **357** (2005), no. 8, 3083–3103, DOI 10.1090/S0002-9947-04-03672-4. MR2135736

[IKT09] Adrian Ioana, Alexander S. Kechris, and Todor Tsankov, *Subequivalence relations and positive-definite functions*, Groups Geom. Dyn. **3** (2009), no. 4, 579–625, DOI 10.4171/GGD/71. MR2529949

[Ioa10] Adrian Ioana, *Relative property (T) for the subequivalence relations induced by the action of* $\mathrm{SL}_2(\mathbb{Z})$ *on* $\mathbb{T}^2$, Adv. Math. **224** (2010), no. 4, 1589–1617, DOI 10.1016/j.aim.2010.01.021. MR2646305

[Ioa11] Adrian Ioana, *Orbit inequivalent actions for groups containing a copy of* $\mathbb{F}_2$, Invent. Math. **185** (2011), no. 1, 55–73, DOI 10.1007/s00222-010-0301-8. MR2810796

[Kid08] Yoshikata Kida, *Orbit equivalence rigidity for ergodic actions of the mapping class group*, Geom. Dedicata **131** (2008), 99–109, DOI 10.1007/s10711-007-9219-8. MR2369194

[Kid10] Yoshikata Kida, *Measure equivalence rigidity of the mapping class group*, Ann. of Math. (2) **171** (2010), no. 3, 1851–1901, DOI 10.4007/annals.2010.171.1851. MR2680399

[MRV13] Niels Meesschaert, Sven Raum, and Stefaan Vaes, *Stable orbit equivalence of Bernoulli actions of free groups and isomorphism of some of their factor actions*, Expo. Math. **31** (2013), no. 3, 274–294, DOI 10.1016/j.exmath.2012.08.012. MR3108102

[OP10] Narutaka Ozawa and Sorin Popa, *On a class of* II_1 *factors with at most one Cartan subalgebra*, Ann. of Math. (2) **172** (2010), no. 1, 713–749, DOI 10.4007/annals.2010.172.713. MR2680430

[OW80] Donald S. Ornstein and Benjamin Weiss, *Ergodic theory of amenable group actions. I. The Rohlin lemma*, Bull. Amer. Math. Soc. (N.S.) **2** (1980), no. 1, 161–164, DOI 10.1090/S0273-0979-1980-14702-3. MR551753

[Pop06a] Sorin Popa, *On a class of type* II_1 *factors with Betti numbers invariants*, Ann. of Math. (2) **163** (2006), no. 3, 809–899, DOI 10.4007/annals.2006.163.809. MR2215135

[Pop06b] Sorin Popa, *Strong rigidity of* II_1 *factors arising from malleable actions of w-rigid groups. II*, Invent. Math. **165** (2006), no. 2, 409–451, DOI 10.1007/s00222-006-0502-3. MR2231962

[Pop08] Sorin Popa, *On the superrigidity of malleable actions with spectral gap*, J. Amer. Math. Soc. **21** (2008), no. 4, 981–1000, DOI 10.1090/S0894-0347-07-00578-4. MR2425177

Department of Mathematics, University of Texas at Austin, Austin, Texas 78712

Contemporary Mathematics
Volume **719**, 2018
https://doi.org/10.1090/conm/719/14474

Shift-coupling of random rooted graphs and networks

Ali Khezeli

ABSTRACT. In this paper, we present a result similar to the shift-coupling result of Thorisson (1996) in the context of random graphs and networks. The result is that a given random rooted network can be obtained by changing the root of another given one if and only if the distributions of the two agree on the invariant sigma-field. Several applications of the result are presented for the case of unimodular networks. In particular, it is shown that the distribution of a unimodular network is uniquely determined by its restriction to the invariant sigma-filed. Also, the theorem is applied to the existence of an invariant transport kernel that balances between two given (discrete) measures on the vertices. An application is the existence of a so called *extra head scheme* for the Bernoulli process on an infinite unimodular graph. Moreover, a construction is presented for balancing transport kernels that is a generalization of the Gale-Shapley stable matching algorithm in bipartite graphs. Another application is on a general method that covers the situations where some vertices and edges are added to a unimodular network and then, to make it unimodular, the probability measure is biased and then a new root is selected. It is proved that this method provides all possible *unimodularizations* in these situations. Finally, analogous existing results for stationary point processes and unimodular networks are discussed in detail.

1. Introduction

This paper deals with random rooted graphs, which are possibly infinite, but finite-degree connected graphs with a distinguished vertex called the *root*. Roughly speaking, each vertex and edge of a graph can be equipped with marks to form a network. Unimodular random rooted networks have been of great interest in the last two decades. They satisfy a formulation of the heuristic property that all vertices are *equally likely* to be the root, although there may be infinitely many vertices. The formulation, called *the mass transport principle*, will be recalled in Section 2. This concept is introduced in [**4**] and developed further in [**1**] to generalize some properties of Cayley graphs to more general classes of graphs and random graphs. It also arises in the study of limits of sequences of finite graphs, which is the novel work of [**2**], and also in stationary point processes. Many concepts and results in stationary point processes have analogues in the context of unimodular networks. This analogy will be addressed many times in this paper.

2010 *Mathematics Subject Classification.* Primary 60C05; Secondary 60K99, 05C80.
Key words and phrases. Shift-coupling, random graphs, random networks, unimodularity, invariant transports, stable transports, network extension, unimodularization.

To introduce the idea of this work, we get help from the following general construction method. Let $[\boldsymbol{G}, \boldsymbol{o}]$ be a given (non-unimodular) random rooted network, where $\boldsymbol{o}$ stands for the root. In some examples in the literature, a unimodular network is constructed from $[\boldsymbol{G}, \boldsymbol{o}]$ by the following two steps: *Bias* the probability measure by an appropriate function and then choose a new random root with an appropriate distribution on the vertices of $\boldsymbol{G}$. Denote the resulting random rooted network by $[\boldsymbol{G}', \boldsymbol{o}']$. Explicit examples of such constructions in the literature will be recalled in Section 5. One may intuitively accept that $[\boldsymbol{G}', \boldsymbol{o}']$ is *equivalent* to $[\boldsymbol{G}, \boldsymbol{o}]$ if we disregard the root ([**1**]), or that $[\boldsymbol{G}', \boldsymbol{o}']$ and $[\boldsymbol{G}, \boldsymbol{o}]$ *have the same non-rooted networks*. However, to state this in a mathematically precise way, one should answer the following question.

QUESTION 1. *When do two given (not necessarily unimodular) random rooted networks have the same non-rooted networks?*

Note that the question is not limited to the setting of the above example. In general, no special relation is assumed between the two random rooted networks and they might be given only by two probability distributions. The answer to this question is not straightforward since the space of non-rooted networks is non-standard. Several definitions of *unrooted-equivalence* are provided in Section 3 as answers to this question, where some of the definitions are shown to be equivalent. It will be shown that in the above example, $[\boldsymbol{G}', \boldsymbol{o}']$ and $[\boldsymbol{G}, \boldsymbol{o}]$ are *weakly* unrooted-equivalent, to be defined later. The *strong* sense in our definition is that (the distribution of) the second one can be obtained from the first by a *root-change* (note that there is a biasing in the above definition of $[\boldsymbol{G}', \boldsymbol{o}']$ before changing the root). Another definition is that the two random rooted networks agree on the *invariant sigma-field*; i.e., any event that does not depend on the root occurs with equal probabilities. Some other definitions will also be given (Definition 7).

The main theorem (Theorem 1) in this work is that the last two definitions mentioned above are equivalent; namely, if two random rooted networks agree on the invariant sigma-field, then they can be obtained from each other by a root-change. This theorem, in its spirit, is similar to a well known result by Thorisson [**20**] that studies *shift-coupling* of random elements in a space equipped with a group action.

In Section 4, we discuss applications of the main theorem in the unimodular case. Theorem 3 says that the distribution of a unimodular network is uniquely determined by the distribution of its non-rooted network, or equivalently, by its restriction to the invariant sigma-field. Theorem 4 deals with *invariant balancing transport kernels*, which are transport kernels that transport a given measure to another given one. In the context of stationary random measures and point processes, this concept has been studied by many authors recently. In this context, under suitable assumptions, the existence of a (random) balancing transport kernel that is invariant under translations is implied by the result of [**20**] (proved in the general case in [**15**]). Based on this abstract result, several constructions have been provided, starting from [**16**] and [**11**], where the latter provides a transport kernel balancing between (a multiple of) the Lebesgue measure and the counting measure of the Poisson point process. Here, in the context of unimodular networks, we consider two discrete measures on the vertices of the random network. In Theorem 4, it will be proved that, roughly speaking, a balancing transport kernel between them exists if and only if the measures have equal *sample intensities*; i.e., the measures of the root have the same expectation conditioned on the invariant sigma-field.

A construction of such transport kernels is discussed in Section 6 (Theorem 7) based on the construction of stable transports in [**9**], which is by itself based on [**11**]. It is a generalization of the Gale-Shapley stable matching algorithm in bipartite graphs [**8**].

In Section 5, we describe a general method for constructing unimodular networks. In some of the examples in the literature, such a network is constructed by the following steps: Adding some vertices and edges to another unimodular network (called a *network extension* here), then biasing the probability measure and finally applying a root-change. These examples are unified in the method presented in Theorem 5. It is also proved in Theorem 6 that this method gives all possible ways to *unimodularize* the extension (to be defined more precisely later).

Many of the definitions and results in this paper have analogues in the context of point processes and random measures, which are discussed in Section 8.

This paper is structured as follows. The definition and basic properties of unimodular random networks are given in Section 2. The definition of *unrooted-equivalence* and the main shift-coupling theorem are presented in Section 3. The applications of the theorem to the unimodular case are studied in Section 4. Section 5 deals with extensions of unimodular networks. Section 6 presents a construction of balancing transport kernels using stable transports. The proofs of some results are moved to Section 7 to help to focus on the main thread of the paper. Finally, Section 8 reviews the analogous results in the context of point processes.

2. Random Rooted Graphs and Networks

In this section, we recall the concepts of random networks and unimodularity mainly from [**1**]. A **network** is a (multi-) graph $G = (V, E)$ equipped with a complete separable metric space Ξ, called the **mark space** and with two maps from V and $\{(v, e) : v \in V, e \in E, v \sim e\}$ to Ξ, where the symbol $\sim$ is used for adjacency of vertices or edges. The image of v (resp. (v, e)) in Ξ is called its **mark**. The degree of a vertex v is denoted by $d(v)$ and the graph-distance of vertices v and w is denoted by $d(u, v)$. The symbol $N_r(v)$ is used for the closed ball centered at v with radius r; i.e., the set of vertices with distance at most r to v.

In this paper, all networks are assumed to be locally finite; that is, the degrees of every vertex is assumed to be finite. Moreover, a network is assumed to be connected except when explicitly mentioned. An **isomorphism** between two networks is a graph isomorphism that also preserves the marks. A **rooted network** is a pair (G, o) in which G is a network and o is a distinguished vertex of G called the **root**. An **isomorphism** of rooted networks is a network isomorphism that takes the root of one to that of the other. Let $\mathcal{G}$ denote the set of isomorphism classes of connected networks and $\mathcal{G}_*$ the set of isomorphism classes of connected rooted networks. The set $\mathcal{G}_{**}$ is defined similarly for doubly-rooted networks; i.e., those with a pair of distinguished vertices. The isomorphism class of a network G (resp. (G, o) or (G, o, v)) is denoted by $[G]$ (resp. $[G, o]$ or $[G, o, v]$).

The sets $\mathcal{G}_*$ and $\mathcal{G}_{**}$ can be equipped with natural metrics that make them a complete separable metric space and equip them with the corresponding Borel sigma-fields. The distance of two rooted networks is defined based on the similarity of finite neighborhoods of their roots. See [**1**] for the precise definition. There are two natural projections $\pi_1, \pi_2 : \mathcal{G}_{**} \to \mathcal{G}_*$ obtained by forgetting the second and the first root respectively. These projections are continuous and measurable.

In contrast, there is no useful metric on $\mathcal{G}$. However, as will be defined in Definition 4, the natural projection $\pi : \mathcal{G}_* \to \mathcal{G}$ of forgetting the root induces a sigma-field on $\mathcal{G}$. This sigma-field is extensively used in this paper although it does not make $\mathcal{G}$ a standard space.

DEFINITION 1. A **random rooted network** is a random element in $\mathcal{G}_*$ and is represented in either of the following ways.

- A probability measure μ on $\mathcal{G}_*$.
- A measurable function from some probability space to $\mathcal{G}_*$ that is denoted by bold symbols $[\boldsymbol{G}, \boldsymbol{o}]$. Here, $\boldsymbol{G}$ and $\boldsymbol{o}$ represent the network and the root respectively.

Note that the whole symbol $[\boldsymbol{G}, \boldsymbol{o}]$ represents one random object, which is a random equivalence class of rooted networks. Therefore, any formula using $\boldsymbol{G}$ and $\boldsymbol{o}$ should be well defined for equivalence classes of rooted networks; i.e., should be invariant under rooted isomorphisms. Moreover, bold symbols are used only in the random case.

The relation between the two representations is expressed by the equation $\mu(A) = \mathbb{P}[[\boldsymbol{G}, \boldsymbol{o}] \in A]$ for events $A \subseteq \mathcal{G}_*$; i.e., μ is the distribution of the random object. These representations are mostly treated equally in this paper. Therefore, all definitions and results expressed for random rooted networks $[\boldsymbol{G}, \boldsymbol{o}]$ also make sense for probability measures on $\mathcal{G}_*$.

For a measurable function $g : \mathcal{G}_{**} \to \mathbb{R}^{\geq 0}$, a network G and $u, v \in V(G)$, let

$$g_G(u, v) := g[G, u, v],$$

where brackets $[\cdot]$ are used as a short form of $([\cdot])$. Also, for $o \in V(G)$, let

$$\begin{aligned} g_G^+(o) &:= \sum_{v \in V(G)} g[G, o, v], \\ g_G^-(o) &:= \sum_{v \in V(G)} g[G, v, o]. \end{aligned}$$

DEFINITION 2. A random rooted network $[\boldsymbol{G}, \boldsymbol{o}]$ is **unimodular** if for all measurable functions $g : \mathcal{G}_{**} \to \mathbb{R}^{\geq 0}$,

$$\mathbb{E}\left[g_{\boldsymbol{G}}^+(\boldsymbol{o})\right] = \mathbb{E}\left[g_{\boldsymbol{G}}^-(\boldsymbol{o})\right], \tag{2.1}$$

where the expectations may be finite or infinite. The term **unimodular network** is used as an abbreviation for unimodular random rooted network. A probability measure on $\mathcal{G}_*$ is called **unimodular** when, by considering it as a random rooted network, one gets a unimodular network.

REMARK 1. For a function g as above, $g_G(\cdot, \cdot)$ can be regarded as a function on $V(G) \times V(G)$ (or a transport kernel on $V(G)$) defined for all networks G. One can interpret $g_G(u, v)$ as the amount of mass that is transported from u to v. Using this intuition, $g_G^+(o)$ (resp. $g_G^-(o)$) can be seen as the amount of mass that goes out of (resp. comes into) o and (2.1) expresses some conservation of mass in expectation. It is referred to as the **mass transport principle** in the literature. With this analogy, a measurable function $g : \mathcal{G}_{**} \to \mathbb{R}^{\geq 0}$ is also called an **invariant transport kernel** in this paper.

The **invariant sigma-field** I on $\mathcal{G}_*$ is the family of events in $\mathcal{G}_*$ that are invariant under changing the root; i.e., events $A \subseteq \mathcal{G}_*$ such that for every rooted network (G, o) and every $v \in V(G)$, if $[G, o] \in A$, then $[G, v] \in A$. Events in I are also called **invariant events** here. A unimodular network $[\boldsymbol{G}, \boldsymbol{o}]$ is called **ergodic** (also called *extremal* in [**1**]) if any invariant event has probability 0 or 1.

A measurable function on $w : \mathcal{G}_* \to \mathbb{R}$ is I-measurable if and only if it **doesn't depend on the root**; i.e., for every rooted network (G, o) and every $v \in V(G)$, one has $w[G, o] = w[G, v]$. Also, if $[\boldsymbol{G}, \boldsymbol{o}]$ is a random rooted network, we say $w[\boldsymbol{G}, \boldsymbol{o}]$ **doesn't depend on the root almost surely** if almost surely, for all $v \in V(\boldsymbol{G})$, one has $w[\boldsymbol{G}, \boldsymbol{o}] = w[\boldsymbol{G}, v]$.

The following definition is borrowed from [**3**].

DEFINITION 3. A **covariant subset** (of the vertices) is a function S which associates to each network G a set $S_G \subseteq V(G)$ such that $\{[G, o] : o \in S_G\}$ is a well-defined and measurable subset of $\mathcal{G}_*$. The same symbol is used for the subnetwork induced by S_G (i.e., the restriction of G to S_G) for all networks G. The latter is called a **covariant subnetwork**. In the applications in this paper, it is clear from context whether S_G is a subset of $V(G)$ or a subnetwork of G.

Note that for the set $\{[G, o] : o \in S_G\}$ to be well defined, S should be *covariant* under network isomorphisms, that is, for all isomorphisms $\rho : G \to G'$, one should have $\rho(S_G) = S_{G'}$. In this paper, the term $1_{S_G}(o)$ is frequently used as a function of $[G, o] \in \mathcal{G}_*$ (it is equal to 1 when $o \in S_G$ and equal to zero otherwise). The measurability of this function is assumed in Definition 3.

In fact, covariant subsets are in one-to-one correspondence with measurable subsets of $\mathcal{G}_*$. For a measurable subset $A \subseteq \mathcal{G}_*$, its corresponding covariant subset is $S_G := \{v \in V(G) : [G, v] \in A\}$. This can be used to define a variety of covariant subsets.

As an example, the set of vertices of a given degree is a covariant subset. Also, the set of vertices with a given mark is a covariant subset (recall the definition of networks above).

The following lemma is borrowed from [**1**] (see also [**3**]).

LEMMA 1. *Let $[\boldsymbol{G}, \boldsymbol{o}]$ be a unimodular network and S be a covariant subset of the vertices. Then $\mathbb{P}[S_{\boldsymbol{G}} \neq \emptyset] > 0$ if and only if $\mathbb{P}[\boldsymbol{o} \in S_{\boldsymbol{G}}] > 0$. Equivalently, $S_{\boldsymbol{G}} = V(\boldsymbol{G})$ a.s. if and only if $\boldsymbol{o} \in S_{\boldsymbol{G}}$ a.s.*

PROOF. Let $g_G(u, v) := 1_{S_G}(v)$. For $o \in V(G)$, one has $g_G^+(o) = \#S_G$ and $g_G^-(o) = 1_{S_G}(o)\#V(G)$. Now, the first claim easily follows from (2.1). The second claim is then implied by considering the complement of the covariant subset. □

3. Shift-Coupling of Random Rooted Networks

In this section, different formulations of *unrooted-equivalence* are defined and the main theorem of this paper is presented, which studies the implications between these formulations. The proofs of most of the results are moved to Section 7 to help to focus on the main thread. The reader can either see the proofs first or proceed to the next results with no problem.

3.1. Definitions.

The following definitions are needed for stating the main definition (Definition 7).

DEFINITION 4. The projection $\pi : \mathcal{G}_* \to \mathcal{G}$ defined by $\pi[G, o] := [G]$ induces a sigma-field, namely J, on $\mathcal{G}$ as follows.

$$J := \{B \subseteq \mathcal{G} : \pi^{-1}(B) \text{ is measurable}\}.$$

A **random non-rooted network** is a random element in, or a probability measure on $(\mathcal{G}, J)$, although it does not form a standard probability space (Proposition 1 below). If $[\boldsymbol{G}, \boldsymbol{o}]$ is a random rooted network with distribution μ, the symbol $[\boldsymbol{G}]$ is used for its corresponding random non-rooted network whose distribution is $\pi_*\mu$. It can also be seen as a natural coupling of μ and $\pi_*\mu$.

Non-standardness of $(\mathcal{G}, J)$ is stated in the following proposition. It is essentially an easy result in theory of *smooth Borel equivalence relations*. See the notes in Subsection 3.4.

PROPOSITION 1. *The measurable space $(\mathcal{G}, J)$ is not a standard Borel space. More precisely, there is no metric on $\mathcal{G}$ that makes it a Polish space whose Borel sigma-field is J.*

Due to non-standardness, several classical tools of probability theory may fail for random non-rooted networks; e.g., conditional expectation. However, it poses no problem for the arguments in this paper; e.g., equality of distributions, pull-back and push-forward of distributions, etc.

Note also that the map π^{-1} corresponds J bijectively to the invariant sigma-field I on $\mathcal{G}_*$. Therefore, probability measures on $(\mathcal{G}, J)$ are in one-to-one correspondence with probability measures on $(\mathcal{G}_*, I)$.

CONVENTION 1. If f is an I-measurable function on $\mathcal{G}_*$, then the same symbol f is used for the J-measurable function $f \circ \pi^{-1}$ defined on $\mathcal{G}$.

Note that in the above convention, for every $x \in \mathcal{G}$, f is constant on $\pi^{-1}(x)$. Therefore $f \circ \pi^{-1}$ is indeed well defined and J-measurable. In the (few) instances in this paper, according to the context, this convention causes no ambiguity.

DEFINITION 5. Let μ be a probability measure on a measurable space X and $w : X \to \mathbb{R}^{\geq 0}$ be a measurable function. Assume $0 < c := \int_X w(x)d\mu(x) < \infty$. By **biasing μ by w** we mean the measure, namely ν, on X defined by

$$\nu(A) := \frac{1}{c}\int_X w(x)1_A(x)d\mu(x).$$

The choice of c implies that ν is a probability measure. It is the unique probability measure on X whose Radon-Nikodym derivative w.r.t. μ is proportional to w.

In particular, let $X := \mathcal{G}_*$ and μ be the distribution of a random rooted network $[\boldsymbol{G}, \boldsymbol{o}]$. Denote by $[\boldsymbol{G}', \boldsymbol{o}']$ a random rooted network with distribution ν defined above. For every measurable function $f : \mathcal{G}_* \to \mathbb{R}^{\geq 0}$, one gets

$$\mathbb{E}\left[f[\boldsymbol{G}', \boldsymbol{o}']\right] = \frac{1}{\mathbb{E}\left[w[\boldsymbol{G}, \boldsymbol{o}]\right]}\mathbb{E}\left[f[\boldsymbol{G}, \boldsymbol{o}]w[\boldsymbol{G}, \boldsymbol{o}]\right].$$

A similar equation holds for biasing the distribution of a random non-rooted network.

It can be seen that biasing μ by w is equal to μ if and only if w is almost surely constant w.r.t. μ; i.e., for some constant c one has $w = c$, μ-a.s. Note that $\int_X w(x)d\mu(x)$ is not assumed to be equal to one. As an example, for an event $B \subseteq X$, conditioning μ on B is just biasing μ by the indicator function 1_B.

LEMMA 2. *By biasing the distribution of a random rooted network $[\boldsymbol{G}, \boldsymbol{o}]$ by a function w, the distribution of $[\boldsymbol{G}]$ becomes biased by $\mathbb{E}\left[w[\boldsymbol{G}, \boldsymbol{o}]\,|I\right]$, where the latter, which is I-measurable, is considered as a function of $[\boldsymbol{G}]$ (see Convention 1).*

This lemma is straightforward and we skip its proof.

DEFINITION 6. Let $[\boldsymbol{G}, \boldsymbol{o}]$ be a (not necessarily unimodular) random rooted network and $T : \mathcal{G}_{**} \to \mathbb{R}^{\geq 0}$ be a measurable function. Assume $T^+_{\boldsymbol{G}}(\boldsymbol{o}) = 1$ a.s. Conditioned on $[\boldsymbol{G}, \boldsymbol{o}]$, choose a new root in $V(\boldsymbol{G})$ with distribution $T_{\boldsymbol{G}}(\boldsymbol{o}, \cdot)$; i.e., consider the following probability measure on $\mathcal{G}_*$.

$$A \mapsto \mathbb{E}\left[\sum_{v \in V(\boldsymbol{G})} T_{\boldsymbol{G}}(\boldsymbol{o}, v) 1_A[\boldsymbol{G}, v]\right]. \tag{3.1}$$

Any random rooted network with this distribution is called the **root-change of $[\boldsymbol{G}, \boldsymbol{o}]$ by kernel T.**

LEMMA 3. *If $[\boldsymbol{G}', \boldsymbol{o}']$ is a root-change of $[\boldsymbol{G}, \boldsymbol{o}]$, then $[\boldsymbol{G}, \boldsymbol{o}]$ is also a root-change of $[\boldsymbol{G}', \boldsymbol{o}']$.*

In fact, the proof of this lemma (given in Section 7) can be used to define the *time reversal* of a root-change similar to that of a Markov chain.

The following simple example helps to illustrate the definitions.

EXAMPLE 1. Let $[\boldsymbol{G}, \boldsymbol{o}]$ be a random rooted graph with at least 2 vertices a.s. Let $[\boldsymbol{G}', \boldsymbol{o}']$ be another random rooted graph with the following distribution:

$$Q(A) := \frac{1}{\mathbb{E}\left[d(\boldsymbol{o})\right]} \mathbb{E}\left[\sum_{v : v \sim \boldsymbol{o}} 1_A[\boldsymbol{G}, v]\right].$$

It is straightforward that Q is indeed a probability measure on $\mathcal{G}_*$. One can rewrite this formula as

$$\frac{1}{\mathbb{E}\left[d(\boldsymbol{o})\right]} \mathbb{E}\left[d(\boldsymbol{o}) \sum_{v \in V(\boldsymbol{G})} T_{\boldsymbol{G}}(\boldsymbol{o}, v) 1_A[\boldsymbol{G}, v]\right],$$

where $T_G(u, v) := \frac{1}{d(u)}$ if $v \sim u$ and $T_G(u, v) = 0$ otherwise. We have $T^+_{\boldsymbol{G}}(\boldsymbol{o}) = 1$ a.s. Therefore, by Definitions 5 and 6, we can say that $[\boldsymbol{G}', \boldsymbol{o}']$ is obtained from $[\boldsymbol{G}, \boldsymbol{o}]$ by a biasing (by the degree $d(\boldsymbol{o})$ of the root) and then a root-change (by changing the root to a neighbor of the root chosen uniformly at random).

We are now ready to present the main definition.

DEFINITION 7. Let $[\boldsymbol{G}_1, \boldsymbol{o}_1]$ and $[\boldsymbol{G}_2, \boldsymbol{o}_2]$ be (not necessarily unimodular) random rooted networks. The following conditions are different definitions for $[\boldsymbol{G}_1, \boldsymbol{o}_1]$ and $[\boldsymbol{G}_2, \boldsymbol{o}_2]$ to be **unrooted-equivalent** (or to **have the same non-rooted networks**).

- **(B):** The distribution of each one is obtained from the other by a *biasing* and then a *root-change.*
- **(R):** The distribution of $[\boldsymbol{G}_2, \boldsymbol{o}_2]$ is obtained from $[\boldsymbol{G}_1, \boldsymbol{o}_1]$ by a root-change.
- **(C):** There is a coupling of them (i.e., a probability measure on $\mathcal{G}_* \times \mathcal{G}_*$ whose marginals are identical with the distributions of $[\boldsymbol{G}_i, \boldsymbol{o}_i]$'s) which is concentrated on the set of pairs of rooted networks with the same non-rooted networks; i.e., $\{([G_1, o_1], [G_2, o_2]) : [G_1] = [G_2]\}$.
- **(D):** There is a random doubly-rooted network $[\boldsymbol{G}, \boldsymbol{o}, \boldsymbol{o}']$ such that $[\boldsymbol{G}, \boldsymbol{o}]$ and $[\boldsymbol{G}, \boldsymbol{o}']$ have the same distributions as $[\boldsymbol{G}_1, \boldsymbol{o}_1]$ and $[\boldsymbol{G}_2, \boldsymbol{o}_2]$ respectively.

(F): By forgetting the roots, the random non-rooted networks $[\boldsymbol{G}_1]$ and $[\boldsymbol{G}_2]$ have the same distribution on $(\mathcal{G}, J)$. Equivalently, the distributions of $[\boldsymbol{G}_1, \boldsymbol{o}_1]$ and $[\boldsymbol{G}_2, \boldsymbol{o}_2]$ agree on the invariant sigma-field I.

It should be noted that the set mentioned in Condition (C) is indeed a measurable subset of $\mathcal{G}_* \times \mathcal{G}_*$ (let $A_{m,n}$ be the event that $N_n(o_1)$ is rooted-isometric to $N_n(v)$ for some $v \in N_m(o_2)$. It can be seen that $A_{m,n}$ is indeed measurable and the desired set is $\cup_m \cap_n A_{m,n}$).

As mentioned in the introduction, the definition with Condition (B) is used heuristically in some examples in the literature, some of which will be mentioned in Section 5.

3.2. Main Theorems. Here, we study the implications between the conditions in Definition 7. At first sight, Condition (F) may seem weaker than the other ones, because the other conditions assume the existence of a third object. But this is not the case as shown below.

THEOREM 1 (Shift-Coupling). *Let $[\boldsymbol{G}, \boldsymbol{o}]$ and $[\boldsymbol{G}', \boldsymbol{o}']$ be (not necessarily unimodular) random rooted networks. Then, $[\boldsymbol{G}', \boldsymbol{o}']$ can be obtained from $[\boldsymbol{G}, \boldsymbol{o}]$ by a root-change if and only if their distributions agree on the invariant sigma-field. In other words, conditions (F) and (R) are equivalent.*

Most results of this paper are based on the above Theorem. Also, the chosen name *shift-coupling* is justified in the notes in Subsection 3.4. This result is the main part in the following implications.

THEOREM 2. *Conditions (R), (C), (D) and (F) are equivalent and imply Condition (B).*

It should be noted that Condition (B) does not imply the other conditions (see Remark 2 below). Theorem 2 allows us to define the following.

DEFINITION 8. Under the assumptions of Definition 7, $[\boldsymbol{G}_1, \boldsymbol{o}_1]$ and $[\boldsymbol{G}_2, \boldsymbol{o}_2]$ are **weakly unrooted-equivalent** if Condition (B) holds and **strongly unrooted-equivalent** if the other equivalent conditions hold.

REMARK 2. According to Condition (F), when two random rooted networks are weakly unrooted-equivalent, the distributions of the corresponding non-rooted networks may be different (but are always mutually absolutely continuous by Lemma 2). This difference can be seen clearly in Example 7.

3.3. Some Applications. The following propositions are presented here as corollaries of Theorem 1. More important applications of the theorem will be presented in the next sections.

PROPOSITION 2. *Let $[\boldsymbol{G}, \boldsymbol{o}]$ be a (not necessarily unimodular) random rooted network and S be a covariant subset (Definition 3) such that $\mathbb{P}[\boldsymbol{o} \in S_{\boldsymbol{G}}] > 0$. Denote by $[\boldsymbol{G}', \boldsymbol{o}']$ the random rooted network obtained by conditioning $[\boldsymbol{G}, \boldsymbol{o}]$ on $\boldsymbol{o} \in S_{\boldsymbol{G}}$. Then, the following are equivalent.*

(i) *$[\boldsymbol{G}', \boldsymbol{o}']$ can be obtained from $[\boldsymbol{G}, \boldsymbol{o}]$ by a root-change.*
(ii) *$\mathbb{P}[\boldsymbol{o} \in S_{\boldsymbol{G}} | I]$ is almost surely constant.*

PROOF. The distribution of $[\boldsymbol{G}', \boldsymbol{o}']$ is obtained from that of $[\boldsymbol{G}, \boldsymbol{o}]$ by biasing by the function $1_{S_{\boldsymbol{G}}}(\boldsymbol{o})$. Lemma 2 implies that the distribution of $[\boldsymbol{G}']$ is obtained from that of $[\boldsymbol{G}]$ by biasing by $\mathbb{P}[\boldsymbol{o} \in S_{\boldsymbol{G}} | I]$ (considered as a function on $\mathcal{G}$).

First, assume the bias function $\mathbb{P}[\boldsymbol{o} \in S_{\boldsymbol{G}} \,|\, I]$ is almost surely constant. It follows that $[\boldsymbol{G}']$ and $[\boldsymbol{G}]$ are identically distributed; i.e., the distributions of $[\boldsymbol{G}, \boldsymbol{o}]$ and $[\boldsymbol{G}', \boldsymbol{o}']$ agree on the invariant sigma-field. Thus, Theorem 1 implies that $[\boldsymbol{G}', \boldsymbol{o}']$ can be obtained from $[\boldsymbol{G}, \boldsymbol{o}]$ by a root-change.

Conversely, assume $[\boldsymbol{G}', \boldsymbol{o}']$ can be obtained from $[\boldsymbol{G}, \boldsymbol{o}]$ by a root-change. Theorem 1 implies that their distributions agree on the invariant sigma-field. In other words, $[\boldsymbol{G}']$ and $[\boldsymbol{G}]$ have the same distribution. Since the former is obtained by biasing the latter by $\mathbb{P}[\boldsymbol{o} \in S_{\boldsymbol{G}} \,|\, I]$, it follows that the bias function is almost surely constant and the claim is proved. □

PROPOSITION 3 (Extra Head Scheme). *Let $[\boldsymbol{G}, \boldsymbol{o}]$ be a unimodular graph. Add i.i.d. marks in $\{0, 1\}$ to the vertices with Bernoulli distribution with parameter $0 < p \leq 1$. If $[\boldsymbol{G}, \boldsymbol{o}]$ is infinite a.s. then there exists a root-change that when applied to $[\boldsymbol{G}, \boldsymbol{o}]$, the result is the same (in distribution) as $[\boldsymbol{G}, \boldsymbol{o}]$ except that the mark of the root is forced to be 1.*

The condition of being infinite is necessary in this proposition as explained in Remark 3. See also [**5**] for the precise definition of adding i.i.d. marks to the vertices. The name *extra head scheme* is borrowed from an analogous definition in [**13**] as will be explained in Section 8.

PROOF OF PROPOSITION 3. Note that the desired random rooted network can be obtained by conditioning $[\boldsymbol{G}, \boldsymbol{o}]$ on $m(\boldsymbol{o}) = 1$, where $m(\cdot)$ denotes the marks of the vertices. Therefore, by Proposition 2, it is enough to prove that $\mathbb{P}[m(\boldsymbol{o}) = 1 \,|\, I]$ is almost surely constant. Let $A \in I$ be an invariant event. By Lemma 4 below, $\mathbb{P}[A \,|\, [\boldsymbol{G}, \boldsymbol{o}]]$ is $\{0, 1\}$-valued and does not depend on the root a.s. Therefore, conditioned on $[\boldsymbol{G}, \boldsymbol{o}]$, A is independent of any random variable including $m(\boldsymbol{o})$. Thus,

$$\mathbb{E}[m(\boldsymbol{o})1_A] = \mathbb{E}[\mathbb{E}[m(\boldsymbol{o}) \,|\, [\boldsymbol{G}, \boldsymbol{o}]]\,\mathbb{E}[1_A \,|\, [\boldsymbol{G}, \boldsymbol{o}]]] = \mathbb{E}[p\mathbb{E}[1_A \,|\, [\boldsymbol{G}, \boldsymbol{o}]]] = \mathbb{E}[p1_A].$$

This equation for all $A \in I$ implies that $\mathbb{E}[m(\boldsymbol{o}) \,|\, I] = p$ a.s. So $\mathbb{P}[m(\boldsymbol{o}) = 1 \,|\, I] = p$ a.s. and the claim is proved. □

The following lemma is used in the proof of Proposition 3 and is interesting in its own. It is similar to the ergodicity of the Bernoulli point process on $\mathbb{Z}^d$ or the Poisson point process in $\mathbb{R}^d$ (see Section 8).

LEMMA 4. *Let $[\boldsymbol{G}, \boldsymbol{o}]$ be a unimodular graph and $[\boldsymbol{G}', \boldsymbol{o}']$ be a random network obtained by adding i.i.d. marks to the vertices of $[\boldsymbol{G}, \boldsymbol{o}]$. If $[\boldsymbol{G}, \boldsymbol{o}]$ is ergodic and almost surely infinite, then so is $[\boldsymbol{G}', \boldsymbol{o}']$. More generally, if $[\boldsymbol{G}, \boldsymbol{o}]$ is infinite a.s., then for any invariant event $A \in I$,*

$$\mathbb{P}\left[[\boldsymbol{G}', \boldsymbol{o}'] \in A \,|\, [\boldsymbol{G}, \boldsymbol{o}]\right] \in \{0, 1\}, \quad a.s. \tag{3.2}$$

and the left hand side does not depend on the root a.s.

Note that in the statement of the lemma, the natural coupling of $[\boldsymbol{G}, \boldsymbol{o}]$ and $[\boldsymbol{G}', \boldsymbol{o}']$ is considered to enable us to condition $[\boldsymbol{G}', \boldsymbol{o}']$ on $[\boldsymbol{G}, \boldsymbol{o}]$. The proof is presented in Section 7.

REMARK 3. The claims of Lemma 4 and Proposition 3 are false for any finite unimodular network. Note that in this case, conditioned on $[\boldsymbol{G}, \boldsymbol{o}]$, with positive probability the marks of all vertices are 0. This contradicts (3.2). Also, the same property holds in any root-change of the network, contradicting the claim of Proposition 3.

3.4. Notes. The name *shift-coupling* for Theorem 1 is borrowed from the analogous result of [**20**]. This result studies when two random elements in a space equipped with some group action have a coupling such that the second one is obtained from the first by a shift corresponding to a random element of the group, called a shift-coupling in the literature. Here, instead of a group action, we have root-changes as in Condition (C) of Definition 7, which don't form a group. In Section 7, a proof of Theorem 1 is presented by mimicking that of [**20**]. A second proof is also presented using the result of [**20**]. With this proof, one can generalize Theorem 1 to the context of Borel equivalence relations as follows.

The following definitions are borrowed from [**7**]. An equivalence relation R on a Polish space X is a **countable Borel equivalence relation** if when considered as a subset of $X \times X$, it is a Borel subset and each equivalence class is countable. The **R-invariant sigma-field** on X consists of Borel subsets of X which are formed by unions of R-equivalence classes. In the following result, a Borel automorphism $F : X \to X$ is called **R-stabilizing** if $F(x)Rx$ for each $x \in X$.

THEOREM 1'. *Let R be a countable Borel equivalence relation on X and Y_1 and Y_2 be random elements in X. Then the following are equivalent.*

(i) *The distributions of Y_1 and Y_2 agree on the R-invariant sigma-field.*
(ii) *There exists a coupling of the distributions of Y_1 and Y_2 that is supported on R (i.e., assigns full measure to R).*
(iii) *There exists a random R-stabilizing Borel automorphism F such that $F(Y_1)$ has the same distribution as Y_2.*

In fact, in the last statement, F can be chosen to be supported on countably many automorphisms. As mentioned above, the proof of this theorem is similar to one of the proofs given for Theorem 1 and is skipped here.

A Borel equivalence relation R is **smooth** if the quotient space X/R with the induced Borel structure is a standard Borel space. Therefore, Proposition 1 just claims that the equivalence relation on $\mathcal{G}_*$ induced by π (see the proof of Theorem 1) is not smooth, which is implied by Corollary 1.3 of [**10**]. A direct proof is also presented in Section 7.

4. The Unimodular Case and Balancing Transport Kernels

In this section, some applications of Theorem 1 are presented for the case of unimodular networks. The main results are theorems 3 and 4 whose proofs are postponed to the end of the section after presenting some minor results.

THEOREM 3 (Uniqueness). *The distribution of a unimodular network $[\boldsymbol{G}, \boldsymbol{o}]$ is uniquely determined by its restriction to the invariant sigma-field (or equivalently, by the distribution of the non-rooted network $[\boldsymbol{G}]$). In other words, if two unimodular networks are strongly unrooted-equivalent, then they are identically distributed.*

Theorem 3 is a precise formulation of a comment in [**1**] saying that 'intuitively, the distribution of the root is forced given the distribution of the unrooted network'. Note also that if we replace 'strongly' with 'weakly' in this theorem, the claim no longer holds. This case will be considered in Lemma 7 and Proposition 4 below.

THEOREM 4 (Balancing Transport Kernel). *Let $[\boldsymbol{G},\boldsymbol{o}]$ be a unimodular network and $w_i : \mathcal{G}_* \to \mathbb{R}^{\geq 0}$ be measurable functions for $i = 1, 2$. Assume $\mathbb{E}[w_1 \mid I] < \infty$ a.s. Then, the following are equivalent.*

(i) *There is an invariant transport kernel that almost surely balances between the functions $w_1[\boldsymbol{G},\cdot]$ and $w_2[\boldsymbol{G},\cdot]$ on the vertices; i.e., a measurable function $T : \mathcal{G}_{**} \to \mathbb{R}^{\geq 0}$ such that almost surely, $T^+_{\boldsymbol{G}}(v) = w_1[\boldsymbol{G}, v]$ and $T^-_{\boldsymbol{G}}(v) = w_2[\boldsymbol{G}, v]$ for all $v \in V(\boldsymbol{G})$.*

(ii) *One has*

$$\mathbb{E}[w_1[\boldsymbol{G},\boldsymbol{o}] \mid I] = \mathbb{E}[w_2[\boldsymbol{G},\boldsymbol{o}] \mid I]. \tag{4.1}$$

Theorem 4 is analogous to similar results for stationary point processes and random measures ([**13**] and [**15**]). This analogy will be explained in Section 8.

REMARK 4. A result similar to Proposition 2 holds with the assumptions of Theorem 4. For $i = 1, 2$, consider biasing the distribution of a (not necessarily unimodular) random rooted network $[\boldsymbol{G},\boldsymbol{o}]$ by a function w_i. Then, the resulting random rooted networks are always weakly unrooted-equivalent, but this holds strongly if and only if the ratio $\mathbb{E}[w_1[\boldsymbol{G},\boldsymbol{o}] \mid I] / \mathbb{E}[w_2[\boldsymbol{G},\boldsymbol{o}] \mid I]$ is almost surely constant. However, the existence of a balancing transport kernel as in Theorem 4 is only proved for the unimodular case.

Before proving the above theorems, we present some other minor results in the unimodular case.

LEMMA 5. *Let $[\boldsymbol{G}_1,\boldsymbol{o}_1]$ be a unimodular network and $[\boldsymbol{G}_2,\boldsymbol{o}_2]$ be an arbitrary random rooted network.*

(i) *If $[\boldsymbol{G}_2,\boldsymbol{o}_2]$ is a root-change of $[\boldsymbol{G}_1,\boldsymbol{o}_1]$ by kernel T, then it can also be obtained by biasing $[\boldsymbol{G}_1,\boldsymbol{o}_1]$ by the function $[G, o] \mapsto T^-_G(o)$.*

(ii) *If $[\boldsymbol{G}_2,\boldsymbol{o}_2]$ is weakly unrooted-equivalent to $[\boldsymbol{G}_1,\boldsymbol{o}_1]$, then the distribution of $[\boldsymbol{G}_2,\boldsymbol{o}_2]$ is obtained from that of $[\boldsymbol{G}_1,\boldsymbol{o}_1]$ by only a biasing (i.e., is absolutely continuous w.r.t. the distribution of $[\boldsymbol{G}_1,\boldsymbol{o}_1]$).*

PROOF. (i). Given an event $A \subseteq \mathcal{G}_*$, define $g[G, o, v] := T_G(o, v)1_A[G, v]$. By (3.1) and unimodularity of $[\boldsymbol{G}_1,\boldsymbol{o}_1]$, one gets

$$\begin{aligned}
\mathbb{P}[[\boldsymbol{G}_2,\boldsymbol{o}_2] \in A] &= \mathbb{E}\left[g^+_{\boldsymbol{G}_1}(\boldsymbol{o}_1)\right] = \mathbb{E}\left[g^-_{\boldsymbol{G}_1}(\boldsymbol{o}_1)\right] \\
&= \mathbb{E}\left[1_A[\boldsymbol{G}_1,\boldsymbol{o}_1] \sum_{v \in V(\boldsymbol{G}_1)} T_{\boldsymbol{G}_1}(v, \boldsymbol{o}_1)\right] \\
&= \mathbb{E}\left[1_A[\boldsymbol{G}_1,\boldsymbol{o}_1] T^-_{\boldsymbol{G}_1}(\boldsymbol{o}_1)\right].
\end{aligned}$$

By letting $A := \mathcal{G}_*$, one gets $\mathbb{E}\left[T^-_{\boldsymbol{G}_1}(\boldsymbol{o}_1)\right] = 1$. Therefore, the above equation means that the distribution of $[\boldsymbol{G}_2,\boldsymbol{o}_2]$ is obtained from that of $[\boldsymbol{G}_1,\boldsymbol{o}_1]$ by the desired biasing (see Definition 5).

(ii). By part (i) and Definition 8, $[\boldsymbol{G}_2,\boldsymbol{o}_2]$ is obtained by biasing $[\boldsymbol{G}_1,\boldsymbol{o}_1]$ by a composition of two biasings, say by functions w_1 and w_2. It is easy to show that the result is just biasing by $w_1 w_2$ and the claim is proved. □

LEMMA 6. *Let $[\boldsymbol{G},\boldsymbol{o}]$ be a unimodular network and $w : \mathcal{G}_* \to \mathbb{R}^{\geq 0}$ be a measurable function. Then, biasing $[\boldsymbol{G},\boldsymbol{o}]$ by w gives a unimodular probability measure if and only if w doesn't depend on the root a.s.*

PROOF. Let $[\boldsymbol{G}', \boldsymbol{o}']$ be a random rooted network whose distribution is obtained by biasing that of $[\boldsymbol{G}, \boldsymbol{o}]$ by w. Let $m := \mathbb{E}[w[\boldsymbol{G}, \boldsymbol{o}]]$. For a measurable function $g : \mathcal{G}_{**} \to \mathbb{R}^{\geq 0}$, one has

$$\mathbb{E}\left[g^+_{\boldsymbol{G}'}(\boldsymbol{o}')\right] = \frac{1}{m}\mathbb{E}\left[w[\boldsymbol{G}, \boldsymbol{o}]g^+_{\boldsymbol{G}}(\boldsymbol{o})\right] = \frac{1}{m}\mathbb{E}\left[h^+_{\boldsymbol{G}}(\boldsymbol{o})\right],$$

where $h_G(o, v) := w[G, o]g_G(o, v)$. By unimodularity of $[\boldsymbol{G}, \boldsymbol{o}]$, one obtains

$$\mathbb{E}\left[g^+_{\boldsymbol{G}'}(\boldsymbol{o}')\right] = \frac{1}{m}\mathbb{E}\left[h^-_{\boldsymbol{G}}(\boldsymbol{o})\right] = \frac{1}{m}\mathbb{E}\left[\sum_{v \in V(\boldsymbol{G})} w[\boldsymbol{G}, v]g_{\boldsymbol{G}}(v, \boldsymbol{o})\right].$$

On the other hand,

$$\mathbb{E}\left[g^-_{\boldsymbol{G}'}(\boldsymbol{o}')\right] = \frac{1}{m}\mathbb{E}\left[w[\boldsymbol{G}, \boldsymbol{o}]g^-_{\boldsymbol{G}}(\boldsymbol{o})\right] = \frac{1}{m}\mathbb{E}\left[\sum_{v \in V(\boldsymbol{G})} w[\boldsymbol{G}, \boldsymbol{o}]g_{\boldsymbol{G}}(v, \boldsymbol{o})\right].$$

Therefore, $[\boldsymbol{G}', \boldsymbol{o}']$ is unimodular if and only if

$$\forall g : \mathbb{E}\left[\sum_{v \in V(\boldsymbol{G})} \left(w[\boldsymbol{G}, v] - w[\boldsymbol{G}, \boldsymbol{o}]\right) g_{\boldsymbol{G}}(v, \boldsymbol{o})\right] = 0. \tag{4.2}$$

First, suppose that almost surely, $w[\boldsymbol{G}, v] = w[\boldsymbol{G}, \boldsymbol{o}]$ for all $v \in V(\boldsymbol{G})$. This implies that (4.2) holds and thus, $[\boldsymbol{G}', \boldsymbol{o}']$ is unimodular. Conversely, assume $[\boldsymbol{G}', \boldsymbol{o}']$ is unimodular. By substitute $g_G(v, o)$ with the positive and negative parts of $w[G, v] - w[G, o]$ respectively, (4.2) gives that almost surely, $w[\boldsymbol{G}, v] = w[\boldsymbol{G}, \boldsymbol{o}]$ for all $v \in V(\boldsymbol{G})$. So, the claim is proved. □

LEMMA 7. *Let $[\boldsymbol{G}_1, \boldsymbol{o}_1]$ and $[\boldsymbol{G}_2, \boldsymbol{o}_2]$ be random rooted networks that are weakly unrooted-equivalent. If both are unimodular, then the distribution of $[\boldsymbol{G}_2, \boldsymbol{o}_2]$ can be obtained by biasing that of $[\boldsymbol{G}_1, \boldsymbol{o}_1]$ by a function that doesn't depend on the root and is almost surely positive.*

PROOF. Since both are unimodular, by lemmas 5 and 6, $[\boldsymbol{G}_2, \boldsymbol{o}_2]$ is obtained from $[\boldsymbol{G}_1, \boldsymbol{o}_1]$ by biasing by a function $w : \mathcal{G}_* \to \mathbb{R}^{\geq 0}$ that doesn't depend on the root. As a result, the distribution of $[\boldsymbol{G}_2, \boldsymbol{o}_2]$ is absolutely continuous w.r.t. that of $[\boldsymbol{G}_1, \boldsymbol{o}_1]$. The same holds by swapping the roles of $[\boldsymbol{G}_1, \boldsymbol{o}_1]$ and $[\boldsymbol{G}_2, \boldsymbol{o}_2]$. Therefore, the Radon-Nikodym derivative, which is proportional to $w[\boldsymbol{G}_1, \boldsymbol{o}_1]$, is positive almost surely. This proves the claim. □

PROPOSITION 4. *Let $[\boldsymbol{G}_1, \boldsymbol{o}_1]$ and $[\boldsymbol{G}_2, \boldsymbol{o}_2]$ be random rooted networks which are weakly unrooted-equivalent. If at least one of them is an ergodic unimodular network, then they are also strongly unrooted-equivalent.*

PROOF. Assume $[\boldsymbol{G}_1, \boldsymbol{o}_1]$ is an ergodic unimodular network. Lemma 5 implies that $[\boldsymbol{G}_2, \boldsymbol{o}_2]$ can be obtained from $[\boldsymbol{G}_1, \boldsymbol{o}_1]$ by biasing by a measurable function $w : \mathcal{G}_* \to \mathbb{R}^{\geq 0}$. Lemma 2 implies that the distribution of $[\boldsymbol{G}_2]$ is obtained from that of $[\boldsymbol{G}_1]$ by biasing by $\mathbb{E}\left[w[\boldsymbol{G}_1, \boldsymbol{o}_1] \,|\, I\right]$. On the other hand, since $[\boldsymbol{G}_1, \boldsymbol{o}_1]$ is ergodic, the I-measurable function $\mathbb{E}\left[w[\boldsymbol{G}_1, \boldsymbol{o}_1] \,|\, I\right]$ is almost surely constant. It follows that $[\boldsymbol{G}_2]$ and $[\boldsymbol{G}_1]$ have the same distribution, which shows that $[\boldsymbol{G}_2, \boldsymbol{o}_2]$ is strongly unrooted-equivalent to $[\boldsymbol{G}_1, \boldsymbol{o}_1]$. □

We are now ready to prove the main theorems of this section.

PROOF OF THEOREM 3. A proof of the result can be obtained by using uniqueness of the ergodic decomposition ([**6**] and [**21**]). However, another proof is given below which is based on Theorem 1 and the above lemmas.

Let $[\boldsymbol{G}_1, \boldsymbol{o}_1]$ and $[\boldsymbol{G}_2, \boldsymbol{o}_2]$ be unimodular networks such that their distributions agree on the invariant sigma-field. Therefore, they are strongly unrooted-equivalent (Condition (F)). The same holds weakly by Theorem 2. Thus, Lemma 7 implies that $[\boldsymbol{G}_2, \boldsymbol{o}_2]$ is obtained by biasing the distribution of $[\boldsymbol{G}_1, \boldsymbol{o}_1]$ by a measurable function $w : \mathcal{G}_* \to \mathbb{R}^{\geq 0}$ that doesn't depend on the root. It is enough to show that $w[\boldsymbol{G}_1, \boldsymbol{o}_1]$ is almost surely constant.

By Lemma 2, the distribution of $[\boldsymbol{G}_2]$ is obtained from that of $[\boldsymbol{G}_1]$ by biasing by $\mathbb{E}\left[w[\boldsymbol{G}_1, \boldsymbol{o}_1] \,|I\right]$. Since the latter distributions are equal by assumption, it follows that $\mathbb{E}\left[w[\boldsymbol{G}_1, \boldsymbol{o}_1] \,|I\right]$ is almost surely constant. On the other hand, since w doesn't depend on the root, it is I-measurable and thus, $\mathbb{E}\left[w[\boldsymbol{G}_1, \boldsymbol{o}_1] \,|I\right] = w[\boldsymbol{G}_1, \boldsymbol{o}_1]$ a.s. It follows that the bias function $w[\boldsymbol{G}_1, \boldsymbol{o}_1]$ is almost surely constant. Therefore, $[\boldsymbol{G}_2, \boldsymbol{o}_2]$ and $[\boldsymbol{G}_1, \boldsymbol{o}_1]$ are identically distributed. $\square$

PROOF OF THEOREM 4. (i)$\Rightarrow$ (ii). Let $A \in I$ be an invariant event. Define $g[G, v, z] := T_G(v, z) 1_A[G, v] = T_G(v, z) 1_A[G, z]$. By the assumption, one gets that almost surely, $g^+_{\boldsymbol{G}}(\boldsymbol{o}) = w_1[\boldsymbol{G}, \boldsymbol{o}] 1_A[\boldsymbol{G}, \boldsymbol{o}]$ and $g^-_{\boldsymbol{G}}(\boldsymbol{o}) = w_2[\boldsymbol{G}, \boldsymbol{o}] 1_A[\boldsymbol{G}, \boldsymbol{o}]$. By unimodularity, one gets

$$\mathbb{E}\left[w_1[\boldsymbol{G}, \boldsymbol{o}] 1_A[\boldsymbol{G}, \boldsymbol{o}]\right] = \mathbb{E}\left[g^+_{\boldsymbol{G}}(\boldsymbol{o})\right] = \mathbb{E}\left[g^-_{\boldsymbol{G}}(\boldsymbol{o})\right] = \mathbb{E}\left[w_2[\boldsymbol{G}, \boldsymbol{o}] 1_A[\boldsymbol{G}, \boldsymbol{o}]\right].$$

By considering this for all $A \in I$, one obtains (4.1).

(ii) $\Rightarrow$ (i). For $i = 1, 2$, let $[\boldsymbol{G}_i, \boldsymbol{o}_i]$ be a random rooted network obtained by biasing $[\boldsymbol{G}, \boldsymbol{o}]$ by w_i. Assumption (4.1) and Lemma 2 imply that $[\boldsymbol{G}_1]$ has the same distribution as $[\boldsymbol{G}_2]$. In other words, $[\boldsymbol{G}_1, \boldsymbol{o}_1]$ and $[\boldsymbol{G}_2, \boldsymbol{o}_2]$ are strongly unrooted-equivalent (Condition (F)). By Theorem 1, $[\boldsymbol{G}_2, \boldsymbol{o}_2]$ can be obtained from $[\boldsymbol{G}_1, \boldsymbol{o}_1]$ by a root-change; i.e., there is a measurable function $t : \mathcal{G}_{**} \to \mathbb{R}^{\geq 0}$ such that $t^+_{\boldsymbol{G}_1}(\boldsymbol{o}_1) = 1$ a.s. and

$$\mathbb{E}\left[h[\boldsymbol{G}_2, \boldsymbol{o}_2]\right] = \mathbb{E}\left[\sum_{v \in V(\boldsymbol{G}_1)} t_{\boldsymbol{G}_1}(\boldsymbol{o}_1, v) h[\boldsymbol{G}_1, v]\right]$$

for any measurable function $h : \mathcal{G}_* \to \mathbb{R}^{\geq 0}$. Fix h arbitrarily. By the definition of $[\boldsymbol{G}_i, \boldsymbol{o}_i]$, one obtains

$$\mathbb{E}\left[w_2[\boldsymbol{G}, \boldsymbol{o}] h[\boldsymbol{G}, \boldsymbol{o}]\right] = \mathbb{E}\left[\sum_{v \in V(\boldsymbol{G})} t_{\boldsymbol{G}}(\boldsymbol{o}, v) w_1[\boldsymbol{G}, \boldsymbol{o}] h[\boldsymbol{G}, v]\right], \tag{4.3}$$

where the equation $\mathbb{E}\left[w_1[\boldsymbol{G}, \boldsymbol{o}]\right] = \mathbb{E}\left[w_2[\boldsymbol{G}, \boldsymbol{o}]\right]$ is used (which holds by (4.1)) to cancel out the denominators. Define an invariant transport kernel T by $T_G(o, v) := t_G(o, v) w_1(G, o)$. By unimodularity, one has

$$\begin{aligned}
\mathbb{E}\left[\sum_{v \in V(\boldsymbol{G})} t_{\boldsymbol{G}}(\boldsymbol{o}, v) w_1[\boldsymbol{G}, \boldsymbol{o}] h[\boldsymbol{G}, v]\right] &= \mathbb{E}\left[\sum_{v \in V(\boldsymbol{G})} T_{\boldsymbol{G}}(\boldsymbol{o}, v) h[\boldsymbol{G}, v]\right] \\
&= \mathbb{E}\left[\sum_{v \in V(\boldsymbol{G})} T_{\boldsymbol{G}}(v, \boldsymbol{o}) h[\boldsymbol{G}, \boldsymbol{o}]\right]
\end{aligned}$$

$$= \mathbb{E}\left[T^-_{\boldsymbol{G}}(\boldsymbol{o})h[\boldsymbol{G},\boldsymbol{o}]\right].$$

So, (4.3) implies that $\mathbb{E}\left[w_2[\boldsymbol{G},\boldsymbol{o}]h[\boldsymbol{G},\boldsymbol{o}]\right] = \mathbb{E}\left[T^-_{\boldsymbol{G}}(\boldsymbol{o})h[\boldsymbol{G},\boldsymbol{o}]\right]$. Since this holds for any h, it follows that $T^-_{\boldsymbol{G}}(\boldsymbol{o}) = w_2[\boldsymbol{G},\boldsymbol{o}]$ a.s. On the other hand, by $t^+_{\boldsymbol{G}_1}(\boldsymbol{o}_1) = 1$ a.s., one gets that $T^+_{\boldsymbol{G}}(\boldsymbol{o}) = w_1[\boldsymbol{G},\boldsymbol{o}]$ a.s. Therefore, Lemma 1 implies that the same holds for all vertices; i.e., almost surely, for all $v \in V(\boldsymbol{G})$, one has $T^+_{\boldsymbol{G}}(v) = w_1[\boldsymbol{G},v]$ and $T^-_{\boldsymbol{G}}(v) = w_2[\boldsymbol{G},v]$. So the theorem is proved. □

5. Network Extension and Unimodularization

In this section, the method of *network extension* is introduced and the shift-coupling theorem is applied to it. This method unifies some of the examples in the literature to construct unimodular networks. First, in Subsection 5.1 we study *unimodularizations* of a random non-rooted network in general. Then, network extension is studied in Subsection 5.2.

5.1. Unimodularizations of a Non-Rooted Network.

DEFINITION 9. Let μ_0 be a probability measure on $(\mathcal{G}, J)$ (or similarly, on $(\mathcal{G}_*, I)$). We say that a random rooted network $[\boldsymbol{G}',\boldsymbol{o}']$ is **unrooted-equivalent to** μ_0,

- **strongly** if the distribution of $[\boldsymbol{G}']$ is identical to μ_0.
- **weakly** if the distribution of $[\boldsymbol{G}']$ and μ_0 are mutually absolutely continuous.

If in addition $[\boldsymbol{G}',\boldsymbol{o}']$ is unimodular, we say it is a **(weak or strong) unimodularization of** μ_0 and μ_0 **can be unimodularized**.

Heuristically, unimodularization means to choose a random root for a given random non-rooted network to obtain a unimodular network.

To see why the weak sense is ever defined here, it will turn out that some well known examples in the literature are weak unimodularizations (see examples 8 and 9 of Subsection 5.2). Moreover, the notions of weak and strong here are analogous to the previous notions as described in the following lemma.

LEMMA 8. *In the case μ_0 is the distribution of $[\boldsymbol{G}]$, where $[\boldsymbol{G},\boldsymbol{o}]$ is a random rooted network, Definition 9 is reduced to Definition 8.*

This lemma is straightforward and we skip its proof.

PROPOSITION 5. *Under the assumptions of Definition 9, if μ_0 can be unimodularized (either weakly or strongly), then there is a unique strong unimodularization of μ_0.*

PROOF. Suppose $[\boldsymbol{G}',\boldsymbol{o}']$ is a weak unimodularization of μ_0. Let $w' : \mathcal{G} \to \mathbb{R}^{\geq 0}$ be the Radon-Nikodym derivative of μ_0 w.r.t. the distribution of $[\boldsymbol{G}']$. Let $[\boldsymbol{G},\boldsymbol{o}]$ be the random rooted network obtained by biasing $[\boldsymbol{G}',\boldsymbol{o}']$ by $w := w' \circ \pi$. Lemma 6 implies that $[\boldsymbol{G},\boldsymbol{o}]$ is unimodular. Lemma 2 implies that the distribution of $[\boldsymbol{G}]$ is equal to μ_0, which means that $[\boldsymbol{G},\boldsymbol{o}]$ is a strong unimodularization of μ_0. Now, Theorem 3 implies that this is the unique strong unimodularization of μ_0. □

Note that some probability measures on $(\mathcal{G}, J)$ (i.e., some random non-rooted networks) cannot be unimodularized; e.g., a deterministic semi-infinite path.

EXAMPLE 2 (Planar Dual I). For a plane graph G, denote by G^* its dual graph. Let $[\boldsymbol{G}, \boldsymbol{o}]$ be a unimodular plane graph (see Example 9.6 of [**1**] for how to regard a plane graph as a network and define its dual). With no need to select a vertex of the dual graph as a root, $[\boldsymbol{G}^*]$ makes sense as a random non-rooted network. In [**1**], a unimodular network is constructed based on the dual graph: First, bias the distribution of $[\boldsymbol{G}, \boldsymbol{o}]$ by F, where $F[G, o] := \sum_{f \sim o} 1/\deg(f)$. Then, choose a face incident to $\boldsymbol{o}$ as the root of the dual graph by choosing face f with probability proportional to $1/\deg(f)$. See Example 9.6 of [**1**] for the precise formula. It can be seen that in our language, this is a weak unimodularization of $[\boldsymbol{G}^*]$. As an example, one can see that if $\boldsymbol{G}$ is the triangular lattice with probability p and the square lattice with probability $q := 1 - p$, then the above construction is the hexagonal lattice with probability $2p/(2p+q)$ and the square lattice with probability $q/(2p+q)$, which does not have the same distribution as $[\boldsymbol{G}^*]$. In the general case, the strong unimodularization of $[\boldsymbol{G}^*]$ will be discussed in Example 9.

EXAMPLE 3 (Subnetwork). Let $[\boldsymbol{G}, \boldsymbol{o}]$ be a unimodular network and S be a covariant subnetwork (Definition 3). Assume $S_{\boldsymbol{G}}$ is nonempty and connected a.s. Therefore, $[S_{\boldsymbol{G}}]$ is a random non-rooted network; i.e., a random element in $(\mathcal{G}, \mathcal{J})$ (note that $S_{\boldsymbol{G}}$ doesn't need to contain $\boldsymbol{o}$ and no root is chosen for $S_{\boldsymbol{G}}$). Condition $[S_{\boldsymbol{G}}, \boldsymbol{o}]$ on the event $\boldsymbol{o} \in S_{\boldsymbol{G}}$; i.e., consider the probability measure $\mathcal{P}'[A] := \mathbb{P}\left[[S_{\boldsymbol{G}}, \boldsymbol{o}] \in A \,|\, \boldsymbol{o} \in S_{\boldsymbol{G}}\right]$ on $\mathcal{G}_*$. Considering $\mathcal{P}'$ as a random rooted network, we claim that it is a weak unimodularization of (the distribution of) $[S_{\boldsymbol{G}}]$. Let $\mu := \pi_* \mathcal{P}'$ be the distribution of the non-rooted network under $\mathcal{P}'$. By taking conditional expectation w.r.t. $\mathcal{I}$, one obtains for any measurable function $g : (\mathcal{G}, \mathcal{J}) \to \mathbb{R}^{\geq 0}$ that $\mathbb{E}\left[g[S_{\boldsymbol{G}}] 1_{S_{\boldsymbol{G}}}(\boldsymbol{o})\right] = \mathbb{E}\left[g[S_{\boldsymbol{G}}] \mathbb{P}\left[\boldsymbol{o} \in S_{\boldsymbol{G}} \,|\, \mathcal{I}\right]\right]$. Therefore,

$$\mathbb{E}\left[g[S_{\boldsymbol{G}}] \,|\, \boldsymbol{o} \in S_{\boldsymbol{G}}\right] = \mathbb{E}\left[g[S_{\boldsymbol{G}}] \frac{\mathbb{P}\left[\boldsymbol{o} \in S_{\boldsymbol{G}} \,|\, \mathcal{I}\right]}{\mathbb{P}\left[\boldsymbol{o} \in S_{\boldsymbol{G}}\right]}\right].$$

In other words, μ is just biasing the distribution of $[S_{\boldsymbol{G}}]$ by $\mathbb{P}\left[\boldsymbol{o} \in S_{\boldsymbol{G}} \,|\, \mathcal{I}\right] / \mathbb{P}\left[\boldsymbol{o} \in S_{\boldsymbol{G}}\right]$, where the latter is considered as a function on $(\mathcal{G}, \mathcal{J})$. Similar to Lemma 1, one can deduce from $S_{\boldsymbol{G}} \neq \emptyset$ a.s. that $\mathbb{P}\left[\boldsymbol{o} \in S_{\boldsymbol{G}} \,|\, \mathcal{I}\right] > 0$ a.s., thus, μ and the distribution of $[S_{\boldsymbol{G}}]$ are mutually absolutely continuous. On the other hand, it is easy to use (2.1) directly to see that $\mathcal{P}'$ is unimodular. Thus, $\mathcal{P}'$ is a weak unimodularization of (the distribution of) $[S_{\boldsymbol{G}}]$.

To obtain a strong unimodularization of $[S_{\boldsymbol{G}}]$, one can bias the distribution of $[\boldsymbol{G}, \boldsymbol{o}]$ by $1_{S_{\boldsymbol{G}}}(\boldsymbol{o}) / \mathbb{P}\left[\boldsymbol{o} \in S_{\boldsymbol{G}} \,|\, \mathcal{I}\right]$ and then consider the subnetwork rooted at $\boldsymbol{o}$ (see Lemma 2). Here, the denominator $\mathbb{P}\left[\boldsymbol{o} \in S_{\boldsymbol{G}} \,|\, \mathcal{I}\right]$ can be regarded as the *sample intensity* of $S_{\boldsymbol{G}}$, which is a random variable and a function of $[\boldsymbol{G}]$.

5.2. Unimodularizations of a Network Extension. In some examples in the literature, given a unimodular network $[\boldsymbol{G}_0, \boldsymbol{o}_0]$, another (not necessarily unimodular) random rooted network $[\boldsymbol{G}, \boldsymbol{o}]$ is obtained by adding some vertices and edges to the original network, called an *extension* here (Definition 10). Then, by biasing the probability measure and changing the root, another unimodular network is constructed. In this subsection, first a general method is presented that covers such examples and helps to construct new unimodular networks. Then, using the previous theorems, it is shown that this method gives all *unimodularizations* of $[\boldsymbol{G}]$ in the sense given in Subsection 5.1. A number of basic examples are provided as applications of the definitions and results, although the examples are not new.

The method presented here needs that the original network can be reconstructed from the extension, as explained in the following definition. In applications, to ensure the reconstruction is possible, one may add extra marks to the newly added vertices and edges (e.g., see Example 6). Nevertheless, after a new unimodular network is successfully constructed using the method, one may forget the extra marks and unimodularity will be preserved.

DEFINITION 10. Let $[\boldsymbol{G}_0, \boldsymbol{o}_0]$ be a unimodular network. An **extension** of $[\boldsymbol{G}_0, \boldsymbol{o}_0]$ is a pair $([\boldsymbol{G}, \boldsymbol{o}], S)$, where $[\boldsymbol{G}, \boldsymbol{o}]$ is a (not necessarily unimodular) random rooted network and S is a covariant subnetwork with the conditions that $\boldsymbol{o} \in S_{\boldsymbol{G}}$ a.s., $S_{\boldsymbol{G}}$ is connected a.s. and $[S_{\boldsymbol{G}}, \boldsymbol{o}]$ has the same distribution as $[\boldsymbol{G}_0, \boldsymbol{o}_0]$. It is called a **covariant extension** if

$$\mathbb{E}\left[\sum_{v \in S_{\boldsymbol{G}}} g[\boldsymbol{G}, \boldsymbol{o}, v]\right] = \mathbb{E}\left[\sum_{v \in S_{\boldsymbol{G}}} g[\boldsymbol{G}, v, \boldsymbol{o}]\right], \forall g \tag{5.1}$$

where g runs over all measurable functions $g : \mathcal{G}_{**} \to \mathbb{R}^{\geq 0}$. Here, $[\boldsymbol{G}, \boldsymbol{o}]$ is allowed to have a larger mark space than $[\boldsymbol{G}_0, \boldsymbol{o}_0]$.

Note that by Lemma 1, $[\boldsymbol{G}, \boldsymbol{o}]$ is non-unimodular except when $S_{\boldsymbol{G}} = \boldsymbol{G}$ a.s. Note also that $[\boldsymbol{G}, \boldsymbol{o}]$ is not necessarily a function of $[\boldsymbol{G}_0, \boldsymbol{o}_0]$; i.e., the newly added vertices and edges might be random. Moreover, (5.1) is stronger than unimodularity of $[S_{\boldsymbol{G}}, \boldsymbol{o}]$ (compare it with (2.1) for $[S_{\boldsymbol{G}}, \boldsymbol{o}]$). We are interested in covariant extensions only, since the results in this section only hold in the covariant case. See the following simple examples for more clarification of (5.1).

A more explicit definition of covariant extensions will be given below in Proposition 7. In this definition, (5.1) is automatically satisfied. The intuition behind Proposition 7 is more applicable than Definition 10 in practice, but it is postponed since formalizing the statements is technical, which is discussed before and after the proposition.

EXAMPLE 4. Let G_0 be the one-dimensional lattice with vertex set $\mathbb{Z}$. Let G be the graph obtained by adding a new vertex v_{2n} for any even number $2n \in \mathbb{Z}$ and connecting it to the vertex $2n$. For the networks that have a unique bi-infinite path, let S be the subnetwork that represents that path. Now, $[S_G, 0]$ has the same distribution as $[G_0, 0]$, but (5.1) does not hold (e.g., let $g(u, v)$ be zero except when $d(u) = 2$ and $d(v) = 3$). So, $([G, 0], S)$ is a non-covariant extension of $[G_0, 0]$.

However, if one chooses $\boldsymbol{o}$ uniformly at random in $\{0, 1\}$, then it can be seen that (5.1) holds and $[G, \boldsymbol{o}]$ is a covariant extension of $[G_0, 0]$. Moreover, by choosing $\boldsymbol{o}'$ uniformly at random in $\{0, 1, v_0\}$, $[G, \boldsymbol{o}']$ is unimodular and is a strong unimodularization of $[G]$. (Definition 9).

EXAMPLE 5. Let $[\boldsymbol{G}', \boldsymbol{o}']$ be a unimodular network and S be a covariant subnetwork such that $S_{\boldsymbol{G}'}$ is nonempty and connected a.s. Let $[\boldsymbol{G}, \boldsymbol{o}]$ be the random rooted network obtained by conditioning $[\boldsymbol{G}', \boldsymbol{o}']$ on $\boldsymbol{o}' \in S_{\boldsymbol{G}'}$ (see Example 3). It can be seen that $([\boldsymbol{G}, \boldsymbol{o}], S)$ is a covariant extension of $[S_{\boldsymbol{G}}, \boldsymbol{o}]$ and by Example 3, $[\boldsymbol{G}', \boldsymbol{o}']$ is a weak unimodularization of $[\boldsymbol{G}]$. This holds strongly if and only if the sample intensity $\mathbb{P}[\boldsymbol{o}' \in S_{\boldsymbol{G}'} | I]$ of S in $\boldsymbol{G}'$ is almost surely constant.

We are now ready to state the results of this section. All proofs are postponed to the end of the subsection.

LEMMA 9. *Let* $([\boldsymbol{G},\boldsymbol{o}],S)$ *be a covariant extension of a unimodular network. If* $[\boldsymbol{G}]$ *can be unimodularized, then there is a unique strong unimodularization of* $[\boldsymbol{G}]$ *and it can be obtained by applying a root-change to* $[\boldsymbol{G},\boldsymbol{o}]$.

Theorem 5 provides a general method to construct unimodularizations of a given covariant extension. Moreover, part (i) of the theorem gives a criteria for verifying existence or non-existence of a unimodularization.

THEOREM 5 (Unimodularization of an Extension). *Let* $([\boldsymbol{G},\boldsymbol{o}],S)$ *be a covariant extension of a unimodular network. Assume* $T:\mathcal{G}_{**}\to\mathbb{R}^{\geq 0}$ *is a measurable function such that* $T_{\boldsymbol{G}}$ *is almost surely a Markovian transport kernel from* $V(\boldsymbol{G})$ *to* $S_{\boldsymbol{G}}$*; i.e., almost surely, for all* $v\in V(\boldsymbol{G})$, $T^+_{\boldsymbol{G}}(v)=1$ *and* $T_{\boldsymbol{G}}(v,\cdot)\equiv 0$ *on* $V(\boldsymbol{G})\setminus S_{\boldsymbol{G}}$. *Let* $M:=M([\boldsymbol{G},\boldsymbol{o}]):=T^-_{\boldsymbol{G}}(\boldsymbol{o})$. *Then,*

(i) $[\boldsymbol{G}]$ *can be unimodularized if and only if* $\mathbb{E}[M\,|I]<\infty$ *a.s.*

(ii) *If* $\mathbb{E}[M]<\infty$, *then the following probability measure gives a weak unimodularization of* $[\boldsymbol{G}]$.

$$\mathcal{P}_T[A]:=\frac{1}{\mathbb{E}[M]}\mathbb{E}\left[\sum_{v\in V(\boldsymbol{G})}T_{\boldsymbol{G}}(v,\boldsymbol{o})1_A[\boldsymbol{G},v]\right]$$

(iii) *If* $\mathbb{E}[M\,|I]<\infty$ *a.s., then the following probability measure gives the unique strong unimodularization of* $[\boldsymbol{G}]$.

$$\mathcal{P}'_T[A]:=\mathbb{E}\left[\frac{1}{\mathbb{E}[M\,|I]}\sum_{v\in V(\boldsymbol{G})}T_{\boldsymbol{G}}(v,\boldsymbol{o})1_A[\boldsymbol{G},v]\right]$$

REMARK 5. The probability measure $\mathcal{P}_T$ (resp. $\mathcal{P}'_T$) in Theorem 5 can be described as biasing the distribution of $[\boldsymbol{G},\boldsymbol{o}]$ by M (resp. $M/\mathbb{E}[M\,|I]$) and then changing the root to a random vertex with distribution $\frac{1}{M}T_{\boldsymbol{G}}(\cdot,\boldsymbol{o})$.

As an example, it can be seen that the unimodularization of Example 4 can be obtained by the constructions in Theorem 5. More examples are provided at the end of this section.

COROLLARY 1. *Under the assumptions of Theorem* 5, *if* $\mathbb{P}[[\boldsymbol{G},\boldsymbol{o}]\in A]\in\{0,1\}$ *for all invariant events* $A\in I$, *then* $\mathcal{P}_T$ *and* $\mathcal{P}'_T$ *are equal and ergodic. Moreover,* $[\boldsymbol{G}]$ *has a unique weak unimodularization.*

In Theorem 5, the function T is assumed to be given. In the following theorem, the existence of T is studied using Theorem 1.

THEOREM 6 (Existence of T). *Let* $([\boldsymbol{G},\boldsymbol{o}],S)$ *be a covariant extension of a unimodular network. If* $[\boldsymbol{G}]$ *can be unimodularized, then*

(i) *There exists a function* $T:\mathcal{G}_{**}\to\mathbb{R}^{\geq 0}$ *satisfying the assumptions in Theorem* 5 *such that* $\mathcal{P}'_T$ *exists.*

(ii) *Any unimodularization of* $[\boldsymbol{G}]$ *can be obtained by biasing* $\mathcal{P}'_T$ *of the previous part by a function that doesn't depend on the root.*

(iii) T *can be chosen such that almost surely,* $T^-_{\boldsymbol{G}}(\cdot)$ *is constant on* $S_{\boldsymbol{G}}$ *(and depends only on* $[\boldsymbol{G}]$*).*

REMARK 6 (Comparison of $\mathcal{P}_T$ vs $\mathcal{P}'_T$). The unimodularization $\mathcal{P}_T$ in Theorem 5 has some advantages comparing to $\mathcal{P}'_T$ and the other unimodularizations.

One advantage is that it is easier to use since there is no division in the term under expectation and no conditional expectation w.r.t. I. Another is that the numerator in the definition of $\mathcal{P}_T[\cdot]$ (which is $\mathbb{E}[M] \times \mathcal{P}_T[\cdot]$) is a linear function of the distribution of $[\boldsymbol{G}, \boldsymbol{o}]$. Another one is shown in the following proposition. In contrast, an advantage of $\mathcal{P}'_T$ is that it is the strong unimodularization and thus, it doesn't change the distribution of the underlying non-rooted network (see Condition (F) in Definition 7 and Example 7). Moreover, in some cases $\mathcal{P}'_T$ is defined but $\mathcal{P}_T$ is not, as shown in Example 7.

PROPOSITION 6. *Let $([\boldsymbol{G}, \boldsymbol{o}], S)$ be a covariant extension of a unimodular network.*

(i) *There is at most one unimodularization $[\boldsymbol{G}', \boldsymbol{o}']$ of $[\boldsymbol{G}]$ with the property that $[\boldsymbol{G}', \boldsymbol{o}']$ conditioned on $\boldsymbol{o}' \in S_{\boldsymbol{G}'}$ has the same distribution as $[\boldsymbol{G}, \boldsymbol{o}]$.*
(ii) *Under the assumptions of Theorem 5, the unimodularization in the previous part exists if and only if $\mathbb{E}[M] < \infty$ and is given by $\mathcal{P}_T$, which is a weak unimodularization.*

COROLLARY 2. *The probability measures $\mathcal{P}_T$ and $\mathcal{P}'_T$ of Theorem 5 (and their existence) do not depend on the choice of the function T.*

REMARK 7. For a covariant extension $([\boldsymbol{G}, \boldsymbol{o}], S)$ of a unimodular network, if $[\boldsymbol{G}]$ has at least two different unimodularizations, then by taking convex combinations of the probability measures one finds infinitely many unimodularizations. So, Corollary 2 implies that there are unimodularizations which cannot be obtained by parts (ii) and (iii) of Theorem 5 (note that a further biasing is enough as described in part (ii) of Theorem 6).

REMARK 8. There is a special case of the construction in Theorem 5 which is sometimes easier to imagine and use, if possible. In the theorem, assume that for all vertices v, the function $T_G(v, \cdot)$ is concentrated on a single vertex, namely $\tau_G(v)$. Such a mapping τ can be called an **allocation** (or a **vertex-shift** in [**3**]). Then, the assumptions for T, apart from measurability, are reduced to $\tau_{\boldsymbol{G}}(\cdot) \in S_{\boldsymbol{G}}$ a.s. Also, M is equal to $\#\tau_{\boldsymbol{G}}^{-1}(\boldsymbol{o})$. However, In the setting of Theorem 6, existence of such an allocation is not guaranteed in general.

REMARK 9. In the setting of Theorem 5, assume $\mathbb{E}[M] = \infty$. By removing the denominator in the definition of $\mathcal{P}_T$, one can still define a measure on $\mathcal{G}_*$ which is not a probability measure. However, it can be seen that this measure is sigma-finite and satisfies the unimodularity condition (2.1).

The following are some basic examples for applications of the definitions and results, although they are not new.

EXAMPLE 6. Let (G_1, o_1) and (G_2, o_2) be the Euclidean lattices in dimension $d = 1$ and $d = 2$ respectively rooted at the origin. One has $V(G_1) = \mathbb{Z}$ and $V(G_2) = \mathbb{Z}^2$. Although $[G_2, o_2]$ is unimodular and can be obtained by adding some vertices and edges to $[G_1, o_1]$, it is not regarded as an extension of $[G_1, o_1]$ in Definition 10 because $[G_1, o_1]$ cannot be recovered from $[G_2, o_2]$ as a subnetwork (not depending on the root). However, let (G, o) be the network obtained from (G_2, o_2) by adding mark 1 to the vertices on the x axis and mark 0 to the other vertices. Assume also that every vertex in G_1 has mark 1. By letting P_G be the set of vertices with mark 1 in G, $([G, o], P)$ is an extension of $[G_1, o_1]$. It is easy to see that (5.1) holds and this extension is covariant. But $[G]$ cannot be unimodularized as shown below. For $v \in V(G)$, let $\tau_G(v)$ be the closest vertex in P_G to v. It can be seen that the assumptions in Remark 8 hold for τ. Since $\#\tau_G^{-1}(o) = \infty$, part (i) of Theorem 5 implies that $[G]$ cannot be unimodularized.

EXAMPLE 7. Let $[G_0, o_0]$ be the graph with a single vertex and L be a random odd number in $\mathbb{N}$. Let $\boldsymbol{G}$ be a path with length L and $\boldsymbol{o}$ be the middle vertex. Then, $[\boldsymbol{G}, \boldsymbol{o}]$ is a covariant extension of $[G_0, o_0]$. By sending unit mass from each vertex to $\boldsymbol{o}$, one gets $M = L$ in Theorem 5 and $\mathbb{E}[M\,|I] = L < \infty$. Therefore, $[\boldsymbol{G}]$ can always be unimodularized. Indeed, $\mathcal{P}'_T$ is just changing the root to a uniformly at random vertex in $[\boldsymbol{G}]$. However, $\mathcal{P}_T$ is defined only when $\mathbb{E}[L] < \infty$. In this case, $\mathcal{P}_T$ is equivalent to choosing a path whose probability of having length k is $k\mathbb{P}[L = k]/\mathbb{E}[L]$ (i.e., the size-biased version of L) and then, choosing a uniformly at random root in the path. Note that the distribution of the underlying non-rooted network is changed under $\mathcal{P}_T$, but unchanged under $\mathcal{P}'_T$.

The following examples shows that examples 9.6 and 9.8 of [**1**] are weak unimodularizations and can be obtained by the method of Theorem 5. Strong unimodularizations of the examples are also discussed.

EXAMPLE 8 (Vertex/Edge Replacement). Let $[\boldsymbol{G}_0, \boldsymbol{o}_0]$ be a unimodular network. In Example 9.6 of [**1**], it is shown how to attach to each edge uv of $\boldsymbol{G}_0$ a doubly-rooted network $L(u, v)$ that depends on the marks of u and v. The two roots of $L(u, v)$ are identified with u and v. Here, to make sure $\boldsymbol{G}_0$ can be reconstructed, add some extra mark to the other added vertices and let $\boldsymbol{G}$ be the resulting network. Now, it can be seen that $[\boldsymbol{G}, \boldsymbol{o}_0]$ gives a covariant extension of $[\boldsymbol{G}_0, \boldsymbol{o}_0]$. So, we may use Theorem 5 as follows. Send unit mass from each vertex of $\boldsymbol{G}_0$ to itself. Also, for each edge uv and each vertex in $L(u, v)$ other than the roots, send mass $\frac{1}{2}$ to each of u and v. It can be seen that this satisfies the assumptions of Theorem 5 and one has $M = 1 + \frac{1}{2}\sum_{v\sim \boldsymbol{o}}(\#V(L(\boldsymbol{o}, v)) - 2)$. Therefore, by Theorem 5, $[\boldsymbol{G}]$ can be unimodularized if and only if $\mathbb{E}[M\,|I] < \infty$ and $\mathcal{P}'_T$ gives the strong unimodularization of $[\boldsymbol{G}]$. Moreover, the probability measure constructed in [**1**] (defined when $\mathbb{E}[M] < \infty$) is identical with $\mathcal{P}_T$, which is a weak unimodularization.

With similar arguments, one can append a rooted network to each vertex of $\boldsymbol{G}_0$ and use the method of Theorem 5 to obtain another unimodular network.

EXAMPLE 9 (Planar Dual II). Let G_0 be a plane graph. Consider the union of G_0 and its dual G_0^* and add an edge vf for each vertex v and each of its adjacent faces $f \sim v$. To make sure G_0 can be recovered, add an extra mark to the vertices of G_0^*. Call the resulting network $G_0^\dagger$ as a function of G_0. Let S be a covariant subnetwork such that $S_{G_0^\dagger} = G_0$ for all possible cases of G_0 in this construction.

Let $[\boldsymbol{G}_0, \boldsymbol{o}_0]$ be a unimodular plane graph (see Example 2) and consider the random rooted network $[\boldsymbol{G}, \boldsymbol{o}] := [\boldsymbol{G}_0^\dagger, \boldsymbol{o}_0]$ as constructed above rooted at $\boldsymbol{o} := \boldsymbol{o}_0$. It can be seen that $([\boldsymbol{G}, \boldsymbol{o}], S)$ is a covariant extension of $[\boldsymbol{G}_0, \boldsymbol{o}_0]$. We will use Theorem 5. Send unit mass from each vertex $v \in V(\boldsymbol{G}_0)$ to itself. From each face f, send mass $\frac{1}{\deg(f)}$ to each of its adjacent vertices in $\boldsymbol{G}_0$, where $\deg(f)$ is the number of vertices of f. It can be seen that this satisfies the assumptions of Theorem 5 and $M = 1 + \sum_{f\sim \boldsymbol{o}} 1/\deg(f)$. Theorem 5 implies that $[\boldsymbol{G}]$ can be unimodularized if and only if $\mathbb{E}[M\,|I] < \infty$. If in addition $\mathbb{E}[M] < \infty$, then the following gives a weak unimodularization of $[\boldsymbol{G}]$.

$$\mathcal{P}_T[A] := \frac{1}{\mathbb{E}\left[M\right]}\mathbb{E}\left[1_A[\boldsymbol{G},\boldsymbol{o}] + \sum_{f\sim \boldsymbol{o}} \frac{1}{\deg(f)} 1_A[\boldsymbol{G}, f]\right].$$

By passing to $\boldsymbol{G}_0^*$ as a subnetwork of $\boldsymbol{G}$ as in Example 3, one can obtain a weak unimodularization of the dual random non-rooted graph $[\boldsymbol{G}_0^*]$. This is obtained by replacing M by $M-1$ in the above formula and deleting the term $1_A[\boldsymbol{G},\boldsymbol{o}]$. This is identical to the one constructed in Example 9.6 of [**1**]. Similarly, the following is the strong unimodularization of $[\boldsymbol{G}_0^*]$.

$$A \mapsto \mathbb{E}\left[\frac{1}{\mathbb{E}\left[M-1\,|I\right]} \sum_{f\sim \boldsymbol{o}_0} \frac{1}{\deg(f)} 1_A[\boldsymbol{G}_0^*, f]\right].$$

Similar to this example, one can add a new vertex for each edge-crossing of $\boldsymbol{G}_0$ with its dual. With similar arguments, the method of Theorem 5 gives the unimodularization of this new network given in Example 9.6 of [**1**].

In the following, a more explicit construction of covariant extensions is discussed. The reader can verify that all covariant extensions in this paper are special cases of this construction. The construction is stated in the Proposition 7 below. Before stating the result, the following definition is needed. Given a deterministic rooted network (G_0, o_0), a **super-network** of (G_0, o_0) is a rooted network (G, o) such that $V(G_0)$ is a subset of $V(G)$, $o = o_0$ and G_0 is the subnetwork of G induced by the subset $V(G_0) \subseteq V(G)$. Also, to make sure that G_0 can be determined given G, we assume an extra mark is assigned to the vertices in $V(G)\setminus V(G_0)$.

To help the reader, the proposition is stated in a heuristic way and is made precise after the proposition. Also, the proof is skipped for brevity.

PROPOSITION 7. *Assume* $\boldsymbol{\xi}$ *is a map that assigns to each deterministic rooted network* (G_0, o_0), *a* random super-network $\boldsymbol{\xi}(G_0, o_0)$ *of* (G_0, o_0). *Assume that*

(i) *its distribution does not depend on the root,*
(ii) *it is compatible with isomorphisms*
(iii) *and it satisfies suitable measurability conditions.*

Then, if $[\boldsymbol{G}_0, \boldsymbol{o}_0]$ *is a unimodular network, then* $\boldsymbol{\xi}(\boldsymbol{G}_0, \boldsymbol{o}_0)$ *is a covariant extension of* $[\boldsymbol{G}_0, \boldsymbol{o}_0]$ *(Definition* 10*).*

The definitions and conditions are made precise below. Special care is needed due to automorphisms of networks. The reader can skip the details in first reading.

Two super-networks (G, o) and (G', o') of (G_0, o_0) are called equivalent if there is a rooted isomorphism $\rho : (G, o) \to (G', o')$ such that its restriction to G_0 is the identity (note that the underlying graphs are *labeled* here). Denote by $\mathcal{G}(G_0, o_0)$ the set of equivalence classes of super-networks of (G_0, o_0). Similar to the arguments for $\mathcal{G}_*$, one can equip $\mathcal{G}(G_0, o_0)$ with a metric that makes it a complete separable metric space (the precise definition is skipped here). So, one can define a *random super-network* of (G_0, o_0) as a random element in $\mathcal{G}(G_0, o_0)$.

Note that if $v \in V(G_0)$ and (G, o) is a super-network of (G_0, o_0), then (G, v) is a super-network of (G_0, v). This provides a natural bijection $\alpha : \mathcal{G}(G_0, o_0) \to \mathcal{G}(G_0, v)$ which is measurable. Condition (i) of Proposition 7 is formalized as follows: For all (G_0, o_0) and all $v \in V(G_0)$, $\alpha(\boldsymbol{\xi}(G_0, o_0))$ has the same distribution as $\boldsymbol{\xi}(G_0, v)$.

Also, assume $\rho : (G_0, o_0) \to (G'_0, o'_0)$ is a rooted isomorphism and (G, o) is a super-network of (G_0, o_0). A natural super-network of (G'_0, o'_0) can be defined with vertex set $V(G'_0) \sqcup (V(G) \setminus V(G_0))$, where $\sqcup$ means disjoint union. The edge set is defined in the obvious way. This induces a natural bijection $\rho^* : \mathcal{G}(G_0, o_0) \to \mathcal{G}(G'_0, o'_0)$ which is also measurable. Condition (ii) of Proposition 7 is formalized as follows: If $\rho : (G_0, o_0) \to (G'_0, o'_0)$ is a rooted isomorphism, then $\rho^*(\boldsymbol{\xi}(G_0, o_0))$ has the same distribution as $\boldsymbol{\xi}(G'_0, o'_0)$.

Moreover, there is a natural function from $\mathcal{G}(G_0, o_0)$ to $\mathcal{G}_*$ (induced by the map $(G, o) \mapsto [G, o]$). Condition (iii) of Proposition 7 is formalized as follows: For every event $A \subseteq \mathcal{G}_*$, the following function, defined on $\mathcal{G}_*$, is measurable.

$$[G_0, o_0] \mapsto \mathbb{P}\left[[\boldsymbol{\xi}(G_0, o_0)] \in A\right],$$

where $[\boldsymbol{\xi}(G_0, o_0)]$ is the (random) element of $\mathcal{G}_*$ corresponding to $\boldsymbol{\xi}(G_0, o_0)$. Note that Condition (ii) implies that the right hand side depends only on the equivalence class of (G_0, o_0) and hence, the function is well defined.

Finally, $\boldsymbol{\xi}(\boldsymbol{G}_0, \boldsymbol{o}_0)$ (more precisely, $[\boldsymbol{\xi}(\boldsymbol{G}_0, \boldsymbol{o}_0)]$) can be defined by combining the two levels of randomness and by using Condition (ii). Also, there is a natural covariant subnetwork S (Definition 3) such that for all (G_0, o_0) and all super-networks (G, o) of (G_0, o_0), one has $S_G = G_0$. This makes $\xi(\boldsymbol{G}_0, \boldsymbol{o}_0)$ an extension of $[\boldsymbol{G}_0, \boldsymbol{o}_0]$. It remains to prove the proposition, which is skipped in this paper.

In fact, a converse to Proposition 7 also holds; i.e., every covariant extension can be obtained by the construction in Proposition 7. So, this gives an equivalent definition of covariant extensions. This claim is verified by the author and also skipped here.

Now, the proofs of the results of this section are presented. First, we start with the following lemma.

LEMMA 10. *Let $([\boldsymbol{G}, \boldsymbol{o}], S)$ be a covariant extension of a unimodular network. If $[\boldsymbol{G}', \boldsymbol{o}']$ is a unimodularization of $[\boldsymbol{G}]$, then $[\boldsymbol{G}, \boldsymbol{o}]$ is obtained by biasing $[\boldsymbol{G}', \boldsymbol{o}']$ by a function w such that almost surely, $w[\boldsymbol{G}', \cdot]$ is constant on $S_{\boldsymbol{G}'}$ (but may depend on $[\boldsymbol{G}']$) and is zero outside $S_{\boldsymbol{G}'}$.*

PROOF. By Lemma 5, $[\boldsymbol{G}, \boldsymbol{o}]$ can be obtained by biasing $[\boldsymbol{G}', \boldsymbol{o}']$ by a function, say w. Since $\boldsymbol{o} \in S_{\boldsymbol{G}}$ a.s., one obtains that on the event $\boldsymbol{o}' \notin S_{\boldsymbol{G}'}$, one has $w[\boldsymbol{G}', \boldsymbol{o}'] = 0$ a.s. Therefore, similar to Lemma 1, (5.1) implies that $w[\boldsymbol{G}', \cdot]$ is zero outside $S_{\boldsymbol{G}'}$ almost surely. Also, using (5.1) and an argument similar to Lemma 6, one obtains that almost surely, $w[\boldsymbol{G}', \cdot]$ is constant on $S_{\boldsymbol{G}'}$. We skip repeating the arguments for brevity. $\square$

PROOF OF LEMMA 9. Existence and uniqueness of a strong unimodularization $[\boldsymbol{G}', \boldsymbol{o}']$ is directly implied by Proposition 5. By definition, $[\boldsymbol{G}', \boldsymbol{o}']$ is strongly unrooted-equivalent to $[\boldsymbol{G}, \boldsymbol{o}]$. Therefore, Theorem 1 implies that $[\boldsymbol{G}', \boldsymbol{o}']$ can be obtained from $[\boldsymbol{G}, \boldsymbol{o}]$ by a root change. $\square$

PROOF OF THEOREM 5. We prove the parts of the theorem in a different order.

(iii) The assumptions imply that almost surely, for some vertex $v \in S_{\boldsymbol{G}}$, $T^-_{\boldsymbol{G}}(v) > 0$. By (5.1) and an argument similar to Lemma 1, one can obtain that $\mathbb{P}\left[T^-_{\boldsymbol{G}}(\boldsymbol{o}) > 0 \,|\, I\right] > 0$ a.s., hence, $\mathbb{E}\left[M \,|\, I\right] > 0$ a.s. Therefore, $\mathcal{P}'_T$ is well-defined. It can be seen that $\mathcal{P}'_T$ is a probability measure. Now, let $[\widetilde{\boldsymbol{G}}, \widetilde{\boldsymbol{o}}]$ be a random rooted network with distribution $\mathcal{P}'_T$. Let $g : \mathcal{G}_{**} \to \mathbb{R}^{\geq 0}$ be a measurable function.

By the definition of $\mathcal{P}'_T$ and conditioning on I, one gets

$$\begin{aligned}
&\mathbb{E}\left[\sum_{u\in V(\widetilde{\boldsymbol{G}})} g[\widetilde{\boldsymbol{G}},\widetilde{\boldsymbol{o}},u]\right]\\
&= \mathbb{E}\left[\frac{1}{\mathbb{E}[M\,|I]}\mathbb{E}\left[\sum_{v\in V(\boldsymbol{G})}\sum_{u\in V(\boldsymbol{G})} T_{\boldsymbol{G}}(v,\boldsymbol{o})g[\boldsymbol{G},v,u]\,|I\right]\right]\\
&= \mathbb{E}\left[\frac{1}{\mathbb{E}[M\,|I]}\mathbb{E}\left[\sum_{z\in S_{\boldsymbol{G}}}\sum_{v\in V(\boldsymbol{G})}\sum_{u\in V(\boldsymbol{G})} T_{\boldsymbol{G}}(v,\boldsymbol{o})T_{\boldsymbol{G}}(u,z)g[\boldsymbol{G},v,u]\,|I\right]\right]\\
&= \mathbb{E}\left[\frac{1}{\mathbb{E}[M\,|I]}\mathbb{E}\left[\sum_{z\in S_{\boldsymbol{G}}} \hat{g}[\boldsymbol{G},\boldsymbol{o},z]\,|I\right]\right],
\end{aligned}$$

where $\hat{g}[G,o,z] := \sum_{v\in V(G)}\sum_{u\in V(G)} T_G(v,o)T_G(u,z)g[G,v,u]$ and in the second equality, the equation $\sum_{z\in S_{\boldsymbol{G}}} T_{\boldsymbol{G}}(u,z) = 1$ a.s. is used, which holds by the assumptions. It can be seen that $\hat{g}$ is well-defined and measurable. One obtains a similar equation by swapping the two roots in all instances of g and $\hat{g}$. To summarize,

$$\text{(5.2)}\qquad \begin{cases} \mathbb{E}\left[g^+_{\widetilde{\boldsymbol{G}}}(\widetilde{\boldsymbol{o}})\right] &= \mathbb{E}\left[\frac{1}{\mathbb{E}[M|I]}\mathbb{E}\left[\sum_{z\in S_{\boldsymbol{G}}}\hat{g}[\boldsymbol{G},\boldsymbol{o},z]\,|I\right]\right],\\ \mathbb{E}\left[g^-_{\widetilde{\boldsymbol{G}}}(\widetilde{\boldsymbol{o}})\right] &= \mathbb{E}\left[\frac{1}{\mathbb{E}[M|I]}\mathbb{E}\left[\sum_{z\in S_{\boldsymbol{G}}}\hat{g}[\boldsymbol{G},z,\boldsymbol{o}]\,|I\right]\right].\end{cases}$$

Let $A\in I$ be an invariant event. By (5.1), one gets

$$\mathbb{E}\left[\sum_{z\in S_{\boldsymbol{G}}}\hat{g}[\boldsymbol{G},\boldsymbol{o},z]1_A[\boldsymbol{G},\boldsymbol{o}]\right] = \mathbb{E}\left[\sum_{z\in S_{\boldsymbol{G}}}\hat{g}[\boldsymbol{G},z,\boldsymbol{o}]1_A[\boldsymbol{G},\boldsymbol{o}]\right].$$

Since this holds for all $A\in I$, it follows that the right hand sides of (5.2) are equal. Thus, so are the left hand sides. This implies that (2.1) holds for $[\widetilde{\boldsymbol{G}},\widetilde{\boldsymbol{o}}]$, hence, $\mathcal{P}'_T$ is unimodular. To show that $\mathcal{P}'_T$ is a strong unimodularization of $[\boldsymbol{G}]$ (which implies that it is the unique one by Proposition 5), we will prove that $[\widetilde{\boldsymbol{G}}]$ has the same distribution as $[\boldsymbol{G}]$. Let $B\in I$. By the definition of $\mathcal{P}'_T$ and invariance of B under changing root, one gets

$$\begin{aligned}
\mathbb{P}\left[[\widetilde{\boldsymbol{G}},\widetilde{\boldsymbol{o}}]\in B\right] &= \mathbb{E}\left[\frac{1}{\mathbb{E}[M\,|I]}\sum_{v\in V(\boldsymbol{G})} T_{\boldsymbol{G}}(v,\boldsymbol{o})1_B[\boldsymbol{G},v]\right]\\
&= \mathbb{E}\left[\frac{1}{\mathbb{E}[M\,|I]}M1_B[\boldsymbol{G},\boldsymbol{o}]\right]\\
&= \mathbb{E}\left[\frac{1}{\mathbb{E}[M\,|I]}\mathbb{E}[M1_B[\boldsymbol{G},\boldsymbol{o}]\,|I]\right]\\
&= \mathbb{E}[1_B[\boldsymbol{G},\boldsymbol{o}]]\\
&= \mathbb{P}[[\boldsymbol{G},\boldsymbol{o}]\in B].
\end{aligned}$$

Therefore, the distributions of $[\widetilde{\boldsymbol{G}},\widetilde{\boldsymbol{o}}]$ and $[\boldsymbol{G},\boldsymbol{o}]$ agree on I, which proves the claim.

(ii) $\mathcal{P}_T$, whenever defined, is just biasing $\mathcal{P}'_T$ by $\mathbb{E}[M\,|I]$. Since the bias is I-measurable, lemmas 6 and 2 imply that $\mathcal{P}_T$ is unimodular and is unrooted-equivalent to $[\boldsymbol{G},\boldsymbol{o}]$ (not necessarily strongly).

(i) If $\mathbb{E}[M\,|I] < \infty$, then $\mathcal{P}'_T$ is well-defined and is a unimodularization of the extension by part (iii) proved above. Conversely, assume $[\boldsymbol{G}', \boldsymbol{o}']$ is a unimodularization of the extension. By definition, the distributions of $[\boldsymbol{G}']$ and $[\boldsymbol{G}]$ are mutually absolutely continuous. Therefore, by the assumptions on T, almost surely, for all $v \in V(\boldsymbol{G}')$ one has $T^+_{\boldsymbol{G}'}(v) = 1$ and $T_{\boldsymbol{G}'}(v,z) = 0$ for all $z \notin S_{\boldsymbol{G}'}$. Now, $\mathbb{E}\left[T^+_{\boldsymbol{G}'}(\boldsymbol{o}')\right] = 1$. So, unimodularity of $[\boldsymbol{G}', \boldsymbol{o}']$ implies that $\mathbb{E}\left[T^-_{\boldsymbol{G}'}(\boldsymbol{o}')\right] = 1$. In particular, one obtains that $\mathbb{E}\left[T^-_{\boldsymbol{G}'}(\boldsymbol{o}')\,|I\right] < \infty$ a.s. Consider the function w in Lemma 10 and let $c = c([\boldsymbol{G}'])$ be the common value of $w[\boldsymbol{G}', \cdot]$ on the vertices of $S_{\boldsymbol{G}'}$. Lemma 10 implies that $\mathbb{E}\left[T^-_{\boldsymbol{G}}(\boldsymbol{o})\,|I\right]$ is proportional to $\mathbb{E}\left[w[\boldsymbol{G}', \boldsymbol{o}']T^-_{\boldsymbol{G}'}(\boldsymbol{o}')\,|I\right]$, which is equal to $c[\boldsymbol{G}']\mathbb{E}\left[T^-_{\boldsymbol{G}'}(\boldsymbol{o}')\,|I\right]$ (note that $w[\boldsymbol{G}', \boldsymbol{o}']T^-_{\boldsymbol{G}'}(\boldsymbol{o}') = c[\boldsymbol{G}']T^-_{\boldsymbol{G}'}(\boldsymbol{o}')$ a.s.) The latter is shown above to be finite a.s. This implies that $\mathbb{E}[M\,|I] < \infty$ a.s. □

PROOF OF COROLLARY 1. By the strong sense in part (iii) of Theorem 5, one obtains that $\mathcal{P}'_T$ also satisfies the assumption on $[\boldsymbol{G}, \boldsymbol{o}]$, which implies that $\mathcal{P}'_T$ is ergodic. Therefore, Proposition 4 implies that any weak unimodularization of $[\boldsymbol{G}]$ is also strongly unrooted-equivalent to $\mathcal{P}'_T$. Now, Theorem 3 implies that $\mathcal{P}'_T$ is the unique weak unimodularization of $[\boldsymbol{G}]$. □

PROOF OF THEOREM 6. (i) Let $[\boldsymbol{G}', \boldsymbol{o}']$ be a strong unimodularization of the extension. Define $w : \mathcal{G}_* \to \mathbb{R}^{\geq 0}$ such that $w[\boldsymbol{G}', \boldsymbol{o}'] = 1_{S_{\boldsymbol{G}'}}(\boldsymbol{o}')/\mathbb{E}\left[1_{S_{\boldsymbol{G}'}}(\boldsymbol{o}')\,|I\right]$. Since $S_{\boldsymbol{G}} \neq \emptyset$ a.s. and this property doesn't depend on the root, the same holds for $\boldsymbol{G}'$. Similar to Lemma 1 one can deduce that the above denominator is nonzero a.s. and w is well-defined up to an event of measure zero. One has $\mathbb{E}\left[w[\boldsymbol{G}', \boldsymbol{o}']\,|I\right] = 1$ a.s. Therefore, Theorem 4 implies that there is a measurable function $T : \mathcal{G}_{**} \to \mathbb{R}^{\geq 0}$ such that almost surely, for all $v \in V(\boldsymbol{G}')$, one has $T^+_{\boldsymbol{G}'}(v) = 1$ and $T^-_{\boldsymbol{G}'}(v) = w[\boldsymbol{G}', v]$. In particular, almost surely, for all $v \in V(\boldsymbol{G}') \setminus S_{\boldsymbol{G}'}$, one has $T^-_{\boldsymbol{G}'}(v) = 0$. Since $[\boldsymbol{G}]$ has the same distribution as $[\boldsymbol{G}']$, the same holds for $[\boldsymbol{G}]$; i.e., almost surely, for all $v \in V(\boldsymbol{G})$, $T^+_{\boldsymbol{G}}(v) = 1$ and if $v \notin S_{\boldsymbol{G}}$, then $T^-_{\boldsymbol{G}}(v) = 0$. So, the assumptions of Theorem 5 are satisfied by T. To show that $\mathcal{P}'_T$ is defined, it remains to prove that $\mathbb{E}\left[T^-_{\boldsymbol{G}}(\boldsymbol{o})\,|I\right] < \infty$.

Let $\lambda : (\mathcal{G}, J) \to \mathbb{R}^{\geq 0}$ be the measurable function such that $\lambda[\boldsymbol{G}'] = \mathbb{E}\left[1_{S_{\boldsymbol{G}'}}(\boldsymbol{o}')\,|I\right]$. According the the above arguments, almost surely, for all $v \in V(\boldsymbol{G}')$, $T^-_{\boldsymbol{G}'}(v) = \frac{1}{\lambda[\boldsymbol{G}']}1_{S_{\boldsymbol{G}'}}(v)$. Since $[\boldsymbol{G}]$ has the same distribution as $[\boldsymbol{G}']$, the same holds for $[\boldsymbol{G}]$; i.e., almost surely, $\forall v \in V(\boldsymbol{G}), T^-_{\boldsymbol{G}}(v) = \frac{1}{\lambda[\boldsymbol{G}]}1_{S_{\boldsymbol{G}}}(v)$. In particular, since $\boldsymbol{o} \in S_{\boldsymbol{G}}$ a.s., one gets $T^-_{\boldsymbol{G}}(\boldsymbol{o}) = \frac{1}{\lambda[\boldsymbol{G}]}$ a.s. Since this doesn't depend on the root, $\mathbb{E}\left[T^-_{\boldsymbol{G}}(\boldsymbol{o})\,|I\right] = \frac{1}{\lambda[\boldsymbol{G}]} < \infty$ a.s. and the claim is proved.

(ii) Let $[\boldsymbol{G}', \boldsymbol{o}']$ be a unimodularization of $[\boldsymbol{G}]$. Any two unimodularizations are weakly unrooted-equivalent by definition. Therefore, Lemma 7 implies that $[\boldsymbol{G}', \boldsymbol{o}']$ can be obtained by biasing $\mathcal{P}'_T$ by a function that doesn't depend on the root.

(iii) The function T constructed in the proof of part (i) satisfies the desired property. □

PROOF OF PROPOSITION 6. By Theorem 6, one can assume $\mathcal{P}'_T$ is defined for some function T satisfying the assumptions of Theorem 5 without loss of generality. Let $[\boldsymbol{G}', \boldsymbol{o}']$ have distribution $\mathcal{P}'_T$ and $[\widetilde{\boldsymbol{G}}, \widetilde{\boldsymbol{o}}]$ be any unimodularization of the extension. By part (ii) of Theorem 6, $[\widetilde{\boldsymbol{G}}, \widetilde{\boldsymbol{o}}]$ is obtained by biasing $[\boldsymbol{G}', \boldsymbol{o}']$ by a measurable function $w : \mathcal{G}_* \to \mathbb{R}^{\geq 0}$ that does not depend on the root. By a scaling, one may assume $\mathbb{E}\left[w[\boldsymbol{G}',\boldsymbol{o}']\right]=1$. For any event $A\subseteq\mathcal{G}_*$, one has

$$\begin{aligned}\mathbb{P}\left[\widetilde{\boldsymbol{o}}\in S_{\widetilde{\boldsymbol{G}}},[\widetilde{\boldsymbol{G}},\widetilde{\boldsymbol{o}}]\in A\right] &= \mathbb{E}\left[w[\boldsymbol{G}',\boldsymbol{o}']1_{S_{\boldsymbol{G}'}}(\boldsymbol{o}')1_A[\boldsymbol{G}',\boldsymbol{o}']\right]\\ &= \mathbb{E}\left[\frac{1}{\mathbb{E}\left[M\,|I\right]}\sum_{v\in S_{\boldsymbol{G}}}T_{\boldsymbol{G}}(v,\boldsymbol{o})w[\boldsymbol{G},v]1_A[\boldsymbol{G},v]\right]\\ &= \mathbb{E}\left[\frac{1}{\mathbb{E}\left[M\,|I\right]}\sum_{v\in S_{\boldsymbol{G}}}T_{\boldsymbol{G}}(\boldsymbol{o},v)w[\boldsymbol{G},\boldsymbol{o}]1_A[\boldsymbol{G},\boldsymbol{o}]\right]\\ &= \mathbb{E}\left[\frac{1}{\mathbb{E}\left[M\,|I\right]}w[\boldsymbol{G},\boldsymbol{o}]1_A[\boldsymbol{G},\boldsymbol{o}]\right],\end{aligned}$$

where in the third equality (5.1) is used. Therefore,

$$\mathbb{P}\left[[\widetilde{\boldsymbol{G}},\widetilde{\boldsymbol{o}}]\in A\,|\widetilde{\boldsymbol{o}}\in S_{\widetilde{\boldsymbol{G}}}\right]=c\mathbb{E}\left[\frac{w[\boldsymbol{G},\boldsymbol{o}]}{\mathbb{E}\left[M\,|I\right]}1_A[\boldsymbol{G},\boldsymbol{o}]\right],$$

where $c=1/\mathbb{P}\left[\widetilde{\boldsymbol{o}}\in S_{\widetilde{\boldsymbol{G}}}\right]$. Thus, $[\widetilde{\boldsymbol{G}},\widetilde{\boldsymbol{o}}]$ has the desired property if and only if $\frac{w[\boldsymbol{G},\boldsymbol{o}]}{\mathbb{E}[M|I]}$ is almost surely constant. If so, the distribution of $[\widetilde{\boldsymbol{G}},\widetilde{\boldsymbol{o}}]$ is equal to biasing $\mathcal{P}'_T$ by $\mathbb{E}\left[M\,|I\right]$, which is just $\mathcal{P}_T$. As a result, $\mathcal{P}_T$ is defined and thus $\mathbb{E}\left[M\right]<\infty$. So the claim is proved. □

PROOF OF COROLLARY 2. Lemma 9 and Proposition 6 imply the claim. □

6. A Construction Using Stable Transports

In some results in this paper, the existence of specific objects are proved based on Theorem 1, including propositions 2 and 3 and theorems 4 and 6. However, Theorem 1 does not help to construct such root-changes by looking only at a realization of the given networks. In this section, we present an algorithm to construct a balancing transport kernel as described in Theorem 4. Special cases of this algorithms will provide the desired constructions in the other results mentioned above. The algorithm is based on the one in [**9**], which is by itself based on [**11**]. It is a generalization of the Gale-Shapley stable matching algorithm for bipartite graphs [**8**]. In fact, it is similar to the many-to-many stable matching algorithm. We should note that by the terms *construction* and *algorithm* we do not mean a computational algorithm, but an explicit definition using formulas that might be defined iteratively.

Fix a rooted network (G,o) and measurable functions $w_i:\mathcal{G}_*\to\mathbb{R}^{\geq 0}$ for $i=1,2$. We will use two names **sites** and **centers** for the vertices and use Roman letters for centers for better readability. Given a measurable function $T:\mathcal{G}_{**}\to\mathbb{R}^{\geq 0}$, we say a site $x\in V(G)$ sends mass $T_G(x,\xi)$ to the center $\xi\in V(G)$. Here is an overview of Algorithm 1. It will finally produce a measurable function $T:\mathcal{G}_{**}\to\mathbb{R}^{\geq 0}$ such that $T^+_G(\cdot)\leq w_2[G,\cdot]$ and $T^-_G(\cdot)\leq w_1[G,\cdot]$ (the goal is equality which will hold under some conditions). $T_G(x,\xi)$ will be defined as the mass x *applies* to ξ minus the mass ξ *rejects* from x. The algorithm consists of infinitely many stages and each stage has two steps. At stage n, each site x_0 *applies* to the closest possible centers with *weight* $A_n(x_0,\cdot)$. A constraint is chosen for the applications, which is $0\leq A_n(x_0,\cdot)\leq w_2[G,\cdot]$. Then, each center ξ_0 *rejects* some of the weights applied to ξ_0 if the sum of the incoming applications exceeds $w_2[G,\xi_0]$. The amount of rejection is denoted by $R_n(\cdot,\xi_0)$. The functions A_n and R_n at stage n are chosen such that each site prefers to apply to the closest possible centers and each center prefers to reject (if necessary) the applications of the farthest possible sites.

ALGORITHM 1 (Stable Transport). Let (G, o) be a given deterministic rooted network and $w_1, w_2 : \mathcal{G}_* \to \mathbb{R}^{\geq 0}$ be measurable. Let $R_0(x, \xi) = 0$ for all $x, \xi \in V(G)$. For each $n \geq 1$, stage n consists of the following two steps.

(i) For each site x_0, define its *application radius* at stage n by

$$a_n(x_0) := \min\{a \geq 0 : \sum_{\xi \in N_a(x_0)} (w_2[G, \xi] - R_{n-1}(x_0, \xi)) \geq w_1[G, x_0]\}.$$

Define *the n-th application function* by

$$A_n(x_0, \xi) := \begin{cases} w_2[G, \xi] & d(x_0, \xi) < a_n(x_0), \\ cR_{n-1}(x_0, \xi) + (1 - c)w_2[G, \xi] & d(x_0, \xi) = a_n(x_0), \\ 0 & d(x_0, \xi) > a_n(x_0), \end{cases}$$

where in the case $a_n(x_0) < \infty$, $c = c_n(x_0)$ is chosen in $[0, 1]$ such that

$$\sum_{\xi \in V(G)} (A_n[x_0, \xi] - R_{n-1}(x_0, \xi)) = w_1[G, x_0].$$

(ii) For each center ξ_0, define its *rejection radius* at stage n by

$$r_n(\xi_0) := \min\{r \geq 0 : \sum_{x \in N_r(\xi_0)} A_n(x, \xi_0) \geq w_2[G, \xi_0]\}.$$

Define *the n-the rejection function* by

$$R_n(x, \xi_0) := \begin{cases} 0 & d(x, \xi_0) < r_n(\xi_0), \\ c' A_n(x, \xi_0) & d(x, \xi_0) = r_n(\xi_0), \\ A_n(x, \xi_0) & d(x, \xi_0) > r_n(\xi_0), \end{cases}$$

where in the case $r_n(\xi_0) < \infty$, $c' = c'_n(\xi_0)$ is chosen in $[0, 1]$ such that

$$\sum_{x \in V(G)} (A_n[x, \xi_0] - R_n(x, \xi_0)) = w_2[G, \xi_0].$$

Finally, define

$$T_G(x, \xi) := \lim_{n\to\infty} A_n(x, \xi) - \lim_{n\to\infty} R_n(x, \xi).$$

Here are some basic facts about this algorithm. The proofs are similar to **[9]** and are skipped here for brevity. The sequences of functions A_n, R_n and a_n are non-decreasing w.r.t n and r_n is non-increasing. So, the limit function T is well defined. Moreover, $T_G^+(\cdot) \leq w_1[G, \cdot]$ and $T_G^-(\cdot) \leq w_2[G, \cdot]$. Call a site x_0 **exhausted** if $T_G^+(x_0) = w_1[G, x_0]$. Similarly, a center ξ_0 is **sated** if $T_G^-(\xi_0) = w_2[G, \xi_0]$. It is shown below that T is **stable** in a sense similar to **[9]** (and many-to-many stable matchings) defined as follows: There is no site x_0 and center ξ_0 such that both **desire** each other, where site x_0 desires center ξ_0 if $T_G(x_0, \xi_0) < w_2[G, \xi_0]$ and either x_0 is unexhausted or $T_G(x_0, \xi_1) > 0$ for some farther center ξ_1. Similarly, center ξ_0 desires site x_0 if $T_G(x_0, \xi_0) < w_2[G, \xi_0]$ and either ξ_0 is unsated or $T_G(x_1, \xi_0) > 0$ for some farther site x_1. Stronger than stability, the following holds.

LEMMA 11. *In Algorithm 1, if a site x_0 desires a center ξ_0 (defined above), then ξ_0 is sated and doesn't desire x_0. Therefore, T is stable.*

PROOF. By definition, either x_0 is unexhausted or $T_G(x_0, \xi_1) > 0$ for some farther center ξ_1. In both cases, x_0 has applied to some center farther than ξ_1 at some stage. The definition of A_n implies that $A_n(x_0, \xi_0) = w_2[G, \xi_0]$ for large enough n. Therefore, by $T_G(x_0, \xi_0) < w_2[G, \xi_0]$, ξ_0 has rejected a positive fraction of the application of x_0 at some stage. By the definition of the rejection function, ξ_0 is sated from that stage on. Moreover, ξ_0 has fully rejected the applications of the sites farther than x_0. So, ξ_0 doesn't desire x_0 and the claim is proved. □

LEMMA 12. *In Algorithm* 1, *if there is an unexhausted site, then all centers are sated and vice versa.*

PROOF. Assume x_0 is an unexhausted site and ξ_0 is an unsated center. Since ξ_0 is unsated, one obtains $T_G(x_0, \xi_0) \leq T_G^-(\xi_0) < w_2[G, \xi_0]$. Therefore, x_0 desires ξ_0 by definition. This contradicts Lemma 11. □

THEOREM 7 (Construction of a Balancing Transport Kernel). *Let* $[\boldsymbol{G}, \boldsymbol{o}]$ *be a unimodular network and* $w_i : \mathcal{G}_* \to \mathbb{R}^{\geq 0}$ *be measurable functions for* $i = 1, 2$. *If*

$$\mathbb{E}\left[w_1[\boldsymbol{G}, \boldsymbol{o}]\,|I\right] = \mathbb{E}\left[w_2[\boldsymbol{G}, \boldsymbol{o}]\,|I\right] < \infty, \qquad a.s.,$$

then the function T *constructed in Algorithm* 1 *satisfies the claims of Theorem* 4; *i.e.,* $T_{\boldsymbol{G}}^+(\cdot) = w_1[\boldsymbol{G}, \cdot]$ *and* $T_{\boldsymbol{G}}^-(\cdot) = w_2[\boldsymbol{G}, \cdot]$ *a.s.*

Note that the condition $\mathbb{E}\left[w_1[\boldsymbol{G}, \boldsymbol{o}]\,|I\right] = \mathbb{E}\left[w_2[\boldsymbol{G}, \boldsymbol{o}]\,|I\right]$ a.s. is also necessary (see Theorem 4).

PROOF. First, it is easy to see that T defines a measurable function on $\mathcal{G}_{**}$. We should prove there is no unexhausted site and no unsated center a.s. Let A be the event that there is an unexhausted site. If $\mathbb{P}[A] > 0$, then by conditioning on A, one may assume $\mathbb{P}[A] = 1$ (notice that $A \in I$ and thus the assumptions are not changed after conditioning on A). Therefore, by Lemma 12, there is no unsated centers a.s. Lemma 1 implies that $\boldsymbol{o}$ is unexhausted with positive probability but $\boldsymbol{o}$ is sated a.s. Equivalently, $T_{\boldsymbol{G}}^+(\boldsymbol{o}) < w_1[\boldsymbol{G}, \boldsymbol{o}]$ with positive probability but $T_{\boldsymbol{G}}^-(\boldsymbol{o}) = w_2[\boldsymbol{G}, \boldsymbol{o}]$ a.s. It follows that

$$\mathbb{E}\left[T_{\boldsymbol{G}}^+(\boldsymbol{o})\right] < \mathbb{E}\left[w_1[\boldsymbol{G}, \boldsymbol{o}]\right] = \mathbb{E}\left[w_2[\boldsymbol{G}, \boldsymbol{o}]\right] = \mathbb{E}\left[T_{\boldsymbol{G}}^-(\boldsymbol{o})\right].$$

This contradicts (2.1). Therefore, all sites are exhausted a.s. One can prove similarly that all centers are sated a.s. So, the proof is complete. □

The following is an application of Theorem 7 to Proposition 2 (see also Remark 10 below). Unimodularity is a crucial assumption to ensure that Algorithm 1 works here. The author is not aware of any general construction for the non-unimodular case of Proposition 2.

COROLLARY 3. *In the setting of Proposition* 2, *assume* $[\boldsymbol{G}', \boldsymbol{o}']$ *can be obtained from* $[\boldsymbol{G}, \boldsymbol{o}]$ *by a root-change. If* $[\boldsymbol{G}, \boldsymbol{o}]$ *is unimodular, then such root-change is obtained by the function* T *constructed in Algorithm* 1 *for* $w_1[G, o] := 1$ *and* $w_2[G, o] := \frac{1}{\mathbb{P}[\boldsymbol{o} \in S_{\boldsymbol{G}}]} 1_{S_G}(o)$.

PROOF. Let $p := \mathbb{P}[\boldsymbol{o} \in S_{\boldsymbol{G}}]$. By Proposition 2, $\mathbb{P}[\boldsymbol{o} \in S_{\boldsymbol{G}}\,|I] = p$ a.s. It follows that $\mathbb{E}\left[w_2[\boldsymbol{G}, \boldsymbol{o}]\,|I\right] = 1$ a.s. Now, Theorem 7 implies that $T_{\boldsymbol{G}}^+(\boldsymbol{o}) = 1$ and $T_{\boldsymbol{G}}^-(\boldsymbol{o}) = \frac{1}{p} 1_{S_{\boldsymbol{G}}}(\boldsymbol{o})$ a.s. Therefore, Lemma 5 implies that applying the root-change by kernel T to $[\boldsymbol{G}, \boldsymbol{o}]$ gives $[\boldsymbol{G}', \boldsymbol{o}']$. So, the claim is proved. □

By Corollary 3 and the proof of Proposition 3, the following corollary is readily obtained.

COROLLARY 4 (Construction of an Extra Head Scheme). *In the setting of Proposition 3, an extra head scheme is obtained by the function T constructed in Algorithm 1 for $w_1[G,o] := 1$ and $w_2[G,o] := \frac{1}{p}m(o)$.*

COROLLARY 5 (Construction of a Unimodularization of an Extension). *Let $([\boldsymbol{G},\boldsymbol{o}],S)$ be a covariant extension of a unimodular network $[\boldsymbol{G}_0,\boldsymbol{o}_0]$. If $[\boldsymbol{G}]$ can be unimodularized, then there exists a constant λ_G for each non-rooted network G such that the root-change corresponding to the function T constructed in Algorithm 1 for $w_1[G,v] := 1$ and $w_2[G,v] := \frac{1}{\lambda_G}1_{S_G}(v)$ satisfies the assumptions of Theorem 5. Therefore, $\mathcal{P}'_T$ is the strong unimodularization of $[\boldsymbol{G}]$.*

PROOF. Let $[\boldsymbol{G}',\boldsymbol{o}']$ be a strong unimodularization of $[\boldsymbol{G}]$ and let λ satisfy $\lambda_{\boldsymbol{G}'} = \mathbb{P}[\boldsymbol{o}' \in S_{\boldsymbol{G}'} | I]$. By Theorem 7, $T_{\boldsymbol{G}'}(\cdot,\cdot)$ balances between $w_1[\boldsymbol{G}',\cdot]$ and $w_2[\boldsymbol{G}',\cdot]$ a.s. as defined in Theorem 4. Now, the proof of Theorem 6 shows that T satisfies the assumptions of Theorem 5 and $\mathcal{P}'_T$ is defined. □

REMARK 10. One may ask how to construct $\lambda_{\boldsymbol{G}}$ in the above corollary by looking only at a realization of $[\boldsymbol{G},\boldsymbol{o}]$. If $[\boldsymbol{G}',\boldsymbol{o}']$ is a unimodularization of $[\boldsymbol{G}]$, then λ is the sample intensity of S in $\boldsymbol{G}'$ defined in Example 3. One may also ask the same question in Corollary 3 on how to construct $\mathbb{P}[\boldsymbol{o} \in S_{\boldsymbol{G}}]$. Note that averaging on a large ball like $\#(S_{\boldsymbol{G}} \cap N_r(\boldsymbol{o}))/\#N_r(\boldsymbol{o})$ and taking limit does not work in general. In fact, *amenability* is a necessary condition for this to work (see [**1**]). For general unimodular networks, one construction for the sample intensity can be done by *frequency of visits* to $S_{\boldsymbol{G}}$ of the *delayed simple random walk* in $[\boldsymbol{G}]$ (see [**17**] and [**1**] for the details). Another construction is the following.

In Corollary 5, replace λ_G by an arbitrary constant $\lambda > 0$. Then, it can be seen that given any (G,v), the value $T^+_G(v)$ is non-increasing in terms of λ (see [**14**] and also [**9**] and [**11**]). Then, one can let λ_G be the supremum value of λ such that $T^+_G(v) = 1$ for all $v \in V(G)$. It can be proved that this construction works in Corollary 5. The proof is similar to the arguments in [**9**] and [**11**] and is skipped for brevity.

7. Proofs

This sections is devoted to the proofs of some of the results of Section 3.

DEFINITION 11. Let μ be a measure (not necessarily a probability measure) on $\mathcal{G}_*$ and $T : \mathcal{G}_{**} \to \mathbb{R}^{\geq 0}$ be a measurable function. Define the measures $T^\uparrow\mu$ and $T^\to\mu$ on $\mathcal{G}_{**}$ and $\mathcal{G}_*$ respectively by

$$(T^\uparrow\mu)(A) := \int_{\mathcal{G}_*} \sum_{v\in V(G)} T_G(o,v)1_A[G,o,v]d\mu([G,o]),$$

$$(T^\to\mu)(B) := \int_{\mathcal{G}_*} \sum_{v\in V(G)} T_G(o,v)1_B[G,v]d\mu([G,o]),$$

for measurable subsets $A \subseteq \mathcal{G}_{**}$ and $B \subseteq \mathcal{G}_*$. It can be seen that $\pi_{2*}T^\uparrow\mu = T^\to\mu$. Moreover, if $T^+_G(o) = 1$ for μ-a.e. $[G,o]$, then $\pi_{1*}T^\uparrow\mu = \mu$. If in addition, μ is a probability measure, then $T^\to\mu$ is just the root-change of μ by kernel T as in Definition 6. It is also worthy to mention that when $T^+(\cdot)$ is always 1, there is a Markov kernel on $\mathcal{G}_*$ that transports μ to $T^{\rightarrow}\mu$, which is defined by $T'([G,o],[G',o']) := \sum_v T_G(o,v)$, where the sum is over all $v \in V(G)$ such that $[G,v]=[G',o']$.

Proof of Proposition 1. Assume $(\mathcal{G},J)$ is a standard Borel space. Let μ be an arbitrary ergodic unimodular probability measure on $\mathcal{G}_*$ and let $\nu := \pi_*\mu$. Since μ is ergodic, one gets $\nu(A)\in\{0,1\}$ for any event $A\in J$. The assumption of standardness of $(\mathcal{G},J)$ implies that ν is concentrated on one atom. Therefore, there should be a network G such that μ is concentrated on $\{[G,v]: v\in V(G)\}$. But this is clearly false for general μ (see for instance the example in Lemma 4). □

Proof of Lemma 2. Denote by $[\boldsymbol{G}',\boldsymbol{o}']$ the random rooted network obtained by biasing the probability measure by w. Let $B\in J$ be an event in $\mathcal{G}$ and $A:=\pi^{-1}(B)$. One has

$$\begin{aligned}\mathbb{P}\left[[\boldsymbol{G}']\in B\right] &= \mathbb{P}\left[[\boldsymbol{G}',\boldsymbol{o}']\in A\right]\\ &= \frac{1}{c}\mathbb{E}\left[w[\boldsymbol{G},\boldsymbol{o}]1_A[\boldsymbol{G},\boldsymbol{o}]\right]\\ &= \frac{1}{c}\mathbb{E}\left[\mathbb{E}\left[w[\boldsymbol{G},\boldsymbol{o}]\,|I\right]1_A[\boldsymbol{G},\boldsymbol{o}]\right]\\ &= \frac{1}{c}\mathbb{E}\left[\mathbb{E}\left[w[\boldsymbol{G},\boldsymbol{o}]\,|I\right]1_B[\boldsymbol{G}]\right],\end{aligned}$$

where $c:=\mathbb{E}\left[w[\boldsymbol{G},\boldsymbol{o}]\right]$ and in the third equation we have used the fact $A\in I$. Now, the claim is obtained by noting that $\mathbb{E}\left[w[\boldsymbol{G},\boldsymbol{o}]\,|I\right]$ is a function of $[\boldsymbol{G}]$. □

Lemma 13. *Conditions (R), (C) and (D) in Theorem 2 are equivalent.*

Proof. Let $\mathcal{P}_1$ and $\mathcal{P}_2$ be the distributions of $[\boldsymbol{G}_1,\boldsymbol{o}_1]$ and $[\boldsymbol{G}_2,\boldsymbol{o}_2]$ respectively.

(R)⇒ (D). Consider a root-change by kernel T such that $T^{\rightarrow}\mathcal{P}_1=\mathcal{P}_2$. Given $[\boldsymbol{G}_1,\boldsymbol{o}_1]$, choose a second root with distribution $T_{\boldsymbol{G}_1}(\boldsymbol{o}_1,\cdot)$. To be more precise, a random doubly-rooted network with distribution $T^{\uparrow}\mathcal{P}_1$ (Definition 6) has the desired properties.

(D)⇒ (C). Let $[\boldsymbol{G},\boldsymbol{o},\boldsymbol{o}']$ be such a random doubly-rooted network. Then, the random rooted networks $[\boldsymbol{G},\boldsymbol{o}]$ and $[\boldsymbol{G},\boldsymbol{o}']$ provide the desired coupling. More precisely, the desired coupling is obtained by pushing forward the distribution of $[\boldsymbol{G},\boldsymbol{o},\boldsymbol{o}']$ by the map $[G,o,o']\mapsto([G,o],[G,o'])$, which is a well defined measurable function on $\mathcal{G}_{**}$.

(C)⇒ (R). Let μ be such a probability measure on $\mathcal{G}_*\times\mathcal{G}_*$ as assumed. Fix a rooted network (G_1,o_1) and let $\mu_{(G_1,o_1)}$ be the conditional distribution of the second rooted network given that the first rooted network is $[G_1,o_1]$. Note that $\mu_{(G_1,o_1)}$ is defined and is supported on $A:=\{[G_1,v]: v\in V(G_1)\}$ for $\mathcal{P}_1$-a.e. $[G_1,o_1]$. In the (zero-probability) cases where this doesn't hold, let $\mu_{(G_1,o_1)}$ be concentrated on $[G_1,o_1]$. Note that A is a countable set.

For $[G_2,o_2]\in A$, let $S_{[G_2,o_2]}\subseteq V(G_1)$ be the set of the closest vertices v to o_1 such that $[G_1,v]=[G_2,o_2]$. This set is a finite subset of $V(G_1)$. Finally, from the measure $\mu_{(G_1,o_1)}$ on A, one can construct a measure $T_{G_1}(o_1,\cdot)$ on $V(G_1)$ defined by

$$T_{G_1}(o_1,v) := \sum_{[G_2,v_2]\in A}\frac{1}{\#S_{[G_2,v_2]}}\mu_{(G_1,o_1)}([G_2,v_2])1_{S_{[G_2,v_2]}}(v)$$

for $v \in V(G_1)$. It is easy to see that T is an invariant transport kernel and $T^+_{G_1}(o_1) = 1$. Moreover, by choosing a new root in $V(G_1)$ with distribution $T_{G_1}(o_1, \cdot)$, the resulting network has distribution $\mu_{(G_1,o_1)}$. By the definition of $\mu_{(G_1,o_1)}$ and choosing (G_1, o_1) randomly with distribution $\mathcal{P}_1$, one gets that $T^{\rightarrow}\mathcal{P}_1 = \mathcal{P}_2$. So, T gives the desired root-change. □

PROOF OF LEMMA 3. By part (R)⇒(C) of Theorem 2 (proved in Lemma 13 above), there is a coupling of $[\boldsymbol{G}, \boldsymbol{o}]$ and $[\boldsymbol{G}, \boldsymbol{o}']$ supported on $\{([G,o],[G',o']) : [G] = [G']\} \subseteq \mathcal{G}_* \times \mathcal{G}_*$. The latter also holds if one swaps $[\boldsymbol{G}, \boldsymbol{o}]$ and $[\boldsymbol{G}', \boldsymbol{o}]$. Therefore, by part (C)⇒(R) of Theorem 2 (proved in Lemma 13 above), $[\boldsymbol{G}, \boldsymbol{o}]$ is a root-change of $[\boldsymbol{G}', \boldsymbol{o}']$. □

PROOF OF THEOREM 1. Let P_1 and P_1' be the distributions of $[\boldsymbol{G}, \boldsymbol{o}]$ and $[\boldsymbol{G}', \boldsymbol{o}']$ respectively.

(⇒). Assume $[\boldsymbol{G}', \boldsymbol{o}']$ can be obtained from $[\boldsymbol{G}, \boldsymbol{o}]$ by a root-change. Equation (3.1) easily implies that $\mathbb{P}\left[[\boldsymbol{G}', \boldsymbol{o}'] \in A\right] = \mathbb{P}\left[[\boldsymbol{G}, \boldsymbol{o}] \in A\right]$ for any invariant event A. Equivalently, $\mathcal{P}[A] = \mathcal{P}'[A]$ for any $A \in I$, which is the desired property.

(⇐, *First Proof*). The proof mimics that of [**20**]. Here is a summary of the proof. The idea is to find two root-changes for $[\boldsymbol{G}_1, \boldsymbol{o}_1]$ and $[\boldsymbol{G}_2, \boldsymbol{o}_2]$ such that the resulting random rooted networks have the same distribution. Then, one can combine them to find the desired root-change. We start with an arbitrary root-change such that every vertex has positive probability to be chosen. Then, update it step by step as will be described. However, in the next steps probability measures will be replaced by finite measures.

For a network G and $o, v \in V(G)$, let

$$S[G, o, v] := \begin{cases} \frac{1}{\#V(G)}, & \#V(G) < \infty \\ \frac{1}{2^{r+1}\#\partial N_r(o)}, & \#V(G) = \infty \end{cases}$$

where $r := d(o, v)$ and $\partial N_r(o)$ is the (internal) boundary of the ball; i.e., the set of vertices with distance r from o. It can be seen that S is well-defined and measurable on $\mathcal{G}_{**}$. All we need from S is the following property: For every network G and $o, v \in V(G)$,

$$\begin{cases} S^+_G(o) = 1, \\ S_G(o, v) > 0. \end{cases} \tag{7.1}$$

Starting from P_1 and P_1', construct the sequences of finite measures Q_n, Q_n', P_n, P_n' and λ_n for $n \geq 1$ as follows. The first two are one $\mathcal{G}_{**}$ and the other three on $\mathcal{G}_*$. Here, the symbol $\wedge$ is used for the minimum of measures.

- $\lambda_n := (S^{\rightarrow}P_n) \wedge (S^{\rightarrow}P_n')$.
- Q_n is the probability measure constructed in Lemma 14 for λ_n, $S^{\uparrow}P_n$ and $i = 2$. Q_n' is defined similarly by Lemma 14 for λ_n, $S^{\uparrow}P_n'$ and $i = 2$.
- $P_{n+1} := P_n - \pi_{1*}Q_n$ and $P_{n+1}' := P_n' - \pi_{1*}Q_n'$.

By the definition of Q_n and Q_n' and Lemma 14, one has

$$\begin{cases} Q_n \leq S^{\uparrow}P_n, & \pi_{2*}Q_n = \lambda_n, \\ Q_n' \leq S^{\uparrow}P_n', & \pi_{2*}Q_n' = \lambda_n. \end{cases} \tag{7.2}$$

Since $\pi_{1*}S^{\uparrow}P_n = P_n$ and $\pi_{1*}S^{\uparrow}P_n' = P_n'$, this implies inductively that all above measures are non-negative (and justifies validity of using Lemma 14 inductively).

Define

$$P_\infty := \lim_n P_n, \qquad Q := \sum_{n=1}^{\infty} Q_n,$$

$$P'_\infty := \lim_n P'_n, \qquad Q' := \sum_{n=1}^{\infty} Q'_n.$$

The limits are well defined since P_n and P'_n are decreasing sequences and the sums of $\|Q_n\|$ and $\|Q'_n\|$ over n are convergent, where the symbol $\|\cdot\|$ is used for the total mass of a measure (note that $\|Q_n\| = \|P_n\| - \|P_{n+1}\|$). This also shows that Q is a finite measure. Now, one has

$$\left\{ \begin{array}{lll} \pi_{1*}Q & = \sum_n (P_n - P_{n+1}) = & P_1 - P_\infty, \\ \pi_{1*}Q' & = \sum_n (P'_n - P'_{n+1}) = & P'_1 - P'_\infty \end{array} \right. \tag{7.3}$$

and

$$\pi_{2*}Q = \pi_{2*}Q'. \tag{7.4}$$

Let $\lambda_\infty := (S^{\rightarrow}P_\infty) \wedge (S^{\rightarrow}P'_\infty)$. By $P_\infty \leq P_n$ and $P'_\infty \leq P'_n$, it is clear that $\lambda_\infty \leq \lambda_n$. Therefore, by (7.2) we get $\|\lambda_\infty\| \leq \|\lambda_n\| = \|Q_n\|$ for every n. The sum of the right hand side over n is convergent (bounded by $\|Q\|$) and so $\|\lambda_\infty\| = 0$. Therefore, $\lambda_\infty = 0$. This means that the measures $S^{\rightarrow}P_\infty$ and $S^{\rightarrow}P'_\infty$ are mutually singular; i.e., there is an event $A \subseteq \mathcal{G}_*$ such that

$$\left\{ \begin{array}{lll} S^{\rightarrow}P_\infty(A^c) & = & 0, \\ S^{\rightarrow}P'_\infty(A) & = & 0, \end{array} \right. \tag{7.5}$$

Consider the event $B := \{[G,o] : \exists v \in V(G) : [G,v] \in A\} = \pi^{-1}(\pi(A))$ in $\mathcal{G}_*$. By (7.5), (7.1) and the definition of $S^{\rightarrow}P_\infty$ and $S^{\rightarrow}P'_\infty$, one gets

$$\left\{ \begin{array}{lll} P_\infty(B^c) & = & 0, \\ P'_\infty(B) & = & 0. \end{array} \right. \tag{7.6}$$

By part (D)$\Rightarrow$(F) of Theorem 2 (proved in Lemma 13 above), the measures $\pi_{1*}Q$ and $\pi_{2*}Q$ agree on I. The same holds for Q' (and any arbitrary measure on $\mathcal{G}_{**}$). Therefore, (7.4) gives that $\pi_{1*}Q$ and $\pi_{1*}Q'$ agree on I. By (7.3) and the assumption that P_1 and P'_1 agree on I, one gets that P_∞ and P'_∞ also agree on I. Since B is clearly an invariant event, one obtains $P_\infty(B) = P'_\infty(B)$. Now, (7.6) readily implies $P_\infty = P'_\infty = 0$. Now, one has

$$\begin{array}{lll} \pi_{1*}Q & = & P_1, \\ \pi_{1*}Q' & = & P'_1, \\ \pi_{2*}Q & = & \pi_{2*}Q'. \end{array}$$

As a result, Q and Q' are probability measures. By part (D)$\Rightarrow$(R) of Theorem 2 (proved in Lemma 13 above), one finds a root-change, say by kernel T, that transports P_1 to $\alpha := \pi_{2*}Q = \pi_{2*}Q'$. Similarly, by the same argument and Lemma 3, one finds a root-change, say by kernel T', that transports α to $\mathcal{P}_2$. Now, let t be the composition of T and T' defined by $t_G(o,v) := \sum_{z\in V(G)} T_G(o,z)T'_G(z,v)$. It can be seen that t gives a root-change of P_1 (as in Definition 6) and $t^{\rightarrow}P_1 = P'_1$. Therefore, by Definition 6, $[\boldsymbol{G}', \boldsymbol{o}']$ can be obtained from $[\boldsymbol{G}, \boldsymbol{o}]$ by the root-change by kernel t, which completes the proof.

($\Leftarrow$, *Second Proof*). Let R be the equivalence relation on $\mathcal{G}_*$ in which $[G_1, o_1]$ is R-related to $[G_2, o_2]$ if and only if $[G_1] = [G_2]$. Following the definitions in

Subsection 3.4, it can be seen that R is a countable Borel equivalence relation. Therefore, by Theorem 1 of [**7**], there is a countable group H consisting of Borel isomorphisms of $\mathcal{G}_*$ that generates R in the sense that

$$xRy \Leftrightarrow \exists h \in H : y = h(x).$$

Endow H with the discrete topology. It can be seen that the invariant sigma-field under the action of H is equal to the invariant sigma-field I in Definition 4. So, the assumption gives that the distributions of $[\boldsymbol{G}_1, \boldsymbol{o}_1]$ and $[\boldsymbol{G}_2, \boldsymbol{o}_2]$ agree on the H-invariant sigma-field. Thus, by Theorem 1 of [**20**], there is a random element F of H such that $F[\boldsymbol{G}_1, \boldsymbol{o}_1]$ has the same distribution as $[\boldsymbol{G}_2, \boldsymbol{o}_2]$. This provides a coupling of $\mathcal{P}$ and $\mathcal{P}'$ that satisfied Condition (C) of Definition 7. Therefore, by part (C)⇒(R) of Theorem 2 (proved in Lemma 13 above), $[\boldsymbol{G}_2, \boldsymbol{o}_2]$ can be obtained from $[\boldsymbol{G}_1, \boldsymbol{o}_1]$ by a root-change and the claim is proved. □

The following lemma is used in the proof of Theorem 1 above. Note that all measures are assumed to be non-signed in this paper.

LEMMA 14. *Let P and Q be finite measures on $\mathcal{G}_*$ and $\mathcal{G}_{**}$ respectively and $i \in \{1, 2\}$. If $\pi_{i*}Q \geq P$, then there is a measure $Q' \leq Q$ such that $\pi_{i*}Q' = P$.*

PROOF. The claim is a direct consequence of Lemma 1 in [**20**]. □

PROOF OF THEOREM 2. According to Lemma 13 and Theorem 1 proved above, the only remaining part is (R)⇒ (B), which is trivial. □

PROOF OF LEMMA 4. By the natural coupling of the two random rooted networks, one may assume $V(\boldsymbol{G}) = V(\boldsymbol{G}')$ and $\boldsymbol{o} = \boldsymbol{o}'$. First, assume (3.2) is proved and $[\boldsymbol{G}, \boldsymbol{o}]$ is ergodic and infinite a.s. The left hand side of (3.2) is an invariant function of $[\boldsymbol{G}, \boldsymbol{o}]$. Therefore, by ergodicity, it is almost surely constant, hence, by (3.2), it is either 1 a.s. or 0 a.s. It follows that $\mathbb{P}\left[[\boldsymbol{G}', \boldsymbol{o}] \in A\right] \in \{0, 1\}$ and so $[\boldsymbol{G}', \boldsymbol{o}]$ is ergodic. So, it is enough to prove (3.2). Let

$$f(G, o) := \mathbb{P}\left[[\boldsymbol{G}', \boldsymbol{o}] \in A \,|\, [\boldsymbol{G}, \boldsymbol{o}] = [G, o]\right].$$

For any measurable function $g : \mathcal{G}_{**} \to \mathbb{R}^{\geq 0}$,

$$\begin{aligned} \mathbb{E}\left[f[\boldsymbol{G}, \boldsymbol{o}] \sum_{v \in V(\boldsymbol{G})} g_{\boldsymbol{G}}(\boldsymbol{o}, v)\right] &= \mathbb{E}\left[f[\boldsymbol{G}, \boldsymbol{o}] g^+_{\boldsymbol{G}}(\boldsymbol{o})\right] \\ &= \mathbb{E}\left[1_A[\boldsymbol{G}', \boldsymbol{o}] g^+_{\boldsymbol{G}}(\boldsymbol{o})\right] \\ &= \mathbb{E}\left[1_A[\boldsymbol{G}', \boldsymbol{o}] g^-_{\boldsymbol{G}}(\boldsymbol{o})\right] \\ &= \mathbb{E}\left[f[\boldsymbol{G}, \boldsymbol{o}] g^-_{\boldsymbol{G}}(\boldsymbol{o})\right] \\ &= \mathbb{E}\left[\sum_{v \in V(\boldsymbol{G})} f[\boldsymbol{G}, v] g_{\boldsymbol{G}}(\boldsymbol{o}, v)\right], \end{aligned}$$

were in the second and forth equations, conditioning on $[\boldsymbol{G}, \boldsymbol{o}]$ is used, in the third one unimodularity of $[\boldsymbol{G}', \boldsymbol{o}']$ and $A \in I$ are used and in the last equation, unimodularity of $[\boldsymbol{G}, \boldsymbol{o}]$ is used. Therefore $\mathbb{E}\left[\sum_{v \in V(\boldsymbol{G})} (f[\boldsymbol{G}, \boldsymbol{o}] - f[\boldsymbol{G}, v]) g_{\boldsymbol{G}}(\boldsymbol{o}, v)\right] = 0$. By substituting $g_{\boldsymbol{G}}(\boldsymbol{o}, v)$ with the positive and negative parts of $f[\boldsymbol{G}, \boldsymbol{o}] - f[\boldsymbol{G}, v]$ separately, one obtains

$$\forall v \in V(\boldsymbol{G}), f[\boldsymbol{G}, v] = f[\boldsymbol{G}, \boldsymbol{o}], \quad a.s. \tag{7.7}$$

In other words, f does not depend on the root a.s.

Consider the root-change of changing the root to a uniformly at random neighbor of the root and compose it with its *time reversal* given by Lemma 3. This can be explicitly written by

$$T_G(u,v) := \frac{1}{d(u)d(v)} \sum_{w:w\sim u, w\sim v} \left(\sum_{z:z\sim w} \frac{1}{d(z)} \right)^{-1}.$$

It is straightforward that when G is not a single vertex, for all vertices $v \in V(G)$, one has $T_G^+(v) = T_G^-(v) = 1$. Moreover, if u, v have a common neighbor, then $T_G(u,v) > 0$. Let $k \in \mathbb{N}$ and t be the k-fold composition of T with itself. One can see T as the law of (the first step of) a random walk on the vertices that preserves the distribution of $[\boldsymbol{G}, \boldsymbol{o}]$ and t as the law of the k'th step of the random walk. By infiniteness of $\boldsymbol{G}$, it can be seen that almost surely, when $k \to \infty$, $t_{\boldsymbol{G}}(\cdot,\cdot) \to 0$ point-wise. We also have $t_{\boldsymbol{G}}^+(\cdot) = t_{\boldsymbol{G}}^-(\cdot) = 1$ a.s.

For an arbitrary $\epsilon > 0$, there exist $n \in \mathbb{N}$ and an event A_n that depends only on the ball with radius n centered at the root such that $\mathbb{P}\left[[\boldsymbol{G}', \boldsymbol{o}] \in A \Delta A_n\right] < \epsilon$. One has

$$\begin{aligned}
\mathbb{E}\left[f[\boldsymbol{G}, \boldsymbol{o}]\right] &= \mathbb{E}\left[1_A[\boldsymbol{G}', \boldsymbol{o}]\right] \\
&= \mathbb{E}\left[1_A[\boldsymbol{G}', \boldsymbol{o}] \sum_v 1_A[\boldsymbol{G}', v] t_{\boldsymbol{G}}(\boldsymbol{o}, v)\right] \\
&\leq \mathbb{E}\left[1_{A_n}[\boldsymbol{G}', \boldsymbol{o}] \sum_v 1_A[\boldsymbol{G}', v] t_{\boldsymbol{G}}(\boldsymbol{o}, v)\right] + \epsilon \\
&= \mathbb{E}\left[1_A[\boldsymbol{G}', \boldsymbol{o}] \sum_v 1_{A_n}[\boldsymbol{G}', v] t_{\boldsymbol{G}}(v, \boldsymbol{o})\right] + \epsilon \\
&\leq \mathbb{E}\left[1_{A_n}[\boldsymbol{G}', \boldsymbol{o}] \sum_v 1_{A_n}[\boldsymbol{G}', v] t_{\boldsymbol{G}}(v, \boldsymbol{o})\right] + 2\epsilon \\
&\leq \mathbb{E}\left[1_{A_n}[\boldsymbol{G}', \boldsymbol{o}] \sum_{v \notin N_{2n}(\boldsymbol{o})} 1_{A_n}[\boldsymbol{G}', v] t_{\boldsymbol{G}}(v, \boldsymbol{o})\right] + 3\epsilon,
\end{aligned}$$

where to ensure the last inequality holds, by dominated convergence, k can be chosen large enough (depending on ϵ, n, A and A_n) in the definition of t. Now, note that conditioned on $[\boldsymbol{G}, \boldsymbol{o}] = [G, o]$, for $v \notin N_{2n}(o)$, the balls $N_n(o)$ and $N_n(v)$ are disjoint and their marks are independent. Therefore, conditioned on $[\boldsymbol{G}, \boldsymbol{o}]$, the terms $1_{A_n}[\boldsymbol{G}', \boldsymbol{o}]$ and $\sum_{v \notin N_{2n}(\boldsymbol{o})} 1_{A_n}[\boldsymbol{G}', v] t_{\boldsymbol{G}}(v, \boldsymbol{o})$ are independent.

On the other hand, by defining $f_n(G, o) := \mathbb{E}\left[1_{A_n}[\boldsymbol{G}', \boldsymbol{o}] \,|\, [\boldsymbol{G}, \boldsymbol{o}] = [G, o]\right]$, one has for any measurable function $h : \mathcal{G}_* \to \mathbb{R}^{\geq 0}$,

$$\begin{aligned}
&\mathbb{E}\left[h[\boldsymbol{G}, \boldsymbol{o}] \sum_{v \notin N_{2n}(\boldsymbol{o})} 1_{A_n}[\boldsymbol{G}', v] t_{\boldsymbol{G}}(v, \boldsymbol{o})\right] \\
= &\mathbb{E}\left[1_{A_n}[\boldsymbol{G}', \boldsymbol{o}] \sum_{v \notin N_{2n}(\boldsymbol{o})} h[\boldsymbol{G}, v] t_{\boldsymbol{G}}(\boldsymbol{o}, v)\right]
\end{aligned}$$

$$
\begin{aligned}
&= \mathbb{E}\left[f_n[\boldsymbol{G},\boldsymbol{o}] \sum_{v\notin N_{2n}(\boldsymbol{o})} h[\boldsymbol{G},v] t_{\boldsymbol{G}}(\boldsymbol{o},v) \right] \\
&= \mathbb{E}\left[h[\boldsymbol{G},\boldsymbol{o}] \sum_{v\notin N_{2n}(\boldsymbol{o})} f_n[\boldsymbol{G},v] t_{\boldsymbol{G}}(v,\boldsymbol{o}) \right].
\end{aligned}
$$

This implies that

$$
\mathbb{E}\left[\sum_{v\notin N_{2n}(\boldsymbol{o})} 1_{A_n}[\boldsymbol{G}',v] t_{\boldsymbol{G}}(v,\boldsymbol{o}) \,|[\boldsymbol{G},\boldsymbol{o}] \right] = \sum_{v\notin N_{2n}(\boldsymbol{o})} f_n[\boldsymbol{G},v] t_{\boldsymbol{G}}(v,\boldsymbol{o}), \quad a.s.
$$

Therefore, by the above inequalities and the mentioned independence, one gets by conditioning on $[\boldsymbol{G},\boldsymbol{o}]$ that

$$
\begin{aligned}
\mathbb{E}\left[f[\boldsymbol{G},\boldsymbol{o}]\right] &\leq \mathbb{E}\left[f_n[\boldsymbol{G},\boldsymbol{o}] \sum_{v\notin N_{2n}(\boldsymbol{o})} f_n[\boldsymbol{G},v] t_{\boldsymbol{G}}(v,\boldsymbol{o}) \right] + 3\epsilon \\
&\leq \mathbb{E}\left[f_n[\boldsymbol{G},\boldsymbol{o}] \sum_{v\in V(\boldsymbol{G})} f_n[\boldsymbol{G},v] t_{\boldsymbol{G}}(v,\boldsymbol{o}) \right] + 3\epsilon.
\end{aligned}
$$

For any measurable function h on $\mathcal{G}_*$ such that $0\leq h\leq 1$, the fact $\mathbb{P}\left[[\boldsymbol{G}',\boldsymbol{o}]\in A\Delta A_n\right] < \epsilon$ easily implies that $|\mathbb{E}\left[(f[\boldsymbol{G},\boldsymbol{o}]-f_n[\boldsymbol{G},\boldsymbol{o}])h[\boldsymbol{G},\boldsymbol{o}]\right]| < \epsilon$. Using this and unimodularity two times, the above inequality implies

$$
\begin{aligned}
\mathbb{E}\left[f[\boldsymbol{G},\boldsymbol{o}]\right] &\leq \mathbb{E}\left[f[\boldsymbol{G},\boldsymbol{o}] \sum_{v\in V(\boldsymbol{G})} f_n[\boldsymbol{G},v] t_{\boldsymbol{G}}(v,\boldsymbol{o}) \right] + 4\epsilon \\
&= \mathbb{E}\left[f_n[\boldsymbol{G},\boldsymbol{o}] \sum_{v\in V(\boldsymbol{G})} f[\boldsymbol{G},v] t_{\boldsymbol{G}}(\boldsymbol{o},v) \right] + 4\epsilon \\
&\leq \mathbb{E}\left[f[\boldsymbol{G},\boldsymbol{o}] \sum_{v\in V(\boldsymbol{G})} f[\boldsymbol{G},v] t_{\boldsymbol{G}}(\boldsymbol{o},v) \right] + 5\epsilon \\
&= \mathbb{E}\left[f[\boldsymbol{G},\boldsymbol{o}]^2\right] + 5\epsilon,
\end{aligned}
$$

where in the last equation, (7.7) is used. Since $0\leq f\leq 1$ and ϵ is arbitrary, this implies that $f[\boldsymbol{G},\boldsymbol{o}]\in\{0,1\}$ a.s. So, (3.2) is proved and the proof is complete. □

8. Bibliography of Analogous Results for Point Processes

In this section, we discuss a similarity between unimodular networks and stationary point processes (and random measures). Then, some of the concepts and results for random networks in this paper will be related to existing ones for point processes in the literature.

8.1. General Analogies. Let us recall Palm distributions and the mass transport principle for stationary point processes briefly. A stationary point processes is, roughly speaking, a random configuration Φ of points in $\mathbb{R}^d$ such that its distribution is invariant under the translations of $\mathbb{R}^d$. The Palm distribution P_Φ of Φ is defined by $\mathbb{P}_\Phi[A] = \frac{1}{\mathbb{E}[\#\Phi\cap B]}\mathbb{E}\left[\sum_{x\in\Phi\cap B} 1_A(\theta_x(\Phi))\right]$, where B is an arbitrary measurable set and $\theta_x(\Phi)$ is just Φ translated by the vector $-x$. In words, to obtain the Palm distribution, one should *bias the probability measure by $\#\Phi \cap B$ and then move the origin to a uniformly at random point in $\Phi \cap B$.* Notice the similarity of this sentence with the examples in the introduction and Subsection 5.2. Another equivalent definition of the Palm distribution can obtained by clarifying the idea of conditioning Φ to have a point at the origin. The *mass transport principle* for stationary point processes is

$$\mathbb{E}_\Phi\left[\sum_{x\in\Phi} g(\Phi,0,x)\right] = \mathbb{E}_\Phi\left[\sum_{x\in\Phi} g(\Phi,x,0)\right]$$

for any (measurable) function g that is translation-invariant (see [**12**] or [**11**]). A result of Mecke [**18**] gives an extension of this property for stationary random measures (see [**15**]). Notice the similarity of the above mass transport principle with the one in Definition 2. This implies that, roughly speaking, any graph that is constructed from Φ in a translation-invariant manner is unimodular [**1**]. Mecke's formula [**18**] may also look related to (2.1), but all points of the space are taken into account:

$$\mathbb{E}\left[\sum_{x\in\Phi} h(0,x)\right] = \lambda\mathbb{E}_\Phi\left[\int_{\mathbb{R}^d} h(x,0)dx\right]$$

for all measurable functions $h(x,y) = h(\Phi,x,y)$ that are invariant under the translations. Here, λ is the *intensity* of Φ.

A difference of the two concepts is that in point processes, there is a group action (that of translations) for *moving the origin to another point*, but for rooted networks, there is no natural group for *changing the root*. However, like the above similarities, one can transfer some concepts and results for point processes to analogous ones for random networks. We will discuss in the next subsection that some of the results in sections 3, 4 and 6 have analogous results for point processes in the literature. See also [**3**] for other results and examples of this analogy including Mecke's point-stationarity theorem and Neveu's exchange formula.

Finally, as mentioned in [**1**], the invariant sigma-field I is analogous to the sigma-field of invariant events under translations and the notion of ergodic unimodular networks is analogous to ergodic point processes.

8.2. Analogies Regarding the Present Paper. Let $[\boldsymbol{G},\boldsymbol{o}]$ be a unimodular network. A covariant subset S (Definition 3) can be considered analogous to a subprocess (or thinning) of a stationary point process. Then, conditioning the probability measure on $\boldsymbol{o}\in S_{\boldsymbol{G}}$ in Example 3 is analogous to the Palm distribution of a subprocess (similarly, biasing by $1_{S_{\boldsymbol{G}}}(\boldsymbol{o})/\mathbb{P}[\boldsymbol{o}\in S_{\boldsymbol{G}}\,|I]$ in Example 3 is analogous to the *modified Palm distribution* [**15**] of a stationary point process). Even fancier, one can think of $[\boldsymbol{G},\boldsymbol{o}]$ and $S_{\boldsymbol{G}}$ as discrete objects analogous to the space and a point process respectively. With this analogy, Proposition 2 is analogous to a result of [**20**] which states that the Palm distribution of a stationary point process Φ can be obtained from Φ by *moving the origin to a random point of* Φ if and only if the *sample intensity* of Φ, defined by $\mathbb{E}[\#\Phi\cap B\,|I]$ for an arbitrary set B with unit volume, is almost surely constant (here, I is used for the sigma-field of invariant events under translations). This notion of sample intensity is also analogous to that of Proposition 2 and Example 3 which is equal to $\mathbb{E}[\#S_{\boldsymbol{G}}\cap\{\boldsymbol{o}\}\,|I]$.

The special case of the result of [**20**] (mentioned above) for a Poisson point process Φ has been of special interest. The result implies that one can move the origin to a random point of Φ such that (the distribution of) the resulting point process is the same as Φ except that a point is added at the origin (see Slivnyak's theorem in [**19**]). Such a change of origin is introduced by Thorisson [**20**] and is called an *extra head scheme* in [**13**]. The same holds for the Bernoulli point process in $\mathbb{Z}^d$ and is analogous to Proposition 3.

More general to a stationary point process and its Palm version, [**20**] also studies when two (not necessarily stationary) point processes Φ and Ψ can be obtained from each other by changing the origin (and covers even more general cases). This gives a coupling of Φ and Ψ that is called a *shift-coupling* in the literature. It is proved in [**20**] that a shift-coupling exists if and only if the distributions of Φ and Ψ agree on the invariant sigma-field. Theorem 1 in the present paper is its analogous result in the context of random networks.

To obtain a shift-coupling of a stationary point process Φ and its Palm version, one can use a translation-invariant *balancing transport kernel* between (a multiple of) the Lebesgue measure and the counting measure on Φ (see [**13**] and [**15**]). A transport kernel is, roughly speaking, a function $T(x, y)$ depending on Φ that shows how much of the mass at each point x of the space goes to each point y in Φ. It is *balancing* when the sum of the outgoing mass is 1 and the integral of the incoming mass is constant for all points. The existence of such a transport kernel is proved in [**13**] and [**15**] using the result of [**20**] and also by an explicit construction. Analogously, Theorem 4 proves the existence of balancing transport kernels in the context of unimodular networks with similar conditions. Invariant transport kernels are analogous to measurable functions on $\mathcal{G}_{**}$ as mentioned in Remark 1. See also [**15**] for similar results for stationary random measures.

Assume $([\boldsymbol{G}, \boldsymbol{o}], S)$ is a covariant extension of a unimodular network as in Section 5. According to the analogy of subnetworks and Palm distributions mentioned earlier, $[\boldsymbol{G}, \boldsymbol{o}]$ is analogous to a Palm distribution. In fact, (5.1) is analogous to the mass transport principle for *point-stationary* point processes (see [**15**]); which are more general than Palm distributions (when dealing with probability measures, as assumed here). Unimodularization of $[\boldsymbol{G}, \boldsymbol{o}]$ is analogous to reconstructing the stationary version of the point process from the Palm version [**18**]. The unimodularization $\mathcal{P}_T$ in Theorem 5 is analogous to *the inversion formula* of [**18**] for the reconstruction. If the sample intensity of a point process is almost surely constant, one can do the reconstruction by a shift-coupling as mentioned above. This is analogous to Theorem 6.

The existence results of [**15**] and [**20**] are abstract results and cannot be used to construct a balancing transport kernel given realizations of the two point processes. However, several constructions are provided in the literature motivated mainly by [**11**] for point processes, which is motivated by the construction in [**16**] and the stable marriage algorithm. This work is generalized in [**9**] to cover the general case of stationary random measures. Analogously, Theorem 4 doesn't provide a construction of the balancing transport kernel in the context of unimodular networks. Theorem 7 is a construction which is analogous to the one in [**9**].

Acknowledgements

The major part of this work was done in a post doctoral program in IPM (Institute for Research in Fundamental Sciences, Tehran, Iran). I thank them for their support. Also, this work was motivated by the author's joint works with Francois Baccelli and Mir-Omid Haji-Mirsadeghi. I thank them also for their useful comments. I thank Lewis Bowen as well for his comments on Borel equivalence relations, especially for the discussion on Proposition 1.

References

[1] D. Aldous and R. Lyons, *Processes on unimodular random networks*, Electron. J. Probab. **12** (2007), no. 54, 1454–1508, DOI 10.1214/EJP.v12-463. MR2354165

[2] D. Aldous and J. M. Steele, *The objective method: probabilistic combinatorial optimization and local weak convergence*, Probability on discrete structures, Encyclopaedia Math. Sci., vol. 110, Springer, Berlin, 2004, pp. 1–72, DOI 10.1007/978-3-662-09444-0_1. MR2023650

[3] F. Baccelli, M. O. Haji-Mirsadeghi, and A. Khezeli, Eternal Family Trees and Dynamics on Unimodular Random Graphs, in: 'Unimodularity in randomly generated graphs', Edt.: F. Sobieczky, Contemporary Mathematics, Amer. Math. Soc, Providence, RI, (to appear)

[4] I. Benjamini, R. Lyons, Y. Peres, and O. Schramm, *Group-invariant percolation on graphs*, Geom. Funct. Anal. **9** (1999), no. 1, 29–66, DOI 10.1007/s000390050080. MR1675890

[5] I. Benjamini, R. Lyons, and O. Schramm, *Unimodular random trees*, Ergodic Theory Dynam. Systems **35** (2015), no. 2, 359–373, DOI 10.1017/etds.2013.56. MR3316916

[6] R. H. Farrell, *Representation of invariant measures*, Illinois J. Math. **6** (1962), 447–467. MR0150264

[7] J. Feldman and C. C. Moore, *Ergodic equivalence relations, cohomology, and von Neumann algebras. I*, Trans. Amer. Math. Soc. **234** (1977), no. 2, 289–324, DOI 10.2307/1997924. MR0578656

[8] D. Gale and L. S. Shapley, *College Admissions and the Stability of Marriage*, Amer. Math. Monthly **69** (1962), no. 1, 9–15, DOI 10.2307/2312726. MR1531503

[9] M.-O. Haji-Mirsadeghi and A. Khezeli, *Stable transports between stationary random measures*, Electron. J. Probab. **21** (2016), Paper No. 51, 25, DOI 10.1214/16-EJP4237. MR3539645

[10] L. A. Harrington, A. S. Kechris, and A. Louveau, *A Glimm-Effros dichotomy for Borel equivalence relations*, J. Amer. Math. Soc. **3** (1990), no. 4, 903–928, DOI 10.2307/1990906. MR1057041

[11] C. Hoffman, A. E. Holroyd, and Y. Peres, *A stable marriage of Poisson and Lebesgue*, Ann. Probab. **34** (2006), no. 4, 1241–1272, DOI 10.1214/009117906000000098. MR2257646

[12] A. E. Holroyd and Y. Peres, *Trees and matchings from point processes*, Electron. Comm. Probab. **8** (2003), 17–27, DOI 10.1214/ECP.v8-1066. MR1961286

[13] A. E. Holroyd and Y. Peres, *Extra heads and invariant allocations*, Ann. Probab. **33** (2005), no. 1, 31–52, DOI 10.1214/009117904000000603. MR2118858

[14] A. Khezeli (2016). *Mass transport between stationary random measures* (Doctoral dissertation), Sharif University of Technology.

[15] G. Last and H. Thorisson, *Invariant transports of stationary random measures and mass-stationarity*, Ann. Probab. **37** (2009), no. 2, 790–813, DOI 10.1214/08-AOP420. MR2510024

[16] T. M. Liggett, *Tagged particle distributions or how to choose a head at random*, In and out of equilibrium (Mambucaba, 2000), Progr. Probab., vol. 51, Birkhäuser Boston, Boston, MA, 2002, pp. 133–162. MR1901951

[17] R. Lyons and O. Schramm, *Indistinguishability of percolation clusters*, Ann. Probab. **27** (1999), no. 4, 1809–1836, DOI 10.1214/aop/1022677549. MR1742889

[18] J. Mecke, *Stationäre zufällige Masse auf lokalkompakten Abelschen Gruppen* (German), Z. Wahrscheinlichkeitstheorie und Verw. Gebiete **9** (1967), 36–58, DOI 10.1007/BF00535466. MR0228027

[19] R. Schneider and W. Weil, *Stochastic and integral geometry*, Probability and its Applications (New York), Springer-Verlag, Berlin, 2008. MR2455326

[20] H. Thorisson, *Transforming random elements and shifting random fields*, Ann. Probab. **24** (1996), no. 4, 2057–2064, DOI 10.1214/aop/1041903217. MR1415240

[21] V. S. Varadarajan, *Groups of automorphisms of Borel spaces*, Trans. Amer. Math. Soc. **109** (1963), 191–220, DOI 10.2307/1993903. MR0159923

Department of Mathematics, Tarbiat Modares University, P.O. Box 14115-134, Tehran, Iran—and—School of Mathematics, Institute for Research in Fundamental Sciences (IPM), P.O. Box 19395-5746, Tehran, Iran

Email address: `khezeli@modares.ac.ir, alikhezeli@ipm.ir`

Selected Published Titles in This Series

719 **Florian Sobieczky, Editor,** Unimodularity in Randomly Generated Graphs, 2018

716 **Alex Martsinkovsky, Kiyoshi Igusa, and Gordana Todorov, Editors,** Surveys in Representation Theory of Algebras, 2018

715 **Sergio R. López-Permouth, Jae Keol Park, S. Tariq Rizvi, and Cosmin S. Roman, Editors,** Advances in Rings and Modules, 2018

714 **Jens Gerlach Christensen, Susanna Dann, and Matthew Dawson, Editors,** Representation Theory and Harmonic Analysis on Symmetric Spaces, 2018

713 **Naihuan Jing and Kailash C. Misra, Editors,** Representations of Lie Algebras, Quantum Groups and Related Topics, 2018

712 **Nero Budur, Tommaso de Fernex, Roi Docampo, and Kevin Tucker, Editors,** Local and Global Methods in Algebraic Geometry, 2018

711 **Thomas Creutzig and Andrew R. Linshaw, Editors,** Vertex Algebras and Geometry, 2018

710 **Raphaël Danchin, Reinhard Farwig, Jiří Neustupa, and Patrick Penel, Editors,** Mathematical Analysis in Fluid Mechanics, 2018

709 **Fernando Galaz-García, Juan Carlos Pardo Millán, and Pedro Solórzano, Editors,** Contributions of Mexican Mathematicians Abroad in Pure and Applied Mathematics, 2018

708 **Christian Ausoni, Kathryn Hess, Brenda Johnson, Ieke Moerdijk, and Jérôme Scherer, Editors,** An Alpine Bouquet of Algebraic Topology, 2018

707 **Nitya Kitchloo, Mona Merling, Jack Morava, Emily Riehl, and W. Stephen Wilson, Editors,** New Directions in Homotopy Theory, 2018

706 **Yeonhyang Kim, Sivaram K. Narayan, Gabriel Picioroaga, and Eric S. Weber, Editors,** Frames and Harmonic Analysis, 2018

705 **Graham J. Leuschke, Frauke Bleher, Ralf Schiffler, and Dan Zacharia, Editors,** Representations of Algebras, 2018

704 **Alain Escassut, Cristina Perez-Garcia, and Khodr Shamseddine, Editors,** Advances in Ultrametric Analysis, 2018

703 **Andreas Malmendier and Tony Shaska, Editors,** Higher Genus Curves in Mathematical Physics and Arithmetic Geometry, 2018

702 **Mark Grant, Gregory Lupton, and Lucile Vandembroucq, Editors,** Topological Complexity and Related Topics, 2018

701 **Joan-Carles Lario and V. Kumar Murty, Editors,** Number Theory Related to Modular Curves, 2018

700 **Alexandre Girouard, Dmitry Jakobson, Michael Levitin, Nilima Nigam, Iosif Polterovich, and Frédéric Rochon, Editors,** Geometric and Computational Spectral Theory, 2017

699 **Mark L. Agranovsky, Matania Ben-Artzi, Catherine Bénéteau, Lavi Karp, Dmitry Khavinson, Simeon Reich, David Shoikhet, Gilbert Weinstein, and Lawrence Zalcman, Editors,** Complex Analysis and Dynamical Systems VII, 2017

698 **Alexander M. Blokh, Leonid A. Bunimovich, Paul H. Jung, Lex G. Oversteegen, and Yakov G. Sinai, Editors,** Dynamical Systems, Ergodic Theory, and Probability: in Memory of Kolya Chernov, 2017

697 **Fabrizio Broglia, Françoise Delon, Max Dickmann, Danielle Gondard-Cozette, and Victoria Ann Powers, Editors,** Ordered Algebraic Structures and Related Topics, 2017

696 **Ara S. Basmajian, Yair N. Minsky, and Alan W. Reid, Editors,** In the Tradition of Ahlfors–Bers, VII, 2017

For a complete list of titles in this series, visit the AMS Bookstore at **www.ams.org/bookstore/conmseries/**.